Antibiotic Alternatives in Poultry and Fish Feed

Edited by

Mohamed E. Abd El-Hack
Department of Poultry
Faculty of Agriculture, Zagazig University
Zagazig
Egypt

&

Mahmoud Alagawany
Department of Poultry
Faculty of Agriculture, Zagazig University
Zagazig
Egypt

Antibiotic Alternatives in Poultry and Fish Feed

Editors: Mohamed E. Abd El-Hack and Mahmoud Alagawany

ISBN (Online): 978-981-5049-01-5

ISBN (Print): 978-981-5049-02-2

ISBN (Paperback): 978-981-5049-03-9

Published by Bentham Science Publishers Pte. Ltd. Singapore. All Rights Reserved.

First published in 2022.

need for a court order if at any point you breach any terms of this License Agreement. In no event will any delay or failure by Bentham Science Publishers in enforcing your compliance with this License Agreement constitute a waiver of any of its rights.

3. You acknowledge that you have read this License Agreement, and agree to be bound by its terms and conditions. To the extent that any other terms and conditions presented on any website of Bentham Science Publishers conflict with, or are inconsistent with, the terms and conditions set out in this License Agreement, you acknowledge that the terms and conditions set out in this License Agreement shall prevail.

Bentham Science Publishers Pte. Ltd.
80 Robinson Road #02-00
Singapore 068898
Singapore
Email: subscriptions@benthamscience.net

BENTHAM SCIENCE

CONTENTS

FOREWORD i

PREFACE ii

LIST OF CONTRIBUTORS iii

CHAPTER 1 HAZARDS OF USING ANTIBIOTIC GROWTH PROMOTERS IN THE POULTRY INDUSTRY 1

Mahmoud Alagawany, Mohamed E. Abd El-Hack, Muhammad Saeed, Muhammad S. Khan, Asghar A. Kamboh, Faisal Siddique, Ali Raza, Mayada R. Farag and *Samir Mahgoub*

    INTRODUCTION 2

    ANTIBIOTICS AS GROWTH PROMOTERS 3

    LETHAL EFFECTS OF ANTIBIOTICS USED IN POULTRY PRODUCTION 6

    ANTIMICROBIAL RESIDUES IN POULTRY PRODUCTS 7

    PUBLIC HEALTH RISKS RELATED TO ANTIBIOTICS 9

    CONCLUSION AND FUTURE DIRECTION 10

    CONSENT FOR PUBLICATION 11

    CONFLICT OF INTEREST 11

    ACKNOWLEDGEMENTS 11

    REFERENCES 11

CHAPTER 2 HERB AND PLANT-DERIVED SUPPLEMENTS IN POULTRY NUTRITION 19

Muhammad Saeed, Muhammad S. Khan, Rizwana Sultan, Amjad I. Aqib, Muhammad A. Naseer, Iqra Muzammil, Mayada R. Farag, Mohamed E. Abd El-Hack, Alessandro Di Cerbo and *Mahmoud Alagawany*

    INTRODUCTION 20

    PHYTOCHEMICALS 21

    TYPES OF PHYTOCHEMICAL ANTIBIOTIC ALTERNATIVES IN POULTRY ... 22

    SYNERGISTIC EFFECT OF PHYTOCHEMICALS IN POULTRY NUTRITION ... 22

    HERBS AS GROWTH AND HEALTH PROMOTORS 23

    HEALTH RISKS RELATED TO HERBS UTILIZATION IN POULTRY FEED ... 26

    CONCLUSION 27

    CONSENT FOR PUBLICATION 27

    CONFLICT OF INTEREST 27

    ACKNOWLEDGEMENTS 27

    REFERENCES 27

CHAPTER 3 GINGER AS A NATURAL FEED SUPPLEMENT IN POULTRY DIETS 33

Mohamed E. Abd El-Hack, Ayman A. Swelum, Youssef A. Attia, Mohamed Abdo, Ahmed I. Abo-Ahmed, Mahmoud A. Emam and *Mahmoud Alagawany*

    INTRODUCTION 34

    IMPACT OF GINGER OIL 35

    BIOCHEMICAL COMPONENTS OF GINGER EXTRACT 35

    VALUABLE EFFECTS OF GINGER SUPPLEMENTATION IN POULTRY 37

        Impact of Ginger Supplementation on Carcass Weight 37

        Impact of Ginger and its Products on Carcass Characters 40

        Impact of Ginger and its Products on Egg Quality and Production ... 40

        Impact of Ginger on Reproductive Function 41

        Impact of Ginger on Blood Parameters 41

        Impact of Ginger and its Products on Microbial Infections 43

        Impact of Ginger on Egg and Meat Quality 44

The Economic Impact of Ginger 45
CONCLUSION 46
CONSENT FOR PUBLICATION 46
CONFLICT OF INTEREST 46
ACKNOWLEDGEMENTS 46
REFERENCES 46

CHAPTER 4 USE OF CINNAMON AND ITS DERIVATIVES IN POULTRY NUTRITION ... 52
Rana M. Bilal, Faiz ul Hassan, Majed Rafeeq, Mayada R. Farag, Mohamed E. Abd El-
Hack, Mahmoud Madkour and Mahmoud Alagawany
    INTRODUCTION 53
    CHEMICAL COMPOSITION 53
    EFFECTS ON GROWTH PERFORMANCE 56
        Body Weight and Body Weight Gain 56
        Feed Utilization 57
    CARCASS TRAITS 58
    BLOOD PARAMETERS 59
    INTESTINAL MICROBIOTA 59
    CONCLUSION 61
    CONSENT FOR PUBLICATION 61
    CONFLICT OF INTEREST 61
    ACKNOWLEDGEMENTS 61
    REFERENCES 61

CHAPTER 5 CLOVE (SYZYGIUM AROMATICUM) AND ITS DERIVATIVES IN
POULTRY FEED 66
Mahmoud Alagawany, Mohamed E. Abd El-Hack, Muhammad Saeed, Shaaban S.
Elnesr and Mayada R. Farag
    INTRODUCTION 67
    NUTRITIVE VALUE OF CLOVE 68
    BENEFICIAL APPLICATION OF CLOVE AND ITS DERIVATIVES 70
        Antimicrobial Properties 70
        Antibacterial Activity of Clove 70
        Antioxidant Activity of Clove 71
    BENEFICIAL APPLICATION OF CLOVE AND ITS DERIVATIVES IN POULTRY 72
        Effect of Clove and its Derivatives on Poultry Performance 72
        Effect of Clove and its Derivatives on Egg Production and Quality 74
    CONCLUSION 74
    CONSENT FOR PUBLICATION 75
    CONFLICT OF INTEREST 75
    ACKNOWLEDGEMENTS 75
    REFERENCES 75

CHAPTER 6 POMEGRANATE (PUNICA GRANATUM L): BENEFICIAL IMPACTS,
HEALTH BENEFITS AND USES IN POULTRY NUTRITION 80
Youssef A. Attia, Ayman E. Taha, Mohamed E. Abd El-Hack, Mohamed Abdo,
Ahmed I. Abo-Ahmed, Mahmoud A. Emam, Karima El Naggar, Mervat A. Abdel-
Latif, Nader R. Abdelsalam and Mahmoud Alagawany
    INTRODUCTION 81
    PHYTOCHEMICALS IN POMEGRANATE 82
    CONVENTIONAL USES OF POMEGRANATE FRUIT 83
    CHEMICAL ANALYSIS OF POMEGRANATE 84

BIOLOGICAL PROPERTIES AND THERAPEUTIC APPLICATIONS 84
    Antioxidant Activity 85
    Anti-inflammatory Activity 86
    Glucose and Lipid Metabolism Activities 86
    Antimicrobial Activity 86
STUDIES ON POULTRY LIVESTOCK 86
    Direct Additive Effects on Meat 89
CONCLUSION 90
CONSENT FOR PUBLICATION 91
CONFLICT OF INTEREST 91
ACKNOWLEDGEMENTS 91
REFERENCES 91

CHAPTER 7 USE OF CHICORY (CICHORIUM INTYBUS) AND ITS DERIVATIVES IN
POULTRY NUTRITION 98
Muhammad Saeed, Faisal Siddique, Rizwana Sultan, Sabry A.A. El-Sayed, Sarah Y.A.
Ahmed, Mayada R. Farag, Mohamed E. Abd El-Hack, Abdelrazeq M. Shehata and
Mahmoud Alagawany
INTRODUCTION 99
THE DESCRIPTION OF PLANTS AND THEIR CHEMICAL COMPOSITION 100
    Advantageous Results of Chicory with Specific Respect to its Function as a
    Hepatoprotective Agent 102
BENEFICIAL APPLICATIONS OF CHICORY 104
PRACTICAL USAGE IN THE POULTRY SECTOR 105
CONCLUSION AND FUTURE RECOMMENDATIONS 105
CONSENT FOR PUBLICATION 106
CONFLICT OF INTEREST 106
ACKNOWLEDGEMENTS 106
REFERENCES 106

CHAPTER 8 USE OF PSYLLIUM HUSK (PLANTAGO OVATA) IN POULTRY FEEDING
AND POSSIBLE APPLICATION IN ORGANIC PRODUCTION 111
Mahmoud Alagawany, Rana Muhammad Bilal, Fiza Batool, Youssef A. Attia,
Mohamed E. Abd El-Hack, Sameh A. Abdelnour, Mayada R. Farag, Ayman A.
Swelum and *Mahmoud Madkour*
INTRODUCTION 112
    Geographical Source of Psyllium 113
    Biological Benefits of Psyllium Husk 115
    The Anti-cholesterol Activity of Psyllium 115
    Psyllium as a Potent Hypocholesterolemic Agent in Humans 116
    Psyllium Husk as a Potent Hypocholesterolemic Agent in Animal and Poultry 116
    Role of Psyllium in the Therapy of other Human Ailments 117
        Digestive Function and Metabolism 117
    Hemorrhoids 118
    Clinical Effect 118
CONCLUSION 118
CONSENT FOR PUBLICATION 119
CONFLICT OF INTEREST 119
ACKNOWLEDGEMENTS 119
REFERENCES 119

CHAPTER 9 DANDELION HERB: CHEMICAL COMPOSITION AND USE IN POULTRY NUTRITION 124
Mahmoud Alagawany, Mohamed E. Abd El-Hack, Mayada R. Farag, Sameh A. Abdelnour, Kuldeep Dhama, Ayman A. Swelum and Alessandro Di Cerbo
    INTRODUCTION 125
        Structure and Chemical Composition 126
        Beneficial Roles of Dandelion for Health 127
            Effect of Dandelion on Performance, Carcass and Meat Quality 127
        Hepatoprotective and Anti-cancer Activities 128
        Antibacterial and Antiparasitic Activities 129
        Antioxidant and Anti-inflammatory Activities 129
        Hypoglycaemic and Hypolipidemic Effects 130
        Immune System Enhancer 130
        Digestion Stimulant 130
        Effect of Dandelion on Hematological and Biochemical Blood Parameters 131
    CONCLUSION 131
    CONSENT FOR PUBLICATION 131
    CONFLICT OF INTEREST 131
    ACKNOWLEDGEMENTS 131
    REFERENCES 131

CHAPTER 10 PROBIOTICS IN POULTRY NUTRITION AS A NATURAL ALTERNATIVE FOR ANTIBIOTICS 137
Mohamed E. Abd El-Hack, Mahmoud Alagawany, Nahed A. El-Shall, Abdelrazeq M. Shehata, Abdel-Moneim E. Abdel-Moneim and Mohammed A. E. Naiel
    INTRODUCTION 138
    PROBIOTICS TYPES AND SOURCES 138
    MECHANISM OF PROBIOTIC ACTION 141
    ASPECTS OF PROBIOTIC APPLICATIONS IN THE POULTRY 142
        Growth Performance 142
        Antibiotic Alternatives to Counter Infectious Pathogens 144
        Egg Production 146
        Health and Immunity 147
    CONCLUSION 149
    CONSENT FOR PUBLICATION 149
    CONFLICT OF INTEREST 149
    ACKNOWLEDGEMENT 149
    REFERENCES 149

CHAPTER 11 PHYTOGENIC SUBSTANCES: A PROMISING APPROACH TOWARDS SUSTAINABLE AQUACULTURE INDUSTRY 160
Abdelrazeq M. Shehata, Abdel-Moneim E. Abdel-Moneim, Ahmed G. A. Gewida, Mohamed E. Abd El-Hack, Mahmoud Alagawany and Mohammed A. E. Naiel
    INTRODUCTION 161
    POTENTIAL OF PHYTOGENIC FEED ADDITIVES IN AQUACULTURE 164
        Phytogenic Feed Additives as Appetite Stimulators and Growth Promoters 164
        Phytogenic Feed Additives as Immunostimulants 166
        Phytogenic Feed Additives as Natural Antioxidant Agents 168
        Phytogenic Feed Additives as Modulators of Gut Health and Microbiota 169
    PHYTOTHERAPY AS AN ALTERNATIVE FOR TREATING FISH DISEASE 172
        Anti-bacterial Activity 172

Anti-viral Activity 175

Anthelminthic [Monogeneans] Activity 177

Anti-fungal Activity 179

CONCLUDING REMARKS 180

CONSENT FOR PUBLICATION 180

CONFLICT OF INTEREST 180

ACKNOWLEDGMENT 180

REFERENCES 180

CHAPTER 12 THE BENEFICIAL IMPACTS OF ESSENTIAL OILS APPLICATION AGAINST PARASITIC INFESTATION IN FISH FARM 194

Samar S. Negm, Mohamed E. Abd El-Hack, Mahmoud Alagawany, Amlan Kumar Patra and *Mohammed A. E. Naiel*

INTRODUCTION 195

ESSENTIAL OIL RESOURCES, STRUCTURE AND BIOACTIVE MOLECULES 196

THE IMMUNOSTIMULATORY ROLE OF EOS 198

THE ANTIOXIDANT AND PROTECTIVE ROLE OF EO 201

THE ANTIPARASITIC EFFECTS OF EOS 203

CONCLUDING REMARKS 205

CONSENT FOR PUBLICATION 206

CONFLICT OF INTEREST 206

ACKNOWLEDGEMENTS 206

REFERENCES 206

CHAPTER 13 THE ROLE OF ANTIMICROBIAL PEPTIDES (AMPS) IN AQUACULTURE FARMING 215

Mohammed A. E. Naiel, Mohamed E. Abd El-Hack, Amlan Kumar Patra and *Mahmoud Alagawany*

INTRODUCTION 215

Antimicrobial Peptides (AMPs) Types and Structure 216

Resources of Antimicrobial Peptides 217

Antimicrobial Peptides Mechanisms, Advantages and Disadvantages 218

The Application of AMPs Against Fish Diseases 219

Anti-parasitic and Antifungal Activity of AMPs 222

The Activity of AMPs Toward Bacterial Fish Diseases 223

The Antiviral Effects of AMPs 224

CONCLUDING REMARKS 226

CONSENT FOR PUBLICATION 226

CONFLICT OF INTEREST 226

ACKNOWLEDGEMENTS 226

REFERENCES 226

SUBJECT INDEX 236

FOREWORD

I was glad when I got a request from Dr. Mohamed E. Abd El-Hack and Dr. Mahmoud Alagawany to write a brief foreword to their reprint book. For several years, I have admired their incredible work. Furthermore, poultry and fish products' consumers always search for organic products. So, the topic of this book is timely and globally needed for all people who produce or consume poultry and fish products.

Looking through this valuable book, I'm pleased with the authors' talent in writing such a useful book. It is a source of inspiration and information for those who work in poultry and fish production.

In conclusion, Dr. Abd El-Hack and Dr. Alagawany's book is peerless and a work to treasure for anyone interested in poultry and fish production. So, it is expected that readers will enjoy reading it and learn from it. Thank you, Dr. Abd El-Hack and Dr. Alagawany, for producing such a masterwork.

Prof. Dr. Vincenzo Tufarelli
DETO - Section of Veterinary Science and Animal Production University of Bari 'Aldo Moro' s.p. Casamassima km 3
70010 Valenzano BA
Italy

PREFACE

For a long time ago, poultry keepers used to add trace levels of antibiotics to poultry feed to act as growth-promoting agents. This practice caused harmful impacts on poultry products' consumers because of antibiotic resistance. This led the European Union to ban the use of antibiotics in poultry feed. So, this book focuses on the use of antibiotic alternatives in poultry and fish feed. Also, it deals with the different impacts of these natural feed additives in poultry and fish nutrition on growth, production, reproduction and health status. This book contains 13 chapters contributed by 38 experts and scientists of animal, poultry and fish nutrition, poultry and fish physiology, toxicology, pharmacology, and pathology, which highlights the significance of herbal plants and their extracts and derivatives, cold-pressed and essential oils, fruits by-products, immunomodulators, antimicrobial peptides, and probiotics with their role in poultry and fish industry instead of antibiotic growth promoters. This book provides detailed information about using antibiotics in the poultry and fish industry as growth promoters and developing bacterial resistance to antibiotics. All chapters give a holistic approach to how organic feed additives (herbal plants and their extracts, probiotics, peptides, *etc.*) can positively impact animal health and production. Also, the book chapters cover the main poultry species, including broilers, laying hens, quails, geese, ducks, turkey, and fish. This book will be useful for poultry and fish keepers and research in nutrition, pharmacology, and veterinary sciences.

Mohamed E. Abd El Hack
Department of Poultry
Faculty of Agriculture, Zagazig University
Zagazig
Egypt

&

Mahmoud Alagawany
Department of Poultry
Faculty of Agriculture, Zagazig University
Zagazig
Egypt

List of Contributors

Abdel-Moneim E. Abdel-Moneim
Biological Applications Department, Nuclear Research Center, Egyptian Atomic Energy Authority, 13759, Egypt

Abdelrazeq M. Shehata
Department of Dairy Science & Food Technology, Institute of Agricultural Sciences, Banaras Hindu University, Varanasi 221005, India
Department of Animal Production, Faculty of Agriculture, Al-Azhar University, Cairo 11651, Egypt

Ahmed G. A. Gewida
Department of Animal Production, Faculty of Agriculture, Al-Azhar University, Cairo 11651, Egypt

Ahmed I. Abo-Ahmed
Department of Anatomy and Embryology, Faculty of Veterinary Medicine,Benha University, Toukh 13736, Egypt

Alessandro Di Cerbo
School of Biosciences and Veterinary Medicine, University of Camerino, Matelica, Italy

Ali Raza
Center for Animal Sciences, Queensland Alliance for Agriculture and Food Innovation, University of Queensland, Australia

Amjad I. Aqib
Cholistan University of Veterinary and Animal Sciences, Bahawalpur, 63100, Pakistan

Amlan Kumar Patra
Department of Animal Nutrition, West Bengal University of Animal and Fishery Sciences, Kolkata 700037, India

Asghar A. Kamboh
Department of Veterinary Microbiology, Faculty of Animal Husbandry and Veterinary Sciences, Sindh Agriculture University, Tandojam, Pakistan

Ayman A. Swelum
Department of Animal Production, College of Food and Agriculture Sciences, King Saud University, P.O. Box 2460, Riyadh 11451, Saudi Arabia
Department of Theriogenology, Faculty of Veterinary Medicine, Zagazig University, Zagazig 44511, Egypt

Ayman E. Taha
Department of Animal Husbandry and Animal Wealth Development, Faculty of Veterinary Medicine, Alexandria University, Rasheed, 22758 Edfina, Egypt

Faisal Siddique
Cholistan University of Veterinary and Animal Sciences, Bahawalpur, 63100, Pakistan

Faiz ul Hassan
Institute of Animal & Dairy Sciences, Faculty of Animal Husbandry, University of Agriculture, Faisalabad, 38040, Pakistan

Fiza Batool
Department of Forestry, Faculty of Agriculture, The Islamia University of Bahawalpur, Bahawalpur, Pakistan

Iqra Muzammil
Department of Clinical Medicine and Surgery, University of Agriculture, Faisalabad-38000, Pakistan

Karima El Naggar
Department of Nutrition and Clinical Nutrition, Faculty of Veterinary Medicine, Alexandria University, Egypt

Kuldeep Dhama
Division of Pathology, ICAR-Indian Veterinary Research Institute, Izatnagar, Bareilly- 243 122, Uttar Pradesh, India

Mahmoud A. Emam	Department of Histology, Faculty of Veterinary Medicine, Benha University, Toukh 13736, Egypt
Mahmoud Alagawany	Department of Poultry, Faculty of Agriculture, Zagazig University, Zagazig 44511, Egypt
Mahmoud Madkour	Animal Production Department, National Research Centre, Dokki, 12622 Giza, Egypt
Majed Rafeeq	University of Balochistan Quetta, Quetta, Pakistan
Mayada R. Farag	Department of Poultry, Faculty of Agriculture, Zagazig University, Egypt Forensic Medicine and Toxicology Department, Veterinary Medicine Faculty, Zagazig University, Zagazig 44519, Egypt
Mervat A. Abdel-Latif	Department of Nutrition and Veterinary Clinical Nutrition, Faculty of Veterinary Medicine, Damanhour University, Damanhour 22511, Egypt
Mohammed A. E. Naiel	Department of Animal Production, Faculty of Agriculture, Zagazig University, Zagazig 44519, Egypt
Mohamed Abdo	Department of Animal Histology and Anatomy, School of Veterinary Medicine, Badr University in Cairo (BUC), Cairo, Egypt Department of Anatomy and Embryology, Faculty of Veterinary Medicine, University of Sadat City, Sadat City, Egypt
Mohamed E. Abd El-Hack	Department of Poultry, Faculty of Agriculture, Zagazig University, Zagazig 44511, Egypt
Muhammad A. Naseer	Department of Clinical Medicine and Surgery, University of Agriculture, Faisalabad-38000, Pakistan
Muhammad S. Khan	Cholistan University of Veterinary and Animal Sciences, Bahawalpur, 63100, Pakistan
Muhammad Saeed	Cholistan University of Veterinary and Animal Sciences, Bahawalpur, 63100, Pakistan
Nader R. Abdelsalam	Agricultural Botany Department, Faculty of Agriculture (Saba Basha), Alexandria University, 21531 Alexandria, Egypt
Nahed A. El-Shall	Department of Poultry and Fish Diseases, Faculty of Veterinary Medicine, Alexandria University, Edfina, Elbehira 22758, Egypt
Rana M. Bilal	College of Veterinary and Animal Sciences, The Islamia University of Bahawalpur, Bahawalpur, 63100, Pakistan
Rizwana Sultan	Cholistan University of Veterinary and Animal Sciences, Bahawalpur, 63100, Pakistan
Sabry A.A. El-Sayed	Department of Nutrition and Clinical Nutrition, Faculty of Veterinary Medicine, Zagazig University, Zagazig, Egypt
Samar S. Negm	Fish Biology and Ecology Department, Central Lab for Aquaculture Research Abbassa, Agriculture Research Centre, Giza, Egypt
Sameh A. Abdelnour	Animal Production Department, Faculty of Agriculture, Zagazig University, Zagazig 44511, Egypt
Samir Mahgoub	Department of Agricultural Microbiology, Faculty of Agriculture, Zagazig University, Zagazig 44511, Egypt

Sarah Y.A. Ahmed   Department of Microbiology, Faculty of Veterinary Medicine, Zagazig University, Zagazig, Egypt

Shaaban S. Elnesr   Poultry Production Department, Faculty of Agriculture, Fayoum University, Fayoum 63514, Egypt

Youssef A. Attia   Animal and Poultry Production Department, Faculty of Agriculture Damanhour University, Damanhour, Egypt
Department of Agriculture , Faculty of Environmental Sciences, King Abdulaziz University, 21589, Jeddah, Kingdom of Saudi Arabia

CHAPTER 1

Hazards of Using Antibiotic Growth Promoters in the Poultry Industry

Mahmoud Alagawany[1,*], Mohamed E. Abd El-Hack[1,*], Muhammad Saeed[2], Muhammad S. Khan[2,*], Asghar A. Kamboh[3], Faisal Siddique[2], Ali Raza[4], Mayada R. Farag[5] and Samir Mahgoub[6]

[1] *Department of Poultry, Faculty of Agriculture, Zagazig University, Zagazig 44511, Egypt*

[2] *Cholistan University of Veterinary and Animal Sciences Bahawalpur, 63100, Pakistan*

[3] *Department of Veterinary Microbiology, Faculty of Animal Husbandry and Veterinary Sciences, Sindh Agriculture University, Tandojam, Pakistan*

[4] *Center for Animal Sciences, Queensland Alliance for Agriculture and Food Innovation, University of Queensland, Australia*

[5] *Forensic Medicine and Toxicology Department, Veterinary Medicine Faculty, Zagazig University, Zagazig 44519, Egypt*

[6] *Department of Agricultural Microbiology, Faculty of Agriculture, Zagazig University, Zagazig 44511, Egypt*

Abstract: The poultry industry is one of the significant hubs of the livestock industry and the world's largest food industry. In the last 50 years, it has become common to observe poultry antibiotic feeding to treat disease and growth. Antibiotics inhibit the growth of toxic and beneficial microorganisms. They are used as growth promoters when given in adjunctive therapy. The Centers for Disease Control and Prevention (CDC) estimates that fifty million pounds of antibiotics will be produced each year in the USA. Forty percent of the total antibiotics produced will be used in agriculture. 11 million pounds are used for the poultry sector and 24 million for domestic and wild animals. Ciprofloxacin, chloramphenicol, enrofloxacin, oxytetracycline, tylosin, tetracycline, virginiamycin, tilmicos, nitrofuran and sulfamids are used as growth promoters in the poultry industry globally. Antibacterial residues are found in various parts of poultry birds, *e.g.*, kidney, heart, gizzard, liver, chest, thigh muscles, albumin and egg yolk. These residues may directly or indirectly produce many health concerns in human beings, such as toxic effects in the liver, brain, bone marrow, kidney, allergic reaction, mutagenicity, reproductive abnormalities and gastrointestinal tract leading to indigestion. In addition, resistant strains of pathogenic microbes pose an indirect threat to antibacterial residues that can spread to humans and contaminate residual fertilizers used as plant fertilizers. This chapter describes the benefits and contraindications of

* **Corresponding authors Mahmoud Alagawany, Mohamed E. Abd El-Hack & Muhammad S. Khan:** Department of Poultry, Faculty of Agriculture, Zagazig University, Zagazig 44511, Egypt & Cholistan University of Veterinary and Animal Sciences Bahawalpur, 63100, Pakistan; E-mails: drsajjad2@yahoo.com, dr.mohamed.e.abdalhaq@gmail.com & mmalagwany@ zu.edu.eg

antibiotics used as growth promoters and the toxic effects of antimicrobial residues in poultry and humans.

Keywords: Antibiotics, Feed, Growth, Poultry, Resistant.

INTRODUCTION

Growth-enhancing effects of in-feed antibiotics were recognized first in the 1940s when dried mycelia of Streptomyces aureofaciens were fed to animals containing chlortetracycline antibiotics [1]. Later, in 1946, Moore and his coworkers reported the growth-promoting effects of antibiotics in commercial poultry [2] that, opened a new era of antibiotics used in poultry feed. The use of antibiotics as a promoter of the growth in the poultry industry has been going on worldwide for over 50 years [1]. In 1981, the American Council for Agricultural Science and Technology published a report on antibiotics in feeding animals [1]. Though the report did not provide any data that the use of antibiotics in animals causes the emergence of resistant microorganisms that may produce drug-resistant infections in human beings, it started a debate on antibiotics in food animals. In 1986, Sweden was the first country that banned antibiotics for growth promotion [3]. Denmark in 1995 and European Union in 1997 banned the Avoparcin (an antibiotic growth promoter) in food-producing animals. Later, in 1999 European Union Commission banned other antimicrobials for use as growth promoters in farm animals [4].

Plenty of evidence establishes that the use of antimicrobials in farm animals for therapeutic purposes and/or growth promotion causes the creation of antibiotic-resistant bacteria in the environment that ultimately deteriorate the therapeutic options in human medicine [5 - 7]. About 7 million deaths in hospitals have been estimated due to antibiotic-resistant infections [8]. Antibiotics used in food-producing animals like poultry led to the transfer of resistant bacteria to human beings *via* animal food. These bacteria may further transfer resistant genes to the non-pathogenic commensal flora [9]. It is estimated that the antibiotics used in poultry are not completely metabolized in body tissues that accumulate in meat [10] and are also excreted into the environment *via* poultry droppings [11].

When poultry droppings are used as manure in agriculture fields, these antimicrobials enter the soil ecosystem and significantly alter the soil communities [12]. Moreover, when consumed by humans, crops/vegetables cultivated in such fields transmit antimicrobial-resistant genes to them [8, 13]. This chapter attempted to highlight the hazards of using antimicrobial growth promoters in the poultry industry. It will also illuminate the health risk of antibiotic's use and its residues in poultry products concerning the environment and human health.

ANTIBIOTICS AS GROWTH PROMOTERS

The mechanism of action of antibiotic growth promoters in poultry is illustrated (Fig. **1**). Antibiotics are growth-promoting drugs or chemicals that stop the growth of bacteria in low, sub-therapy doses [14, 15]. In recent years, antibiotics as growth promoters have increased dramatically due to increasing consumer demand for fast livestock and animal feed products. Microbial infections can reduce the growth and/ or yield of farm animals raised for food purposes; thus, controlling these infections by adding antibiotics to animal feed has been effective [16]. Antibiotics as growth promoters have several beneficial effects, such as efficient feed digestion in the animal body to achieve better feed-conversion ratios, which allow an animal to grow into a strong and healthy living being. In addition, the use of antibiotics in the feed may also help in controlling zoonotic infectious organisms at an early stage. In addition to the benefits of antibiotics as growth promoters, this practice involves many ecological and ethical concerns [17 - 19]. For example, antibiotics may be used as growth promoters and may not be cost-effective [20].

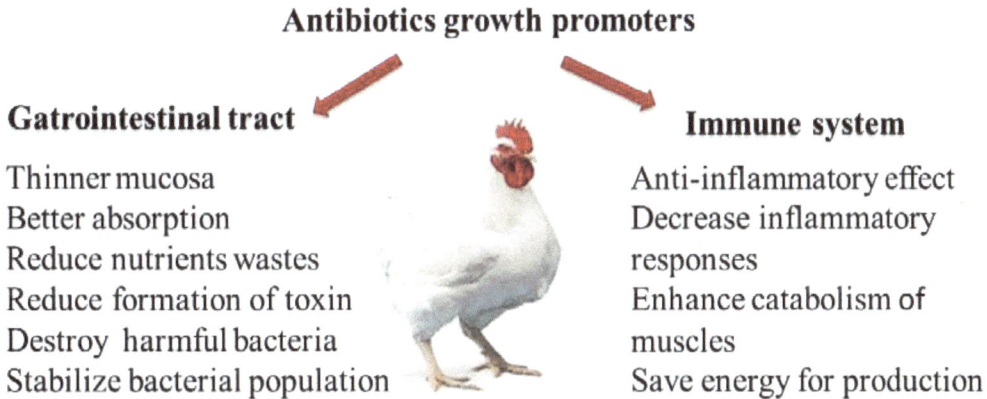

Antibiotics growth promoters

Gatrointestinal tract

Thinner mucosa
Better absorption
Reduce nutrients wastes
Reduce formation of toxin
Destroy harmful bacteria
Stabilize bacterial population

Immune system

Anti-inflammatory effect
Decrease inflammatory
responses
Enhance catabolism of
muscles
Save energy for production

Fig. (1). Beneficial uses of antibiotic growth promoters in poultry.

In the 1950s, antibiotics used in domestic animals were introduced to fulfill the rapidly increasing food demand. Currently, some countries use antibiotics as growth promoters without strict regulation [21], while other regions and countries, such as Europe and America, prohibit their use as agricultural or growth promoters. However, antibiotics at therapeutic doses are allowed for prophylactic and preventive purposes. In fact, the United States (US) and many European countries are the main users of antibiotics in domestic food animals [22], whereas their use is growing rapidly in some developing countries. Antibiotic use is expected to increase 67% by 2030, almost doubling in China, Brazil, India, South Africa, and Russia [23]. The livestock industry is the second-largest consumer of antibiotics after human health care. Although antibiotics are used therapeutically to treat infectious diseases and prevent the spread of disease in poultry, fish, and domestic food animals, forty percent of antimicrobial drugs are used as growth supporters in the livestock sector [17, 19]. Such intensive use raises concerns about using antibiotics in food and food animals. For example, intensive use of antibiotic growth promoters over a while results in increased selection pressure on the local bacterial populations, leading to resistance to these antibiotics [24, 25].

Antibiotics are chemotherapeutic agents used to treat bacterial diseases in humans and animals; however, many antibiotics produced every year are used for other purposes instead of therapeutic [26]. According to the CDC survey, the United States produces more than 50 million pounds of antibiotics each year, out of which 40% of the total antibiotics are used in the agriculture sector. It was estimated that livestock uses 24.6 million pounds of antibiotics each year for non-therapeutic treatment. About 11 million pounds are used for poultry, 9 million pounds for pigs and 3.7 million pounds for cattle [27].

These estimates did not include the therapeutic use of antibiotics to treat sick animals. Thus, most antibiotics in animal agriculture were used as growth promoters. In 2010, thirteen thousand tonnes of antibiotics were administered to animals; however, this was mainly utilized as a growth promoter [28]. In 1940, chlortetracycline antibiotics were first used to promote growth in small doses produced from *Streptomyces aureofaciens* [29]. This resulted in higher growth rates in the chickens fed with feed containing by-products. Since then, antibiotics have been widely used as a worldwide growth promoter in animal farming [30].

The underlying mechanism of the growth-stimulating effect is unknown. Still, antibiotics are believed to suppress the susceptible microbial population in the gut, which utilizes a significant amount of nutrients during fermentation in the gastrointestinal tract. It turns out that microbial yeast in the gut can waste about 6% of the pure energy of pig feed [31].

It was hypothesized that animals raised in unhealthy environments might carry latent infections that stimulate the immune system. Antibiotics can control such latent infections and reduce the production of cytokines in animals, resulting in a subsequent increase in muscle weight. The resulting cytokines lead to the release of certain catabolic hormones, resulting in muscle wasting by controlling the causative agent of the infection [32, 33]. Therefore, lost energy can be redirected to microbial population growth and better controlled and this is achieved by using antibiotics at low doses in the animal feed.

In addition, the effects of growth promoters have been mediated by their antibacterial effects in the following possible ways; (i) protecting essential micro and macronutrients against bacterial damage (ii) improving nutrient absorption by thinning the epithelial layer of the gut (iii) reduced toxic production by enterobacteria which decrease the subclinical gut infections [34].

There has been a serious concern regarding antimicrobial resistance in the food animal bacterial population due to antimicrobial usage as growth promoters. The rate of antimicrobial resistance directly proportional by the usage in the animals, resulting in an exchange of resistant bacteria between animals, their products and the environment [24]. A recent survey of seven European countries showed the direct relationship between antibacterial usage and *Escherichia coli* resistance in poultry, pigs and cattle [35]. On the other hand, the intensive use of antibiotics in human medicine to the developing antibiotic resistance cannot be excluded. For example, there is no link between a human and animal resistant strain in some cases. Currently, some countries prohibit antibiotics in food animals [36]. Most of the antibiotics currently used in food animals are the same as those used for the treatment of humans [24]. These antibiotics are used as anti-infective agents to treat many common pathogenic bacterial infections and other procedures such as major surgery, chemotherapy for tumors, organ transplants and premature babies [37].

It uses both natural and synthetic antibiotics to slow down or stop the growth of bacteria. Bacteria and fungus primarily synthesize antibiotics. Antibiotics are frequently used to treat and preclude human and animal infections. However, scientific verification endorses that the overuse of these composites amplifies the hazard of antibiotic resistance as growth enhancers [38]. The use of antibiotics, combined with firm biosecurity and cleanliness procedures, assists the poultry industry to thrive by counteracting the ill effects of many diseases occurring in birds. Broadly, antibiotics are classified according to chemical nature and mode of action. Bactericidal eradicate bacteria by barring cell wall syntheses, such as cephalosporins, carbapenems, vancomycin, and fluoroquinolones. However, the bacteriostatic slows down the growth patterns by inhibiting protein synthesis or weakening the phagocytic processes [12].

Various antibiotics such as tylosin, tetracycline and virginiamycin are widely used on animal farms in the U.S. Of these, two-thirds of tetracyclines are administered to animals, and 37% are used in the European Union [39 - 41]. European countries do not recommend antibiotics as growth promoters to monitor the European Surveillance of Veterinary Antimicrobial Consumption [42].

Numerous antibiotics have been used to promote the growth of poultry birds, which mainly act by controlling gastrointestinal infections and changes in the intestinal flora [43, 44]. The administration of virginiamycin antibiotic (90-100 ppm) in broiler as a growth enhancer was directly linked with an enriched frequency of different species of *Lactobacillus* genus, which depicted that virginiamycin changes the structure of the intestinal flora in chickens [45]. However, different results were observed when using tylosin antibiotics in feed [46, 47]. Tylosin reduces lactic acid bacteria production. The basic properties reduce bile hydrolase production and increase the presence of bile salts, but also promote phospholipid digestion and energy production and increase the weight of animals [48]. Two necessary antibiotics, virginiamycin and bacitracin, have been extensively used as growth promoters in poultry production. *S. virginiae* produces virginiamycin which inhibits protein synthesis by binding to the 50 S ribosomal subunits. When it's mixed with chicken feed, it decreases the mortality percentage due to infection caused by *Clostridium perfringens* [49].

LETHAL EFFECTS OF ANTIBIOTICS USED IN POULTRY PRODUCTION

As the world's population grows daily, the demand for chicken meat, eggs and their products increases, making poultry a critical food industry in the world [11]. Rapid development causes environmental problems such as soil and water pollution [50]. Every year, several antibiotics are used worldwide to prevent, eliminate and treat poultry and livestock diseases [51]. Antibiotics are used around 2 million tons annually worldwide [52]. The poultry droppings consist of important nutrients (nitrogen, phosphorus), pathogens, hormones, antibiotics, and toxic heavy metals [53]. This suggests a need to study the sources, behavior, fate, risks and control of pollutants from the environment [54, 55].

Many scientists have discovered the remnants of antibiotics and antibiotic-resistant bacteria in the soil and water. Antibiotic-resistant germs carry and exchange genes that survive antibiotics and eventually attack humans [56, 57]. Several groups of multiple antibiotic-resistant bacteria (MARBs) were detected in the soil and plant tissues with poultry manure grown in the soil. These bacteria can colonize deep tissues and pose a potential threat to human health [58].

Several studies have detected antibiotics and other medicines in soil and water [59]. Certain antibiotics in poultry have been used as prophylactic, meta-prophylactic, therapeutic, or dietary supplements or as a drug, particularly in developed countries [60]. These antibiotics can harm humans and the environment in two different ways either directly or indirectly; low levels of antibiotics in animal products can affect endocrine function, metabolism and human development [61]. The presence of antimicrobial residues in water and soil is hazardous for humans and other biotas. The remnants of medicine have a profound effect on the environment, so the world is now thinking of it. Environmental antibiotic residues may be associated with increased toxicity of antibiotic-resistant bacteria *via* activation of antibiotic resistance genes (ARGs) [11]. Therefore, antibiotic-resistant bacteria are not killed by using normal doses of antibiotics and could be lethal to human and animal health [6].

ANTIMICROBIAL RESIDUES IN POULTRY PRODUCTS

Detection of antibiotic residues in poultry products by different techniques is illustrated in Table **1**. Chicken meat is a substitute for beef and mutton in terms of cost and affordability; however, the irrational use of antimicrobial drugs and the lack of adequate biosecurity measures have decreased meat quality [62, 63]. Researchers have identified the presence of different antibiotic residues in the edible tissue of chicken [64 - 66]. For example, the concentration of quinolones, enrofloxacin, oxytetracycline and chloramphenicol was observed at 30.81 µg kg−1, 18.32 ng g−1, 88.217 ng g−1, and 89.33–223.05 µg kg−1, respectively in chicken meat [64, 67, 68]. Furthermore, a high quantity of enrofloxacin was observed in broiler meat [69]. Antimicrobial residues such as 22% enrofloxacin, 20% tetracycline and 34% ciprofloxacin are present in thigh muscles and 26% amoxicillin, 30% ciprofloxacin, 24% tetracycline residues in breast muscles of poultry broiler birds [70]. The existence of many antimicrobial residues such as quinolone, aminoglycoside, β-lactam, sulfonamide and tetracycline residues in breast and thigh muscles of chickens was also reported by Hakem *et al.* [71].

Table 1. Detection of antibiotic residues in poultry products by different techniques.

Literatures	Residue (ppb)	Sample	Antibiotic Found	Detection Method
104	30.81	Chicken	Quinolone	ELISA
105	10-10690	Liver-Poultry	Enrofloxacin	
106	12.64- 226.62	Chicken	Chloramphenicol	
107	42-360	Cured meat	Oxytetracycline	HPLC
108	176.3	Triceps muscle	Tetracycline	HPLC-FL
	405.3	Gluteal muscle		
	96.8	Diaphragm		
	672.40	Kidney		
	651.30	Liver		

(Table 1) cont.....

Literatures	Residue (ppb)	Sample	Antibiotic Found	Detection Method
104	-	Chicken	Enrofloxacin and Tetracycline	LC-MS
109	847.7	Poultry muscle	Doxycycline	LC-MS/MS

Antimicrobial residues have been reported in edible poultry tissues, including the kidney, heart, gizzard, and liver [72 - 75]. The high concentration of levamisole, ciprofloxacin, chloramphenicol, enrofloxacin and oxytetracycline residues was observed in the liver and kidney of broiler as compared to thigh muscles [69]. Chloramphenicol and enrofloxacin levels exceeded the maximum residue levels (MRL) in the liver and kidney of broiler meat [76]. The highest concentrations of amoxicillin (42%), tetracycline (48%), enrofloxacin (40%) and ciprofloxacin (44%) residues were also found in the liver of broiler and layer chicken. On the other hand, 30%, 24%, 34%, and 42% of each antibiotic remained in the kidney [70]. In addition, it was reported that high amounts of residues of nicarbazin were found in the liver of the boiler [77, 78].

Sipramycin antibiotic belongs to the macrolide group and is commonly used in poultry as a growth promoter. Its residues were present in thigh, gizzard, and liver tissues [79]. A higher concentration was reported in the liver (40%), followed by the gizzard (10%) [80]. Another study reported the presence of sulphaquinoxaline and amoxicillin residues in chicken liver and gizzard [81]. In addition, tilmicosin, nitrofuran, sulfamids, tetracycline, furaltadone, nifursol and chloramphenicol residues were also present in the liver and gizzard samples [73, 82 - 84].

Similarly, the accumulation of antimicrobial residues in the various components of the egg is of great concern [85]. After the antibiotic has been absorbed through the blood in the intestine, it enters the bird's ovaries and is deposited in the egg yolk and albumin [86]. Many factors, such as the chemical structure of antimicrobial agents, bird physiology, and the egg structure, were affected by the egg's accumulation and distribution residues [87]. The antimicrobial drug residues were reported to deposit faster with egg yolk and albumin protein [88]. High nitrofuran residues (268.25 ng kg-1) were detected in eggs that treated salmonellosis [74]. Similarly, the presence of residues of many drugs, including furaprol, streptomycin, sulfonamide, amprolium, and tylosin has been confirmed in other studies [62, 63, 89].

When a laying bird is given a high dose of amoxicillin, it is transmitted and accumulates in the yolk and egg white [86]. Cooking, and cooling (40°C) for 10 minutes did not help minimize the effects of these residues. However, boiling at 10 minutes and storing at 4°C-25°C did not help minimize amoxicillin residues' effect [86]. Likewise, when gentamicin was administered subcutaneously or intramuscularly, there was variability in the retention between protein and yolk even after withdrawal of the drug, in contrast to albumin, gentamicin residues accumulated in the egg yolk in significantly higher concentrations (90%) [88]. The egg white albumin contained a high concentration of sulfonamide and chlortetracycline residues compared to egg yolk [89, 90]. Therefore, it has been found that various egg compartments collect antimicrobial residues and eating these contaminated eggs can pose a serious health risk to the consumers [63].

PUBLIC HEALTH RISKS RELATED TO ANTIBIOTICS

In the USA, it has been reported that 80% of all sold antibiotics are used on healthy food-producing animals to promote growth performance. This unfair use results in major health concerns in the human population in the form of antimicrobial resistance in microorganisms [91]. Antibiotics commonly used in food-producing animals include aminoglycosides, tetracyclines, β-lactams, pleuromutilins, sulfonamides, lincosamides, and macrolides. Thus the residues of such antibiotics could be easily detected in animal products, *viz* ., milk and meat [92]. These residues could be in a parent compound or its metabolites [93]. Antimicrobial residues may produce several health consequences in humans, including immunopathological diseases, allergic disorders, cancer-causing effects, hepatotoxicity, bone marrow toxicity, nephropathy, mutagenicity, reproductive disorders, and damage to the damage beneficial bacteria existing in the gastrointestinal tract, mainly to teen-agers leading to gastritis [94, 95]. Antibiotic residues may produce chronic toxicity to several multicellular organisms and break down the gut's fatty acids made by supportive microbes [96]. In addition, resistant strains of antimicrobial microflora pose an indirect risk of antibacterial residues that can spread to humans and contaminate residues used as fertilizers for crops [97, 98]. The harmful effects of antibiotics and hormones are also observed on aquatic animals like [99] reported reduced testicular growth in rainbow trout and the small size of ovary and testis in zebrafish.

Unethical usage of antimicrobials results in the development of resistant bacteria in humans and animals, which cause infections that do not respond to usual antibiotic treatments. Such strong infection persists in the body for a long time and does not recover with common antibiotic options. The growing number of antibiotic-resistant microorganisms derived from animals associated with antibiotic residues, and their subsequent effects on human health, make important claims for the treatment of antimicrobial drugs [100].

Common bacterial organisms, including *Escherichia coli, Salmonella, Listeria monocytogenes* and *Staphylococcus aureus* were recognized as antimicrobial-resistant foodborne pathogens of animal-origin food [101]. Contaminated water used for washing foods and crop cultivation is recognized as the main source of antibiotic resistance and a potential threat to human health [102]. Poultry meat and eggs and pork and beef are considered vital factors for the transmission of antimicrobial resistance genes.

Research showed that resistance genes are transmitted from animals to humans through food, meat and waste [10, 21]. Antimicrobial-resistant genes in soil have been observed to enter the food chain and act as a possible cause of antimicrobial-resistant genes in human organisms *via* polluted crops and groundwater, thus affecting human and animal health. As soon as an antimicrobial-resistant gene enters an animal's body, it can remain in the body for a long time before it is measured. Therefore, antibiotic resistance should be considered a multifaceted public health challenge affecting health professionals in the medical, veterinary, medical and environmental fields. Thus, the "concept of health" was introduced [103].

CONCLUSION AND FUTURE DIRECTION

The use of antibiotics as a growth promoter is a feature of the modern poultry industry, but it is not widespread. Initially, all antibiotics could be used, but some did not promote growth and were very expensive. The use of antibiotics, especially those of medical importance, should be prohibited as growth promoters in livestock. The Food and Agriculture Organization has banned antibiotics as growth enhancers because of their harmful effects on human health. Various side effects, *e.g.*, allergic reaction, poisonous effects in vital organs of humans, inhibition of methane production and increased drug-resistant pathogens, have been observed during feeding as growth promoters. Only prudent use and effective regulation can balance the benefits and risks of using antimicrobials in animal production. However, international regulation of antibiotics in poultry birds will require long-term policies.

To best understand factors that promote the dissemination of resistance genes and to elucidate relationships between producer, environmental, and pathogenic bacteria, new and improved strategies for screening different microbial populations, sampling procedures and metagenomic libraries are prerequisites. Moreover, better algorithms and, therefore, bioinformatic approaches for determining relationships between resistance determinants of various environmental niches will be highly beneficial. Additional genome sequencing data will also help fill the knowledge gaps in intermediate stages and carriers for mobilization. It is expected that these bioinformatic tools will unify information on resistance genes and their products found in thousands of bacterial species isolated from the clinic or the environment as their associated mobile genetic elements and allow this information to be quickly mined by researchers in this field.

CONSENT FOR PUBLICATION

Not applicable.

CONFLICT OF INTEREST

The author declares no conflict of interest, financial or otherwise.

ACKNOWLEDGEMENTS

Declared none.

REFERENCES

[1]   Dibner JJ, Richards JD. Antibiotic growth promoters in agriculture: history and mode of action. Poult Sci 2005; 84(4): 634-43.
        [http://dx.doi.org/10.1093/ps/84.4.634] [PMID: 15844822]

[2]   Moore PR, Evenson A, Luckey TD, McCoy E, Elvehjem CA, Hart EB. Use of sulfasuxidine, streptothricin, and streptomycin in nutritional studies with the chick. J Biol Chem 1946; 165(2): 437-41.
        [http://dx.doi.org/10.1016/S0021-9258(17)41154-9] [PMID: 20276107]

[3]   Aarestrup FM. Effects of termination of AGP use on antimicrobial resistance in food animals. Working papers for the WHO international review panels evaluation. Geneva, Switzerland: World Health Organization 2003; pp. 6-11.

[4]   World Health Organization. Impacts of antimicrobial growth promoter termination in Denmark. Document WHO/CDS/CPE/ZFK/20031. Foulum, Denmark: WHO 2003; pp. 1-57.

[5]   Li Y, Zhu G, Ng WJ, Tan SK. A review on removing pharmaceutical contaminants from wastewater by constructed wetlands: Design, performance and mechanism. Sci Total Environ 2014; 468-469: 908-32.
        [http://dx.doi.org/10.1016/j.scitotenv.2013.09.018] [PMID: 24091118]

[6]   Zhang YJ, Hu HW, Gou M, Wang JT, Chen D, He JZ. Temporal succession of soil antibiotic resistance genes following application of swine, cattle and poultry manures spiked with or without antibiotics. Environ Pollut 2017; 231(Pt 2): 1621-32.
        [http://dx.doi.org/10.1016/j.envpol.2017.09.074] [PMID: 28964602]

[7]   Kamboh AA, Shoaib M, Abro SH, Khan MA, Malhi KK, Yu S. Antimicrobial resistance in Enterobacteriaceae isolated from liver of commercial broilers and backyard chickens. J Appl Poult Res 2018; 27(4): 627-34.
        [http://dx.doi.org/10.3382/japr/pfy045]

[8]   Xie WY, Shen Q, Zhao FJ. Antibiotics and antibiotic resistance from animal manures to soil: a review. Eur J Soil Sci 2018; 69(1): 181-95.
        [http://dx.doi.org/10.1111/ejss.12494]

[9]   Guetiya Wadoum RE, Zambou NF, Anyangwe FF, *et al.* Abusive use of antibiotics in poultry farming in Cameroon and the public health implications. Br Poult Sci 2016; 57(4): 483-93.
        [http://dx.doi.org/10.1080/00071668.2016.1180668] [PMID: 27113432]

[10]   Kumar K, Gupta SC, Chander Y, Singh AK. Antibiotic use in agriculture and its impact on the

terrestrial environment. Adv Agron 2005; 87: 1-54.
[http://dx.doi.org/10.1016/S0065-2113(05)87001-4]

[11]   Muhammad J, Khan S, Su JQ, *et al.* Antibiotics in poultry manure and their associated health issues: a systematic review. J Soils Sediments 2020; 20(1): 486-97.
[http://dx.doi.org/10.1007/s11368-019-02360-0]

[12]   Zhao M, Fang Y, Ma L, *et al.* Synthesis, characterization and *in vitro* antibacterial mechanism study of two Keggin-type polyoxometalates. J Inorg Biochem 2020; 210: 111131.
[http://dx.doi.org/10.1016/j.jinorgbio.2020.111131] [PMID: 32563103]

[13]   Luby E, Ibekwe AM, Zilles J, Pruden A. Molecular methods for assessment of antibiotic resistance in agricultural ecosystems: prospects and challenges. J Environ Qual 2016; 45(2): 441-53.
[http://dx.doi.org/10.2134/jeq2015.07.0367] [PMID: 27065390]

[14]   Cully M. Public health: The politics of antibiotics. Nature 2014; 509(7498): S16-7.
[http://dx.doi.org/10.1038/509S16a] [PMID: 24784425]

[15]   Cardinal KM, da Silva Pires PG, Ribeiro AML. Growth promoter in broiler and pig production. Pubvet 2020; 14: 1-11.
[http://dx.doi.org/10.31533/pubvet.v14n3a532.1-11]

[16]   Rushton J. Antimicrobial use in animals: how to assess the trade-offs. Zoonoses Public Health 2015; 62(s1) (Suppl. 1): 10-21.
[http://dx.doi.org/10.1111/zph.12193] [PMID: 25903492]

[17]   Littmann J, Buyx A, Cars O. Antibiotic resistance: An ethical challenge. Int J Antimicrob Agents 2015; 46(4): 359-61.
[http://dx.doi.org/10.1016/j.ijantimicag.2015.06.010] [PMID: 26242553]

[18]   Finley RL, Collignon P, Larsson DGJ, *et al.* The scourge of antibiotic resistance: the important role of the environment. Clin Infect Dis 2013; 57(5): 704-10.
[http://dx.doi.org/10.1093/cid/cit355] [PMID: 23723195]

[19]   Van TTH, Yidana Z, Smooker PM, Coloe PJ. Antibiotic Use in Food Animals in the World with Focus on Africa: Pluses and Minuses. J Glob Antimicrob Resist 2019; 170-7.

[20]   Durso LM, Cook KL. Impacts of antibiotic use in agriculture: what are the benefits and risks? Curr Opin Microbiol 2014; 19: 37-44.
[http://dx.doi.org/10.1016/j.mib.2014.05.019] [PMID: 24997398]

[21]   Founou LL, Founou RC, Essack SY. Antibiotic resistance in the food chain: a developing country-perspective. Front Microbiol 2016; 7: 1881.
[http://dx.doi.org/10.3389/fmicb.2016.01881] [PMID: 27933044]

[22]   Elliott KA, Kenny C, Madan JA. Global treaty to reduce antimicrobial use in livestock. Center for Global Development. Washington, DC 2017.

[23]   Van Boeckel TP, Brower C, Gilbert M, *et al.* Global trends in antimicrobial use in food animals. Proc Natl Acad Sci USA 2015; 112(18): 5649-54.
[http://dx.doi.org/10.1073/pnas.1503141112] [PMID: 25792457]

[24]   Boamah VE, Agyare C, Odoi H, Dalsgaard A. Practices and factors influencing the use of antibiotics in selected poultry farms in Ghana. J Antimicrob Agents 2: 120.

[25]   Lekshmi M, Ammini P, Kumar S, Varela MF. The food production environment and the development of antimicrobial resistance in human pathogens of animal origin. Microorganisms 2017; 5(1): 11.
[http://dx.doi.org/10.3390/microorganisms5010011] [PMID: 28335438]

[26]   Van TTH, Yidana Z, Smooker PM, Coloe PJ. Antibiotic use in food animals worldwide, with a focus on Africa: Pluses and minuses. J Glob Antimicrob Resist 2020; 20: 170-7.
[http://dx.doi.org/10.1016/j.jgar.2019.07.031] [PMID: 31401170]

[27]   Oliver SP, Murinda SE, Jayarao BM. Impact of antibiotic use in adult dairy cows on antimicrobial

resistance of veterinary and human pathogens: a comprehensive review. Foodborne Pathog Dis 2011; 8(3): 337-55.
[http://dx.doi.org/10.1089/fpd.2010.0730] [PMID: 21133795]

[28]   Spellberg B, Bartlett JG, Gilbert DN. The future of antibiotics and resistance. N Engl J Med 2013; 368(4): 299-302.
[http://dx.doi.org/10.1056/NEJMp1215093] [PMID: 23343059]

[29]   Jukes TH, Williams WL. Nutritional effects of antibiotics. Pharmacol Rev 1953; 5(4): 381-420.
[PMID: 13120335]

[30]   Chattopadhyay MK. Use of antibiotics as feed additives: a burning question. Front Microbiol 2014; 5
[http://dx.doi.org/10.3389/fmicb.2014.00334]

[31]   Jensen B. The impact of feed additives on the microbial ecology of the gut in young pigs. J Anim Feed Sci 1998; 7 (Suppl. 1): 45-64.
[http://dx.doi.org/10.22358/jafs/69955/1998]

[32]   Thomke S, Elwinger K. Growth promotants in feeding pigs and poultry. Ann Zootech 1998; 47(3): 153-67.
[http://dx.doi.org/10.1051/animres:19980301]

[33]   Ghosh S, Padalia J, Ngobeni R, *et al.* Targeting Parasite-Produced Macrophage Migration Inhibitory Factor as an Antivirulence Strategy With Antibiotic–Antibody Combination to Reduce Tissue Damage. J Infect Dis 2020; 221(7): 1185-93.
[http://dx.doi.org/10.1093/infdis/jiz579] [PMID: 31677380]

[34]   Butaye P, Devriese LA, Haesebrouck F. Antimicrobial growth promoters used in animal feed: effects of less well known antibiotics on gram-positive bacteria. Clin Microbiol Rev 2003; 16(2): 175-88.
[http://dx.doi.org/10.1128/CMR.16.2.175-188.2003] [PMID: 12692092]

[35]   Chantziaras I, Boyen F, Callens B, Dewulf J. Correlation between veterinary antimicrobial use and antimicrobial resistance in food-producing animals: a report on seven countries. J Antimicrob Chemother 2014; 69(3): 827-34.
[http://dx.doi.org/10.1093/jac/dkt443] [PMID: 24216767]

[36]   Maron DF, Smith TJS, Nachman KE. Restrictions on antimicrobial use in food animal production: an international regulatory and economic survey. Global Health 2013; 9(1): 48.
[http://dx.doi.org/10.1186/1744-8603-9-48] [PMID: 24131666]

[37]   Laxminarayan R, Duse A, Wattal C, *et al.* Antibiotic resistance—the need for global solutions. Lancet Infect Dis 2013; 13(12): 1057-98.
[http://dx.doi.org/10.1016/S1473-3099(13)70318-9] [PMID: 24252483]

[38]   Zarei-Baygi A, Harb M, Wang P, Stadler LB, Smith AL. Evaluating antibiotic resistance gene correlations with antibiotic exposure conditions in anaerobic membrane bioreactors. Environ Sci Technol 2019; 53(7): 3599-609.
[http://dx.doi.org/10.1021/acs.est.9b00798] [PMID: 30810034]

[39]   Diarra MS, Malouin F. Antibiotics in Canadian poultry productions and anticipated alternatives. Front Microbiol 2014; 5: 282.
[http://dx.doi.org/10.3389/fmicb.2014.00282] [PMID: 24987390]

[40]   Carvalho IT, Santos L. Antibiotics in the aquatic environments: A review of the European scenario. Environ Int 2016; 94: 736-57.
[http://dx.doi.org/10.1016/j.envint.2016.06.025] [PMID: 27425630]

[41]   Gonzalez Ronquillo M, Angeles Hernandez JC. Antibiotic and synthetic growth promoters in animal diets: Review of impact and analytical methods. Food Control 2017; 72: 255-67.
[http://dx.doi.org/10.1016/j.foodcont.2016.03.001]

[42]   ESVAC, Sales of veterinary antimicrobial agents in 30 european countries in 2015. Seventh Esvac Report 2017.

[43]   Torok VA, Allison GE, Percy NJ, Ophel-Keller K, Hughes RJ. Influence of antimicrobial feed additives on broiler commensal posthatch gut microbiota development and performance. Appl Environ Microbiol 2011; 77(10): 3380-90.
[http://dx.doi.org/10.1128/AEM.02300-10] [PMID: 21441326]

[44]   Mehdi Y, Létourneau-Montminy MP, Gaucher ML, *et al.* Use of antibiotics in broiler production: Global impacts and alternatives. Anim Nutr 2018; 4(2): 170-8.
[http://dx.doi.org/10.1016/j.aninu.2018.03.002] [PMID: 30140756]

[45]   Dumonceaux TJ, Hill JE, Hemmingsen SM, Van Kessel AG. Characterization of intestinal microbiota and response to dietary virginiamycin supplementation in the broiler chicken. Appl Environ Microbiol 2006; 72(4): 2815-23.
[http://dx.doi.org/10.1128/AEM.72.4.2815-2823.2006] [PMID: 16597987]

[46]   Danzeisen JL, Kim HB, Isaacson RE, Tu ZJ, Johnson TJ. Modulations of the chicken cecal microbiome and metagenome in response to anticoccidial and growth promoter treatment. PLoS One 2011; 6(11): e27949.
[http://dx.doi.org/10.1371/journal.pone.0027949] [PMID: 22114729]

[47]   Lee KW, Ho Hong Y, Lee SH, *et al.* Effects of anticoccidial and antibiotic growth promoter programs on broiler performance and immune status. Res Vet Sci 2012; 93(2): 721-8.
[http://dx.doi.org/10.1016/j.rvsc.2012.01.001] [PMID: 22301016]

[48]   Lin J, Hunkapiller AA, Layton AC, Chang YJ, Robbins KR. Response of intestinal microbiota to antibiotic growth promoters in chickens. Foodborne Pathog Dis 2013; 10(4): 331-7.
[http://dx.doi.org/10.1089/fpd.2012.1348] [PMID: 23461609]

[49]   Bai J, Qiao X, Ma Y, *et al.* Protection Efficacy of Oral Bait Probiotic Vaccine Constitutively Expressing Tetravalent Toxoids against Clostridium perfringens Exotoxins in Livestock (Rabbits). Vaccines (Basel) 2020; 8(1): 17.
[http://dx.doi.org/10.3390/vaccines8010017] [PMID: 31936328]

[50]   Erian I, Phillips C. Public understanding and attitudes towards meat chicken production and relations to consumption. Animals (Basel) 2017; 7(12): 20.
[http://dx.doi.org/10.3390/ani7030020] [PMID: 28282911]

[51]   Uslu MO, Jasim S, Arvai A, Bewtra J, Biswas N. A survey of occurrence and risk assessment of pharmaceutical substances in the Great Lakes Basin. Ozone Sci Eng 2013; 35(4): 249-62.
[http://dx.doi.org/10.1080/01919512.2013.793595]

[52]   Ahmed ST, Islam MM, Mun HS, Sim HJ, Kim YJ, Yang CJ. Effects ofBacillus amyloliquefaciens as a probiotic strain on growth performance, cecal microflora, and fecal noxious gas emissions of broiler chickens. Poult Sci 2014; 93(8): 1963-71.
[http://dx.doi.org/10.3382/ps.2013-03718] [PMID: 24902704]

[53]   Steinfeld H, Gerber P, Wassenaar TD, Castel V, De Haan C. Livestock's long shadow: environmental issues and options. Food & Agriculture Organization 2006.

[54]   Zhou M, Zhu B, Wang S, Zhu X, Vereecken H, Brüggemann N. Stimulation of N_2O emission by manure application to agricultural soils may largely offset carbon benefits: a global meta-analysis. Glob Change Biol 2017; 23(10): 4068-83.
[http://dx.doi.org/10.1111/gcb.13648] [PMID: 28142211]

[55]   Yin Y, Gu J, Wang X, *et al.* Effects of rhamnolipid and Tween-80 on cellulase activities and metabolic functions of the bacterial community during chicken manure composting. Bioresour Technol 2019; 288: 121507.
[http://dx.doi.org/10.1016/j.biortech.2019.121507] [PMID: 31128544]

[56]   Franklin AM, Aga DS, Cytryn E, *et al.* Antibiotics in Agroecosystems: Introduction to the Special Section. J Environ Qual 2016; 45(2): 377-93.
[http://dx.doi.org/10.2134/jeq2016.01.0023] [PMID: 27065385]

[57]   Kim JH, Kuppusamy S, Kim SY, Kim SC, Kim HT, Lee YB. Occurrence of sulfonamide class of antibiotics resistance in Korean paddy soils under long-term fertilization practices. J Soils Sediments 2017; 17(6): 1618-25.
[http://dx.doi.org/10.1007/s11368-016-1640-x]

[58]   Fang J, Shen Y, Qu D, Han J. Antimicrobial resistance profiles and characteristics of integrons in Escherichia coli strains isolated from a large-scale centralized swine slaughterhouse and its downstream markets in Zhejiang, China. Food Control 2019; 95: 215-22.
[http://dx.doi.org/10.1016/j.foodcont.2018.08.003]

[59]   Camacho-Muñoz D, Martín J, Santos JL, Aparicio I, Alonso E. Effectiveness of conventional and low-cost wastewater treatments in the removal of pharmaceutically active compounds. Water Air Soil Pollut 2012; 223(5): 2611-21.
[http://dx.doi.org/10.1007/s11270-011-1053-9]

[60]   Tasho RP, Cho JY. Veterinary antibiotics in animal waste, its distribution in soil and uptake by plants: A review. Sci Total Environ 2016; 563-564: 366-76.
[http://dx.doi.org/10.1016/j.scitotenv.2016.04.140] [PMID: 27139307]

[61]   Kümmerer K. Antibiotics in the aquatic environment – A review – Part II. Chemosphere 2009; 75(4): 435-41.
[http://dx.doi.org/10.1016/j.chemosphere.2008.12.006] [PMID: 19178931]

[62]   Mehtabuddin A, Ahmad T, Nadeem S, Tanveer Z, Arshad J. Sulfonamide Residues Determination in Commercial Poultry Meat and Eggs. J Anim Plant Sci 2012; 22: 473-8.

[63]   Mund MD, Khan UH, Tahir U, Mustafa BE, Fayyaz A. Antimicrobial drug residues in poultry products and implications on public health: A review. Int J Food Prop 2017; 20(7): 1433-46.
[http://dx.doi.org/10.1080/10942912.2016.1212874]

[64]   Hind EA, Adil SM, El-Rade SA. Screening of Antibiotic Residues in Poultry Liver, Kidney and Muscle in Khartoum State, Sudan. J Appl Indus Sci 2014; 2: 116-22.

[65]   Ali MR, Sikder MMH, Islam MS, Islam MS. Investigation of discriminate and indiscriminate use of doxycycline in broiler: an indoor research on antibiotic doxycycline residue study in edible poultry tissue. Asian J Med Biol Res 2020; 6(1): 1-7.
[http://dx.doi.org/10.3329/ajmbr.v6i1.46472]

[66]   Patrabansh S, Parajuli N, Jha VK. Rapid Detection of Tetracycline Residues in Chicken. Int J Appl Sci Biotechnol 2020; 8(1): 14-20.
[http://dx.doi.org/10.3126/ijasbt.v8i1.27201]

[67]   Er B, Onurdağ FK, Demirhan B, Özgacar SÖ, Öktem AB, Abbasoğlu U. Screening of quinolone antibiotic residues in chicken meat and beef sold in the markets of Ankara, Turkey. Poult Sci 2013; 92(8): 2212-5.
[http://dx.doi.org/10.3382/ps.2013-03072] [PMID: 23873571]

[68]   Ebrahimzadeh Attari V, Mesgari Abbasi M, Abedimanesh N, Ostadrahimi A, Gorbani A. Investigation of enrofloxacin and chloramphenicol residues in broiler chickens carcasses collected from local markets of tabriz, northwestern iran. Health Promot Perspect 2014; 4(2): 151-7.
[PMID: 25648045]

[69]   Amjad H, Iqbal J, Naeem M. Analysis of Some Residual Antibiotics in Muscle, Kidney and Liver Samples of Broiler Chicken by Various Methods. Proceedings of the Pakistan Academy of Sciences 2005; 42: 223-31.

[70]   Sattar S, Hassan MM, Islam SKMA, *et al.* Antibiotic residues in broiler and layer meat in Chittagong district of Bangladesh. Vet World 2014; 7(9): 738-43.
[http://dx.doi.org/10.14202/vetworld.2014.738-743]

[71]   Hakem A, Titouche Y, Houali K, *et al.* Screening of Antibiotics Residues in Poultry Meat by Microbiological Methods. Bull. University of Agricultural Sciences and Veterinary Medicine. Vet

Med 2013; 70: 77-82.

[72]   Aslam B, Kousar N, Javed I, *et al.* Determination of Enrofloxacin Residues in Commercial Broilers Using High Performance Liquid Chromatography. Int J Food Prop 2016; 19(11): 2463-70.
[http://dx.doi.org/10.1080/10942912.2015.1027922]

[73]   Morshdy AE, El-Atabany A, Hussein MA, Abdelaziz S, Ahmed AM. Antibiotic Residues in Poultry Edible Offal. 2nd Conference of Food Safety. 188-95.

[74]   Amiri HM, Tavakoli H, Hashemi G, *et al.* The Occurrence of Residues of Furazolidone Metabolite, 3-Amino-2-Oxazolidone, in Eggs Distributed in Mazandaran Province, Iran. Scimetr 2014; 2(4): e19353.

[75]   Yang Y, Qiu W, Li Y, Liu L. Antibiotic residues in poultry food in Fujian Province of China. Food Addit Contam Part B Surveill 2020; 13(3): 177-84.
[http://dx.doi.org/10.1080/19393210.2020.1751309] [PMID: 32308157]

[76]   Mehdizadeh S, Kazerani HR, Jamshidi A. Screening of Chloramphenicol Residues in Broiler Chickens Slaughtered in an Industrial Poultry Abattoir in Mashhad, Iran. Iran J Vet Sci Technol 2010; 2: 25-32.

[77]   Danaher M, Campbell K, O'Keeffe M, Capurro E, Kennedy G, Elliott CT. Survey of the anticoccidial feed additive nicarbazin (as dinitrocarbanilide residues) in poultry and eggs. Food Addit Contam Part A Chem Anal Control Expo Risk Assess 2008; 25(1): 32-40.
[http://dx.doi.org/10.1080/02652030701552956] [PMID: 17957540]

[78]   Silva JM, Azcárate FJ, Knobel G, Sosa JS, Carrizo DB, Boschetti CE. Multiple response optimization of a QuEChERS extraction and HPLC analysis of diclazuril, nicarbazin and lasalocid in chicken liver. Food Chem 2020; 311: 126014.
[http://dx.doi.org/10.1016/j.foodchem.2019.126014] [PMID: 31864181]

[79]   Awad AM, El-Shall NA, Khalil DS, *et al.* Incidence, Pathotyping, and Antibiotic Susceptibility of Avian Pathogenic *Escherichia coli* among Diseased Broiler Chicks. Pathogens 2020; 9(2): 114.
[http://dx.doi.org/10.3390/pathogens9020114] [PMID: 32059459]

[80]   Amroa FH, Hassan MA, Mahmoud AH. Spiramycin Residues in Chicken Meat and Giblets. Benha Vet Med J 2013; 24: 51-61.

[81]   Hassan MA, Heikal GI, Gad-Ghada A. Determination of Some Antibacterial Residues in Chicken Giblets. Benha Vet Med J 2014; 26: 213-8.

[82]   Shahid MA, Siddique M, Abubakar M, Arshed MJ. Asif M and Ahmad, A. Status of Oxytetracycline Residues in Chicken Meat in Rawalpindi/Islamabad Area of Pakistan. Asian J Polit Sci 2007; 1: 8-15.

[83]   Barbosa J, Freitas A, Moura S, Mourão JL, Noronha da Silveira MI, Ramos F. Detection, accumulation, distribution, and depletion of furaltadone and nifursol residues in poultry muscle, liver, and gizzard. J Agric Food Chem 2011; 59(22): 11927-34.
[http://dx.doi.org/10.1021/jf2029384] [PMID: 22011291]

[84]   Johnston J, Duverna R, Williams M, Kishore R, Yee C, Jarosh J. Investigating the Suitability of Semicarbazide as an Indicator of Preharvest Nitrofurazone Use in Raw Chicken. J Food Prot 2020; 83(8): 1368-73.
[http://dx.doi.org/10.4315/JFP-20-090] [PMID: 32294171]

[85]   Thi Huong-Anh N, Van Chinh D, Thi Tuyet-Hanh T. Antibiotic Residues in Chickens and Farmers' Knowledge of Their Use in Tay Ninh Province, Vietnam, in 2017. Asia Pac J Public Health 2020; 32(2-3): 126-32.
[http://dx.doi.org/10.1177/1010539520909942] [PMID: 32174126]

[86]   Khattab WO, Elderea HB, Salem EG, Gomaa NF. Transmission of Administered Amoxicillin Drug Residues from Laying Chicken to their Commercial Eggs. J Egypt Public Health Assoc 2010; 85(5-6): 297-316.
[PMID: 22054104]

[87]   Cornejo J, Pokrant E, Figueroa F, *et al.* Assessing Antibiotic Residues in Poultry Eggs from Backyard

Production Systems in Chile, First Approach to a Non-Addressed Issue in Farm Animals. Animals (Basel) 2020; 10(6): 1056.
[http://dx.doi.org/10.3390/ani10061056] [PMID: 32575363]

[88] Alm-El-Dein AK, Elhearon ER. Antibiotic Residue in Eggs of Laying Hens Following Injection with Gentamicin. New York Science Journal 2010; 3: 135-40.

[89] Alaboudi A, Basha EA, Musallam I. Chlortetracycline and sulfanilamide residues in table eggs: Prevalence, distribution between yolk and white and effect of refrigeration and heat treatment. Food Control 2013; 33(1): 281-6.
[http://dx.doi.org/10.1016/j.foodcont.2013.03.014]

[90] Abbas K, Maryam M, Mahmod E, Sohila S. Occurrence of Tetracycline Residue in Table Eggs and Genotoxic Effects of Raw and Heated Contaminated Egg Yolks on Hepatic Cells. Iran J Public Health 2020; 7: 1-8.

[91] Ghorbani B, Ghorbani M, Abedi M, Tayebi M. Effect of antibiotics overuse in animal food and its link with public health risk. Int J Sci Res Sci Technol 2016; 2: 46-50.

[92] De Briyne N, Atkinson J, Borriello SP, Pokludová L. Antibiotics used most commonly to treat animals in Europe. Vet Rec 2014; 175(13): 325.
[http://dx.doi.org/10.1136/vr.102462] [PMID: 24899065]

[93] Riviere JE, Sundlof SF. Chemical residue in tissues of food animals. In: Adams HR, Ed. Veterinary Pharmacology and Therapeutics. Iowa: Blackwell Publishing Professional 2001; Vol. 8: pp. 1166-74.

[94] Nisha A. Antibiotic residues – a global health hazard. Vet World 2008; 2(2): 375-7.
[http://dx.doi.org/10.5455/vetworld.2008.375-377]

[95] Nonga HE, Simon C, Karimuribo ED, Mdegela RH. Assessment of antimicrobial usage and residues in commercial chicken eggs from smallholder poultry keepers in Morogoro municipality, Tanzania. Zoonoses Public Health 2010; 57(5): 339-44.
[PMID: 19486498]

[96] Shukla SD, Budden KF, Neal R, Hansbro PM. Microbiome effects on immunity, health and disease in the lung. Clin Transl Immunology 2017; 6(3): e133.
[http://dx.doi.org/10.1038/cti.2017.6] [PMID: 28435675]

[97] Dubois M, Fluchard D, Sior E, Delahaut P. Identification and quantification of five macrolide antibiotics in several tissues, eggs and milk by liquid chromatography–electrospray tandem mass spectrometry. J Chromatogr, Biomed Appl 2001; 753(2): 189-202.
[http://dx.doi.org/10.1016/S0378-4347(00)00542-9] [PMID: 11334331]

[98] Tagg K. Human health, animal health, and ecosystems are interconnected. BMJ 2013; 347(aug14 4): f4979.
[http://dx.doi.org/10.1136/bmj.f4979] [PMID: 23945362]

[99] Van den Belt K, Wester PW, van der Ven LTM, Verheyen R, Witters H. Effects of ethynylestradiol on the reproductive physiology in zebrafish (*Danio rerio*): Time dependency and reversibility. Environ Toxicol Chem 2002; 21(4): 767-75.
[http://dx.doi.org/10.1002/etc.5620210412] [PMID: 11951950]

[100] Akbar A, Anal AK. Prevalence and antibiogram study of Salmonella and Staphylococcus aureus in poultry meat. Asian Pac J Trop Biomed 2013; 3(2): 163-8.
[http://dx.doi.org/10.1016/S2221-1691(13)60043-X] [PMID: 23593598]

[101] Tanih NF, Sekwadi E, Ndip RN, Bessong PO. Detection of pathogenic *Escherichia coli* and *Staphylococcus* aureus from cattle and pigs slaughtered in abattoirs in Vhembe District, South Africa. ScientificWorldJournal 2015; 2015: 1-8.
[http://dx.doi.org/10.1155/2015/195972] [PMID: 25811040]

[102] Berendonk TU, Manaia CM, Merlin C, *et al.* Tackling antibiotic resistance: the environmental framework. Nat Rev Microbiol 2015; 13(5): 310-7.

[http://dx.doi.org/10.1038/nrmicro3439] [PMID: 25817583]

[103] WHO (World Health Organization), Draft global action plan on antimicrobial resistance WHA 68.7. 2015.

[104] Kim DP, Degand G, Douny C, Pierret G, Delahaut P, Ton VD, *et al.* Preliminary evaluation of antimicrobial residue levels in marketed pork and chicken meat in the red river delta region of Vietnam. Food Public Health 2013; 3(6): 267-76.

[105] Sultan IA. Detection of Enrofloxacin in livers of livestock animals obtained from a slaughterhouse in Mosul City. J Vet Sci Technol 2014; 5(2): 1-3.
[http://dx.doi.org/10.4172/2157-7579.1000168]

[106] Yibar A, Cetinkaya F, Soyutemiz GE. ELISA screening and liquid chromatography-tandem mass spectrometry confirmation of chloramphenicol residues in chicken muscle, and the validation of a confirmatory method by liquid chromatography-tandem mass spectrometry. Poult Sci 2011; 90(11): 2619-26.
[http://dx.doi.org/10.3382/ps.2011-01564] [PMID: 22010249]

[107] Senyuva H, Ozden T, Sarica DY. High-performance liquid chromatographic determination of Oxytetracycline residue in cured meat products. Turk J Chem 2000; 24: 395-400.

[108] Mesgari Abbasi M, Nemati M, Babaei H, Ansarin M, Nourdadgar AOS. Solid-Phase extraction and simultaneous determination of tetracycline residues in edible cattle tissues using an HPLC-FL method. Iran J Pharm Res 2012; 11(3): 781-7.
[PMID: 24250505]

[109] Jank L, Martins MT, Arsand JB, *et al.* Liquid chromatography-tandem mass spectrometry multiclass method for 46 antibiotics residues in milk and meat: Development and validation. Food Anal Methods 2017; 10(7): 2152-64.
[http://dx.doi.org/10.1007/s12161-016-0755-4]

CHAPTER 2

Herb and Plant-derived Supplements in Poultry Nutrition

Muhammad Saeed[1], Muhammad S. Khan[1,*], Rizwana Sultan[1], Amjad I. Aqib[1], Muhammad A. Naseer[2], Iqra Muzammil[2], Mayada R. Farag[4], Mohamed E. Abd El-Hack[4,*], Alessandro Di Cerbo[5] and Mahmoud Alagawany[4,*]

[1] *Cholistan University of Veterinary and Animal Sciences, Bahawalpur, 63100, Pakistan*

[2] *Department of Clinical Medicine and Surgery, University of Agriculture, Faisalabad-38000, Pakistan*

[3] *Forensic Medicine and Toxicology Department, Veterinary Medicine Faculty, Zagazig University, Zagazig 44519, Egypt*

[4] *Department of Poultry, Faculty of Agriculture, Zagazig University, Zagazig 44511, Egypt*

[5] *School of Biosciences and Veterinary Medicine, University of Camerino, Matelica, Italy*

Abstract: Modern poultry industry faces the everlasting challenge of the growing demand for high-quality, low-priced food without compromising general hygiene, health, and welfare standards. To exploit optimal growth potential, antibiotic-supplemented feeds were implemented in the past decades. But later on, alternative strategies to trigger the productive characteristics of birds were proposed, including the use of phytochemicals. Phytobiotics are herbs and their derivatives, endowed with many beneficial effects. Herbs and their products enhance feed intake by mitigating intestinal damage, strengthening intestinal integrity, compensating nutritional needs for local and general immune response, reducing the concentration of pathogenic microflora, and preventing local inflammatory response. This form of feed manipulation recently gained interest in the poultry sector due to the lack of side effects, immune system modulation boosting, and stress tolerance. On the other hand, several types of research highlighted the potentially harmful effects of some herbs and their metabolites. This raised concerns among consumers about their safety and implications as feed supplements or medicines. This chapter will provide insights into phytobiotics, their role in immunity and growth, and the possible risks of herbal supplemented feeds in the poultry sector.

Keywords: Feed additives, Growth, Health, Herbs, Immunity, Poultry.

* **Corresponding authors Muhammad S. Khan, Mohamed E. Abd El-Hack & Mahmoud Alagawany:** Cholistan University of Veterinary and Animal Sciences, Bahawalpur, 63100, Pakistan & Department of Poultry, Faculty of Agriculture, Zagazig University, Zagazig 44511, Egypt; E-mails: drsajjad2@yahoo.com, dr.mohamed.e.abdalhaq@gmail.com & mmalagwany@zu.edu.eg

INTRODUCTION

The use of antibiotics as growth promoters in animal feed leads to the inactivation of the immune system and poor antioxidant defense in poultry [3]. So, there is a need for alternative feed additives which can support the animal's growth without any side effects, *e.g.*, the antibiotic resistance of pathogenic microorganisms. For this purpose, the effects of probiotics, prebiotics, acidifiers, enzymes, and herbs on gut microbes have been studied in animals. Current antibiotic growth promoters (AGP) banned in poultry feeds increased the demand for numerous natural substances such as herbal medicines, and a new array of feed supplements for animal and poultry birds, which are useful due to their antimicrobial, antioxidant, and antifungal properties as well as immunomodulatory and anticoccidial effects [1].

Several medicinal herbs such as *Aloe vera*, Fenugreek, Ashwagandha, *Moringa oleifera*, Cinnamon, Tulsi, Garlic, and Pepper can be used as natural feed extracts in poultry. However, the types of effects produced by herbal preparations have not yet been determined. Generally, they are influenced by the concentration of active ingredients and the presence of supplementary substances and linkages with other active constituents of feed [2]. Herbs and herbal extracts have been used in poultry production for years to promote digestion [3, 4]. The multi-beneficial potential of bioactive components of green tea in poultry has also been assessed [5, 6]. These are natural substances, they are considered cost-effective, safe, and environmental friendly without any side effects. So, they are added to feed to improve birds' performance, feed utilization, health maintenance, and environmental stress-related effects. Several studies have reported the use of herbal derivatives in the diets of farm animals and poultry birds to improve the growth rate, gut integrity, nutrients adsorption, antioxidant activity and immunity and decrease diarrhea occurrence [4, 7]. As a result, these products are known as good alternatives to antibiotics to reduce their residues in milk, meat, and eggs [8]. Sometimes biological substances show different results in *in vivo* studies due to animal species, production stage, environmental condition, and features of the plant material used; hence, exhibiting the best results in young birds [9]. Among natural substances, essential oils showed great potential in poultry depending on the composition, type, origin, inclusion level, and the environmental condition of the trial [10 - 13]. For instance, cinnamon and a mixture of thymol and cinnamon have been reported to enhance the performance of broilers [10].

In contrast, cinnamon extract and oil enhanced bile production, reduced toxins, restored electrolytes, and improved digestion [13]. Moreover, curcuma alone or combined with *capsicum* increased resistance against intestinal disturbances such as coccidiosis and necrotic enteritis [11]. Besides essential oils, herbs also improved several functions in bird's body system, acting as sialogogues and allowing a more comfortable and easier swallowing. Extracts from *Salvia officinalis, Rosmarinus officinalis,* and *Thymus vulgaris* and the blend of carvacrol, capsaicin, and cinnamaldehyde enhanced feed digestibility in broilers [14]. At the same time, the main components of *Aloe vera,* including anthraquinones, saccharides, enzymes, vitamins, and low molecular weight compounds [17] are responsible for its immunomodulatory, anti-inflammatory, antiviral, wound healing, antitumor, antifungal, antidiabetic, and antioxidant effects [6, 15, 16, 18, 19].

The current chapter will focus on using herbs and phytochemicals in poultry nutrition as growth promoters and alternative sources to reduce the risk of antibiotic residues in poultry products.

PHYTOCHEMICALS

Phytochemicals, also known as phytobiotics or phytogenics, are natural derivatives of plants whose main bioactive components are polyphenols [11]. They can be added to animal feed to increase production and defend against infections, pests, stress, and physical damage [11]. Phytochemicals exist in solid, dried, and crushed forms or can be used as extracts categorized as oleoresins and essential oils based on the procedure used to obtain active ingredients [12].

However, phytochemicals also include other compounds like flavonoids, alkaloids, tannins, flavanones, and cyanogenic glycosides. The characteristics of flavonoids are broadly studied in animal experiments related to health status, protection against diseases, and food preservation strategies [20]. Flavonoids possess antioxidant, anti-inflammatory, and immunomodulatory properties [22]. In fact, they can increase mucosal and cellular immunity making them suitable to be added to feed to improve the health and immunity of broiler chickens. Phytochemicals exhibiting antioxidant properties have been documented as natural alternatives to artificial feed additives [21]. Moreover, some essential oils like thymol, carvacrol, cinnamaldehyde, and eugenol have also been used separately or combined with increasing animal production [11].

Beyond antioxidant properties, phytochemicals have also shown antimicrobial properties, so they can be added to feed to stabilize the intestinal microbes and decrease toxic microbial metabolites in the intestine. For instance, phytochemicals of *Allium hookeri* have been reported to improve the function of the gut barrier by improving gut proteins expression in the mucosa of young broilers [23]. *Moringa oleifera* leaves have strong prebiotic and antioxidant phytochemicals such as chlorogenic and caffeic acids [24]. Ginger and garlic can be the best substitute for artificial growth-promoting supplements such as antibiotics [25]. Ginger is used as medicine or spice. Its extract can control and detoxify the free radical species and the peroxidation of lipids [26]. Garlic, spice, and native medicine have antiviral, antibacterial, antiparasitic, antifungal, anticholesterolemic, antioxidant, anti-cancerous, and vasodilation features [27].

On the other hand, it is worth noting that stress can impair the immunity of broilers due to the production of reactive oxygen species and free radicals at the cell level are responsible for the weakening of immunity, enhancing liability for infection, and lowering the production [28, 29]. Kamboh *et al.* (2013) studied the ability of dietary antioxidants to improve the harmful effects of stress [30]. For this purpose, genistein and hesperidin have been reported to normalize birds' mucosal and cellular immunity [31].

TYPES OF PHYTOCHEMICAL ANTIBIOTIC ALTERNATIVES IN POULTRY

Health-promoting activities of phytochemicals are mainly due to their prominent ability to enhance host body defense against pathogenic invasion [10]. Some medical herbs, such as dandelion, safflower, and mustard, prevent the growth of tumor cells, activate the innate immune response, and produce antioxidant entities in poultry [32]. Previously, intestinal diseases such as necrotic enteritis and coccidiosis were controlled by in-feed antibiotics, but now there is a need for alternative approaches. Thus, phytochemicals were selected to treat avian diseases [10].

Propyl thiosulfinate oxide and propyl thiosulfinate are garlic metabolites used in poultry. When garlic metabolites were supplemented at 10mg/kg in the feed of broilers, they resulted in an improvement in weight gain and serum antibody titers were increased against profilin, an immunogenic protein of *Eimeria*. This occurs due to the alteration of many genes linked with innate immunity, such as TLR3, TLR5, and NF-κB. At the same time, in uninfected chickens, it enhances the transcripts encoding of IL-4, IFN-γ, and the release of antioxidant enzyme, paraoxonase 2, but reduces transcripts encoding for peroxiredoxin-6 [33].

SYNERGISTIC EFFECT OF PHYTOCHEMICALS IN POULTRY NUTRITION

Various phytochemicals showed synergistic effects in combined form. Firstly, a commercial mixture of phytonutrients (*Capsicum* oleoresin, carvacrol, and cinnamaldehyde) was used to improve innate immunity and to decrease the negative impacts of enteric pathogens in broilers [34]. But later on, this commercial blend was used in various trials to enhance the growth rate and feed efficiency of broilers [12]. Dietary supplementation of the mixture of *Lentinus edodes, Curcuma longa,* and *Capsicum annuum* enhanced weight gain and serum antibody titers against profiling pathogens compared to the control diet or a dietary supplement only on *Capsicum annuum* and *Lentinus edodes* [10].

Similarly, the carvacrol, cinnamaldehyde and *Capsicum* oleoresin increased the levels of transcripts for IFN-γ, IL-6, IL-1β, and IL-15 in the gut lymphocytes of broilers as compared to the birds reared with the *Capsicum/Lentinus* or *Curcuma* diet [33]. Moreover, the same phytochemicals also resulted helpful in stimulating an immune response against experimentally-induced *E. tenella* infection following immunization with profilin [10]. Other phytochemicals prevent innate immune response by targeting pathogen pattern recognition receptors. *Clostridium*-related poultry disease like necrotic enteritis causes significant financial losses worldwide. It has been reported that dietary phytonutrients like *Capsicum* and *Curcuma longa* oleoresins help in increasing body weight and decreasing gut lesion scores in necrotic infected birds as compared to infected birds feeding diet without supplements [35].

HERBS AS GROWTH AND HEALTH PROMOTORS

Due to global concerns over increasing bacterial resistance, antibiotic-resistant pathogenic strains (MRSA, VRSA) and antibiotic residues in the products, there is a dire need to develop alternative strategies for the replacement of antibiotics in food-producing animals, particularly in the poultry sector, to produce a quality protein with minimal or no allopathic products [36 - 38]. Natural growth promoters such as herbal plants and their derivatives have been reported as effective alternatives to antibiotics in food animals. Phytobiotics or herbal feed supplements are gaining research attention and popularity for their gut health and immunity [39]. This form of feed manipulation recently gained interest in the poultry sector with increasing scientific publications since the ban on antibiotic inclusions in poultry feed by regulatory authorities in 1999 [40]. The difference in outcomes of phytobiotics is the consequence of multiple factors like time of harvest, additive preparation method, part of the plant used, extract harvest method, and compatibility with other food constituents [41].

Herbs and their derivatives enhance feed intake by a wide variety of modes of action such as mitigating intestinal damage, strengthening intestinal integrity, compensating nutritional needs for local and general immune response, reducing the concentration of pathogenic microflora, and preventing local inflammatory response. Oppositely, reduction in feed intake by birds can also be achieved *via* the supplementation of synthetic enzymes or plant originated essential oils for the improvement of dietary energy value and to avoid feed refusals in birds due to organoleptic issues associated with medium-chain fatty acids and organic acids supplementation at higher inclusion density, proving to determine general impacts in feed [42]. Improvement in villus height was evident in the birds reared with herbal supplemental diets. However, the depth of crypts and villus to crypt ratio was found to be the same in the control group [37]. The basic way of the beneficial impact of herbal growth promotors can be mainly ascribed to the stabilization, sustainment, and improvement of gastrointestinal integrity and microbiota control by eliminating harmful pathogenic strains [40]. This phenomenon in poultry birds becomes evident by the active improvement of feed intake and the enhanced digestive enzymatic secretions, immune boost up, antipathogenic, antiviral, coccidiostat, antiparasitic, anti-inflammatory, or antioxidant activities [43, 44]. Many plant secondary metabolites, including glucosinolates, isoprene derivatives, and flavonoids, can act as antibiotics and antioxidant entities *in vivo*. All the increase in the metabolic profile of birds leads to improved productive potential and better feed efficiency [37, 44, 45].

The inclusion of herbs and their derivatives in the broiler feed improves the production profile same as antibiotics [37]. Multiple studies showed the beneficial effects of herbal supplements on growth and productive performance in poultry. Supplementation of 1% *Allium hookeri* roots in broiler feed showed improved weight gain at day 14 when compared to the control group (p < 0.05). Up-regulated transcript levels of catalase (CAT), aflatoxin B1 aldehyde reductase (AFAR), and heme oxygenase 1 (HMOX1) were also noted in the jejunum of birds fed on an herbal supplemented diet [46]. The chickens at experimental herbal supplemented diets gained significantly higher final body weight (p < 0.05) compared to control and other treatment groups to conclude that supplementation of hot red pepper 0.5 to 1 g per 100 g feed was able to improve birds' performance [40]. There was increased body weight, improved feed conversion ratio (FCR), higher protein efficiency index (PEI), percent livability, and increased gross profit in birds offered 2% herbal growth promoters [47]. All of the herbal additives can decrease aflatoxin B1's negative impact on chicken's performance, blood indices, and immunity [48]. *In vitro* studies also revealed inhibitory effects of *Selaginella* involved and *Selaginella inaequalifolia* extracts against pathogenic strains of *E. coli* (8 to 13 mm) and *Pseudomonas* (6.5 to 13 mm), exploiting the antibacterial potential of the herbs [49].

Herbal oils and extracts can be incorporated into bird diets as a replacement to antibiotics without compromising the metabolizable energy of diets and the performance or integrity of intestinal mucosa [37]. The egg quality parameters (fatty acids profile, yolk color) are highly affected by herbs and herb product supplementation. In contrast, the sensory properties of an egg are in the least alteration to depict herbs intake in organic layers [39, 50]. Yolk fatty acid profile is highly dependent on forage parts of herb plant and alteration towards the presence of relatively higher polyunsaturated fatty acids, particularly n-3 fatty acids, is evident, which decreases the ratio of n-6/n-3 from 11 to 19 in the control group eggs. Different carotenoids of herbs are reflected primarily in the yolk carotenoids [50]. Shell strength, laying rate, egg mass, plasma glutathione peroxidase, luteinizing hormone, and superoxide dismutase levels were reported to be higher in birds reared on a supplemental herbal diet than in the controls (P < 0.05) [51].

Several herbs and herbal products seek the poultry industry's inline attention as thermoregulatory entities in the tropics. Additionally, the immunogenic boost-up, improved product performance and antioxidant nature of these are of interest to researchers *via* enhancement of immune response and stress tolerance and popularizes usage of herbs amongst poultry producers [52], as shown in Fig. (1).

Fig. (1). Importance of various plant-derived supplements in poultry nutrition (Adopted from Abd El-Hack *et al.*, 2020).

Aloe vera leaves contain CARN 750, which stimulates cytokine release and lymphocytes activation. *Angelica Gigas* roots contain angelan, which modulates cytokines release. *Astragalus membranaceus* roots supplementation increases macrophage count. *Ganoderma lucidum* extract boosts up lymphocyte activation. *Panax ginseng* roots contain panaxadiol, panaxatriol, and ginsan which stimulate interferon-gamma, interleukin – 1, interleukin – 12, tumor necrosis factor-alpha, and lymphocytes activity. *Panax ginseng* leaves contain rhamnogalacturonan-II, which stimulates IL-6 activity and enhances macrophage action. *Scutellaria baicalensis* possesses wogonin, which activates the iNOS mechanism by stimulating TNF-alpha [41, 53]. So, it is suggested to duly implement herbal feed additives in poultry feed as a natural growth promoter.

HEALTH RISKS RELATED TO HERBS UTILIZATION IN POULTRY FEED

Herbs and their derivatives are used and trusted for their health benefits and very minute side effects from ancient times [54, 55]. A general belief is that these are natural curates and always safe to implicate. Most of the experiments also support this belief by indicating clinical safety and better histopathological endurance by the organs or tissues of exposed animals [56]. Genotoxic analysis (chromosome aberration, bacterial reverse mutation test, and micronucleus test) indicated that turmeric and its polysaccharide extract are genotoxically safe and exhibit a tolerable dose of more than 5 g/kg body weight in animal models [57]. The same results were noted for *Ocimum sanctum L.* (Lamiaceae) [58] and *Osmanthus fragrans var. thunbergii* (*O. fragrans*) [59] when tested for genotoxicity.

But research also revealed that some herbs and their metabolites might not be safe [60, 61]. This indicates concerns at the consumer end about their safety and implications as feed supplements or medicine. Toxicity testing evaluates the associated risks and possible harmful effects of herbs inclusion as feed additives. Herbal mutagenicity and toxicity are often assessed in model animals by genotoxicity evaluation tests (bacterial reverse mutation, chromosomal aberration, micronucleus, and acute oral toxicity) [60]. Extensive experience has been collected in recent decades from these testing protocols about herbal plants and their derivatives by feeding model animals (rats, mice, guinea pigs, rabbits) for longer durations, and parameters like feed consumption, body weight, blood chemistry, histopathology, organ weights have been evaluated. Some cases have shown herbal toxicity and mutagenic potential [56]. Abdominal pain and diarrhea can result from ingesting aloe vera [62]. Loss of consciousness, stroke, or death can be seen in the case of bitter orange ingestion [63]. Comfrey can cause diarrhea or liver failure and cancer of the liver when mixed with poultry feed in higher doses [64]. Eucalyptus leaves can harbor Aspergillosis spores (a dangerous fungus that can cause death in various types of birds, including chickens and ducks). Also, full-strength eucalyptus oil is toxic to humans and animals [65]. Foxglove, also known as digitalis, can cause heart failure in animals and humans. An herb called "henbane" can be potentially risky to use in chickens as it causes heart problems, coma, and even death [65, 66]. Horse nettle affects the central nervous system. Monkshood (also called wolfsbane) can cause heart palpitations, respiratory distress, abdominal pain, nausea, vomiting, or even death in birds [67]. Pennyroyal, a member of the mint family, is toxic to birds, resulting in liver failure and eventual death [64, 65]. Tansy can cause diarrhea, liver disease, blindness, the inability to swallow, and even death. Tea tree essential oils cause vomiting, paralysis, seizures, unconsciousness, and even coma [67]. Wormwood (antiparasitic supplement) affects the nervous system and can cause abdominal pain, convulsions, and seizures if ingested in large amounts [62, 68]. Herbal essential oils are extremely concentrated and powerful [13]. These can be easily misused or given in too large amounts in birds, resulting in gut issues and respiratory distress. In worse cases, death may occur [69]. Highly toxic cyanogenic glycosides are present in apricot leaves and pits, triggering seizures, breathing stress, and low blood pressure. Azalea plant toxicity can cause weakness, digestive upset, loss of coordination, and cardiac damage. Unprocessed beans mixture contain hemagglutinin, a highly toxic entity to poultry birds [62, 64, 67, 68]. So, prolonged genotoxicity and oral toxicity are indicated to declare if a herbal product can be considered safe in poultry feed.

CONCLUSION

With the growing concern of antibiotic residues, antimicrobial resistance, and the emergence of resistant pathogenic strains globally, the poultry sector was forced to withdraw from the diffused practice of antibiotic abuse. Therefore, alternative strategies were proposed to boost the productive characteristics of birds by using phytobiotics as antibiotics alternative. Although some herbs are considered safe for animal health, some others and their metabolites have minute side effects. But overall, herbs are used as the best alternative to antibiotics in feed to get good health and maximize poultry's performance.

CONSENT FOR PUBLICATION

Not applicable.

CONFLICT OF INTEREST

The author declares no conflict of interest, financial or otherwise.

ACKNOWLEDGEMENTS

Declared none.

REFERENCES

[1]   Madhupriya V, Shamsudeen P, Manohar GR, Senthilkumar S. Phyto feed additives in poultry nutrition – A review. Int J Sci Environ Technol 2018; 7(3): 815-22.

[2]   Hashemi SR, Davoodi H. Phytogenies as new class of feed additive in poultry industry. J Anim Vet Adv 2010; 9(17): 2295-304.
[http://dx.doi.org/10.3923/javaa.2010.2295.2304]

[3]   Saeed M, *et al.* Chicory (cichorium intybus) herb: Chemical composition, pharmacology, nutritional and healthical applications. Int J Pharmacol 2017; 13(4): 351-60.
[http://dx.doi.org/10.3923/ijp.2017.351.360]

[4]   Saeed M, Naveed M, BiBi J, Ali Kamboh A, Phil L, Chao S. Potential nutraceutical and food additive properties and risks of coffee: a comprehensive overview. Crit Rev Food Sci Nutr 2019; 59(20): 3293-319.
[http://dx.doi.org/10.1080/10408398.2018.1489368] [PMID: 30614268]

[5]   Saeed M, Naveed M, Arif M, *et al.* Green tea (*Camellia sinensis*) and l-theanine: Medicinal values and beneficial applications in humans—A comprehensive review. Biomed Pharmacother 2017; 95: 1260-75.
[http://dx.doi.org/10.1016/j.biopha.2017.09.024] [PMID: 28938517]

[6]   Saeed M, Abd El-Hac ME, Alagawany M, *et al.* Phytochemistry, modes of action and beneficial health applications of green tea (*Camellia sinensis*) in humans and animals. Int J Pharmacol 2017; 13(7): 698-708.
[http://dx.doi.org/10.3923/ijp.2017.698.708]

[7]   Saeed M, Abbas G, Alagawany M, *et al.* Heat stress management in poultry farms: A comprehensive overview. J Therm Biol 2019; 84: 414-25.
[http://dx.doi.org/10.1016/j.jtherbio.2019.07.025] [PMID: 31466781]

[8]   Hernández F, Madrid J, García V, Orengo J, Megías MD. Influence of two plant extracts on broilers performance, digestibility, and digestive organ size. Poult Sci 2004; 83(2): 169-74.
[http://dx.doi.org/10.1093/ps/83.2.169] [PMID: 14979566]

[9]   Zdunczyk Z, Gruzauskas R, Juskiewicz J, *et al.* Growth performance, gastrointestinal tract responses, and meat characteristics of broiler chickens fed a diet containing the natural alkaloid sanguinarine from Macleaya cordata. J Appl Poult Res 2010; 19(4): 393-400.
[http://dx.doi.org/10.3382/japr.2009-00114]

[10]   Lee SH, Lillehoj HS, Jang SI, Kim DK, Ionescu C, Bravo D. Effect of dietary Curcuma, Capsicum, and Lentinus, on enhancing local immunity against Eimeria acervulina infection. J Poult Sci 2010; 47(1): 89-95.
[http://dx.doi.org/10.2141/jpsa.009025]

[11]   Gadde U, Kim WH, Oh ST, Lillehoj HS. Alternatives to antibiotics for maximizing growth performance and feed efficiency in poultry: a review. Anim Health Res Rev 2017; 18(1): 26-45.
[http://dx.doi.org/10.1017/S1466252316000207] [PMID: 28485263]

[12]   Bravo D, Pirgozliev V, Rose SP. A mixture of carvacrol, cinnamaldehyde, and capsicum oleoresin improves energy utilization and growth performance of broiler chickens fed maize-based diet. J Anim Sci 2014; 92(4): 1531-6.
[http://dx.doi.org/10.2527/jas.2013-6244] [PMID: 24496847]

[13]   Saeed M, Kamboh AA, Syed SF, *et al.* Phytochemistry and beneficial impacts of cinnamon (*Cinnamomum zeylanicum*) as a dietary supplement in poultry diets. Worlds Poult Sci J 2018; 74(2): 331-46.
[http://dx.doi.org/10.1017/S0043933918000235]

[14]   Guardia S. Effets de phytobiotiques sur les performances de croissance et l'équilibre du microbiote digestif du poulet de chair. 2012; p. 472.

[15]   Iser M, Martínez Y, Ni H, *et al.* The Effects of *Agave fourcroydes* Powder as a Dietary Supplement on Growth Performance, Gut Morphology, Concentration of IgG, and Hematology Parameters in Broiler Rabbits. BioMed Res Int 2016; 2016: 1-7.

[http://dx.doi.org/10.1155/2016/3414319] [PMID: 27777945]

[16]   Aguilar YM, Yero OM, Navarro MIV, Hurtado CAB, López JAC, Mejía LBG. Effect of squash seed meal (Cucurbita moschata) on broiler performance, sensory meat quality, and blood lipid profile. Rev Bras Cienc Avic 2011; 13(4): 219-26.
[http://dx.doi.org/10.1590/S1516-635X2011000400001]

[17]   Aguilar YM, *et al.* Effect of dietary supplementation with Anacardium occidentale on growth performance and immune and visceral organ weights in replacement laying pullets. J Food Agric Environ 2013; 11(3–4): 1352-7.

[18]   Saeed M, Babazadeh D, Naveed M, *et al. In ovo* delivery of various biological supplements, vaccines and drugs in poultry: current knowledge. J Sci Food Agric 2019; 99(8): 3727-39.
[http://dx.doi.org/10.1002/jsfa.9593] [PMID: 30637739]

[19]   Saeed M, Naveed M, Leskovec J, *et al.* Using Guduchi (Tinospora cordifolia) as an eco-friendly feed supplement in human and poultry nutrition. Poult Sci 2020; 99(2): 801-11.
[http://dx.doi.org/10.1016/j.psj.2019.10.051] [PMID: 32029162]

[20]   Acamovic T, Brooker JD. Biochemistry of plant secondary metabolites and their effects in animals. Proc Nutr Soc 2005; 64(3): 403-12.
[http://dx.doi.org/10.1079/PNS2005449] [PMID: 16048675]

[21]   Kamboh AA. Flavonoids: Health Promoting Phytochemicals for Animal Production-a Review. Journal of Animal Health and Production 2015; 3(1): 6-13.
[http://dx.doi.org/10.14737/journal.jahp/2015/3.1.6.13]

[22]   Glen F. Rall, HHS Public Access. Physiol Behav 2017; 176(1): 139-48.
[http://dx.doi.org/10.1016/j.physbeh.2017.03.040]

[23]   Kim DK, Lillehoj HS, Lee SH, *et al.* Immune effects of dietary anethole on Eimeria acervulina infection. Poult Sci 2013; 92(10): 2625-34.
[http://dx.doi.org/10.3382/ps.2013-03092] [PMID: 24046409]

[24]   Teixeira EMB, Carvalho MRB, Neves VA, Silva MA, Arantes-Pereira L. Chemical characteristics and fractionation of proteins from Moringa oleifera Lam. leaves. Food Chem 2014; 147: 51-4.
[http://dx.doi.org/10.1016/j.foodchem.2013.09.135] [PMID: 24206684]

[25]   Demir E, Sarica Ş, Özcan MA, Suiçmez M. The use of natural feed additives as alternative to an antibiotic growth promoter in broiler diets Verwendung von natürlichen Futterzusätzen im Broilerfutter   als   Alternative   zu   Leistungs-   förderern   mit   Antibiotika-Charakter 2005.https://www.european-poultry-science.com/artikel.dll/m-3-28mk_NDIxNjA4Nw.PDF?UID=97FE7C45A7FD385446ACBD2B3AB9D8923101A43ADE0AB8

[26]   Morakinyo AO, Akindele AJ, Ahmed Z. Modulation of antioxidant enzymes and inflammatory cytokines: Possible mechanism of anti-diabetic effect of ginger extracts. Afr J Biomed Res 2011; 14(3): 195-202.

[27]   Abd El-Hack ME, Alagawany M, Shaheen H, *et al.* Ginger and Its Derivatives as Promising Alternatives to Antibiotics in Poultry Feed. Animals (Basel) 2020; 10(3): 452.
[http://dx.doi.org/10.3390/ani10030452] [PMID: 32182754]

[28]   Kidd MT. Nutritional modulation of immune function in broilers. Poult Sci 2004; 83(4): 650-7.
[http://dx.doi.org/10.1093/ps/83.4.650] [PMID: 15109062]

[29]   Ahmed AA. Hepatocyte Nuclear Factor 4-α, Glucocorticoid Receptor and Heat Shock Protein 70 mRNA Expression during Embryonic Development in Chickens. Journal of Animal Health and Production 2015; 3(3): 54-8.
[http://dx.doi.org/10.14737/journal.jahp/2015/3.3.54.58]

[30]   Kamboh AA, Hang SQ, Bakhetgul M, Zhu W-Y. Effects of genistein and hesperidin on biomarkers of heat stress in broilers under persistent summer stress. Poult Sci 2013; 92(9): 2411-8.
[http://dx.doi.org/10.3382/ps.2012-02960] [PMID: 23960125]

[31]   Kamboh AA, Khan MA, Kaka U, *et al.* Effect of dietary supplementation of phytochemicals on immunity and haematology of growing broiler chickens. Ital J Anim Sci 2018; 17(4): 1038-43.
[http://dx.doi.org/10.1080/1828051X.2018.1438854]

[32]   Lillehoj HS, Kim DK, Bravo DM, Lee SH. Effects of dietary plant-derived phytonutrients on the genome-wide profiles and coccidiosis resistance in the broiler chickens BMC Proc. 5: S34.
[http://dx.doi.org/10.1186/1753-6561-5-S4-S34]

[33]   Kim DK, Lillehoj HS, Lee SH, Lillehoj EP, Bravo D. Improved resistance to *Eimeria acervulina* infection in chickens due to dietary supplementation with garlic metabolites. Br J Nutr 2013; 109(1): 76-88.
[http://dx.doi.org/10.1017/S0007114512000530] [PMID: 22717023]

[34]   Mehana ESE, Khafaga AF, Elblehi SS, *et al.* Biomonitoring of heavy metal pollution using acanthocephalans parasite in ecosystem: Anupdated overview. Animals 2020; 10(5): 811.

[35]   Oh ST, Lillehoj HS. The role of host genetic factors and host immunity in necrotic enteritis. Avian Pathol 2016; 45(3): 313-6.
[http://dx.doi.org/10.1080/03079457.2016.1154503] [PMID: 26957203]

[36]   Vinus RD, Sheoran N, Maan N, Tewatia B. Potential benefits of herbal supplements in poultry feed: A review. Pharma Innov 2018; 7(6): 651-6.

[37]   Petrolli TG, Albino LFT, Rostagno HS, Gomes PC. Herbal extracts in diets for broilers. Revista Brasileira de Zootecnia. 2012; 41: pp. 1683-90.

[38]   Lillehoj H, Liu Y, Calsamiglia S, *et al.* Phytochemicals as antibiotic alternatives to promote growth and enhance host health. Vet Res 2018; 49(1): 76.
[http://dx.doi.org/10.1186/s13567-018-0562-6] [PMID: 30060764]

[39]   Suganya T, Senthilkumar S, Deepa K, Muralidharan J, Gomathi G, Gobiraju S. Herbal feed additives in poultry. Int J Sci Environ Technol 2016; 5(3): 1137-45.

[40]   Puvača N, Lukač D, Stanaćev V, *et al.* Effect ofspice herbs in broiler chicken nutrition on productive performances. Proceedings of XVI International.

[41]   Hashemi SR, Davoodi H. Herbal plants as new immuno-stimulator in poultry industry: a review. Asian J Anim Vet Adv 2012; 7(2): 105-16.
[http://dx.doi.org/10.3923/ajava.2012.105.116]

[42]   van der Aar PJ, Molist F, van der Klis JD. The central role of intestinal health on the effect of feed additives on feed intake in swine and poultry. Anim Feed Sci Technol 2017; 233: 64-75.
[http://dx.doi.org/10.1016/j.anifeedsci.2016.07.019]

[43]   Suriya R, Zulkifli I, Alimon AR. The effect of dietary inclusion of herbs as growth promoter in broiler chickens. J Anim Vet Adv 2012; 11(3): 346-50.
[http://dx.doi.org/10.3923/javaa.2012.346.350]

[44]   Akib MA, Ambar A, Rusman ADP, Abdullah A. Herbal for increasing immunity and weight of poultry. IOP Conf Ser Earth Environ Sci 2019; 247(1): 12056.

[45]   Abaza I. The use of some medicinal plants as feed additives in broiler diets. 2001.

[46]   Lee Y, Lee S, Lee SJ, *et al.* Effects of dietary Allium hookeri root on growth performance and antioxidant activity in young broiler chickens. Res Vet Sci 2018; 118: 345-50.
[http://dx.doi.org/10.1016/j.rvsc.2018.03.007] [PMID: 29635171]

[47]   Mahanta JD, Borgohain B, Sharma M, Sapcota D, Hussain J. Effect of dietary supplementation of herbal growth promoteron performance of commercial broiler chicken. Indian J Anim Res 2016; 51(OF): 1097-100.
[http://dx.doi.org/10.18805/ijar.11420]

[48]   Rashidi N, Khatibjoo A, Taherpour K, Akbari-Gharaei M, Shirzadi H. Effects of Licorice Extract,

Probiotic, Toxin Binder and Poultry Litter Biochar on Performance, blood and Liver tissue indices of Broiler chickens Exposed to Aflatoxin-B1. Poult Sci 2020.
[http://dx.doi.org/10.1016/j.psj.2020.08.034] [PMID: 33142507]

[49]   Duraiswamy H, Nallaiyan S, Nelson J, Samy PR, Johnson M, Varaprasadam I. The effect of extracts of Selaginella involvens and Selaginella inaequalifolia leaves on poultry pathogens. Asian Pac J Trop Med 2010; 3(9): 678-81.
[http://dx.doi.org/10.1016/S1995-7645(10)60164-2]

[50]   Hammershøj M, Johansen NF. Review: The effect of grass and herbs in organic egg production on egg fatty acid composition, egg yolk colour and sensory properties. Livest Sci 2016; 194: 37-43.
[http://dx.doi.org/10.1016/j.livsci.2016.11.001]

[51]   Xiao YQ, Shao D, Sheng ZW, Wang Q, Shi SR. A mixture of daidzein and Chinese herbs increases egg production and eggshell strength as well as blood plasma Ca, P, antioxidative enzymes, and luteinizing hormone levels in post-peak, brown laying hens. Poult Sci 2019; 98(8): 3298-303.
[http://dx.doi.org/10.3382/ps/pez178] [PMID: 30993323]

[52]   Abd El-Hack ME, Abdelnour SA, Taha AE, *et al.* Herbs as thermoregulatory agents in poultry: An overview. Sci Total Environ 2020; 703: 134399.
[http://dx.doi.org/10.1016/j.scitotenv.2019.134399] [PMID: 31757531]

[53]   Tan B, Vanitha J. Immunomodulatory and antimicrobial effects of some traditional chinese medicinal herbs: a review. Curr Med Chem 2004; 11(11): 1423-30.
[http://dx.doi.org/10.2174/0929867043365161] [PMID: 15180575]

[54]   Kim WH, Lillehoj HS. Immunity, immunomodulation, and antibiotic alternatives to maximize the genetic potential of poultry for growth and disease response. Anim Feed Sci Technol 2019; 250: 41-50.
[http://dx.doi.org/10.1016/j.anifeedsci.2018.09.016]

[55]   Elagawany MMII. Multiple beneficial applications and modes of action of herbs in poultry health and production-A review. Sci Alert 2015; 11: 152-76.

[56]   Safety and nutritional assessment of GM plants and derived food and feed: The role of animal feeding trials. Food Chem Toxicol 2008; 46 (Suppl. 1): S2-S70.
[http://dx.doi.org/10.1016/j.fct.2008.02.008] [PMID: 18328408]

[57]   Velusami CC, Boddapati SR, Hongasandra Srinivasa S, *et al.* Safety evaluation of turmeric polysaccharide extract: assessment of mutagenicity and acute oral toxicity. BioMed Res Int 2013; 2013: 1-10.
[http://dx.doi.org/10.1155/2013/158348] [PMID: 24455673]

[58]   Chandrasekaran CV, Srikanth HS, Anand MS, Allan JJ, Viji MMH, Amit A. Evaluation of the mutagenic potential and acute oral toxicity of standardized extract of *Ocimum sanctum* (OciBest™). Hum Exp Toxicol 2013; 32(9): 992-1004.
[http://dx.doi.org/10.1177/0960327112472992] [PMID: 23424203]

[59]   Lu B, Li M, Zhou F, *et al.* The Osmanthus fragrans flower phenylethanoid glycoside-rich extract: Acute and subchronic toxicity studies. J Ethnopharmacol 2016; 187: 205-12.
[http://dx.doi.org/10.1016/j.jep.2016.04.049] [PMID: 27130643]

[60]   Srinivasa Rao B, Chandrasekaran CV, Srikanth HS, *et al.* Mutagenicity and Acute Oral Toxicity Test for Herbal Poultry Feed Supplements. J Toxicol 2018; 2018: 1-12.
[http://dx.doi.org/10.1155/2018/9412167] [PMID: 29861724]

[61]   Mohammadi Gheisar M, Kim IH. Phytobiotics in poultry and swine nutrition – a review. Ital J Anim Sci 2018; 17(1): 92-9.
[http://dx.doi.org/10.1080/1828051X.2017.1350120]

[62]   Dollahite JW, Henson JB. Toxic plants as the etiologic agent of myopathies in animals. Am J Vet Res 1965; 26: 749-52.

[PMID: 14316795]

[63]   Riet-Correa F, Medeiros RMT, Pfister JA, Mendonça FS. Toxic plants affecting the nervous system of ruminants and horses in Brazil. Pesqui Vet Bras 2017; 37(12): 1357-68.
[http://dx.doi.org/10.1590/s0100-736x2017001200001]

[64]   Bruneton J. Toxic plants dangerous to humans and animals. Intercept Limited 1999.

[65]   Guitart R, Croubels S, Caloni F, *et al.* Animal poisoning in Europe. Part 1: Farm livestock and poultry. Vet J 2010; 183(3): 249-54.
[http://dx.doi.org/10.1016/j.tvjl.2009.03.002] [PMID: 19359202]

[66]   Stegelmeier BL. Pyrrolizidine Alkaloid–Containing Toxic Plants (Senecio, Crotalaria, Cynoglossum, Amsinckia, Heliotropium, and Echium spp.). Vet Clin North Am Food Anim Pract 2011; 27(2): 419-428, ix.
[http://dx.doi.org/10.1016/j.cvfa.2011.02.013] [PMID: 21575778]

[67]   Anadón A, Martínez-Larrañaga MR, Ares I, Martínez MA. Poisonous Plants of the Europe. Veterinary Toxicology. Elsevier 2018; pp. 891-909.

[68]   Chesson A, Flachowsky G. Transgenic plants in poultry nutrition. Worlds Poult Sci J 2003; 59(2): 201-7.
[http://dx.doi.org/10.1079/WPS20030012]

[69]   Dang L, Van Damme EJM. Toxic proteins in plants. Phytochemistry 2015; 117: 51-64.
[http://dx.doi.org/10.1016/j.phytochem.2015.05.020] [PMID: 26057229]

CHAPTER 3

Ginger as a Natural Feed Supplement in Poultry Diets

Mohamed E. Abd El-Hack[1,*]**, Ayman A. Swelum**[2,3]**, Youssef A. Attia**[4,5]**, Mohamed Abdo**[6,7]**, Ahmed I. Abo-Ahmed**[8]**, Mahmoud A. Emam**[9] **and Mahmoud Alagawany**[1]

[1] *Department of Poultry, Faculty of Agriculture, Zagazig University, Zagazig 44511, Egypt*

[2] *Department of Animal Production, College of Food and Agriculture Sciences, King Saud University, P.O. Box 2460, Riyadh 11451, Saudi Arabia*

[3] *Department of Theriogenology, Faculty of Veterinary Medicine, Zagazig University, Zagazig 44511, Egypt*

[4] *Animal and Poultry Production Department, Faculty of Agriculture Damanhour University, Damanhour, Egypt*

[5] *Department of Agriculture, Faculty of Environmental Sciences, King Abdulaziz University, 21589, Jeddah, Kingdom of Saudi Arabia*

[6] *Department of Animal Histology and Anatomy, School of Veterinary Medicine, Badr University in Cairo (BUC), Cairo, Egypt*

[7] *Department of Anatomy and Embryology, Faculty of Veterinary Medicine, University of Sadat City – Sadat City, Egypt*

[8] *Department of Anatomy and Embryology, Faculty of Veterinary Medicine, Benha University, Toukh 13736, Egypt*

[9] *Department of Histology, Faculty of Veterinary Medicine, Benha University, Toukh 13736, Egypt*

Abstract: Poultry ventures have progressed quickly over the last three decades. Therefore, curative or growth-promoting antibacterial agents have been utilized extensively. Because of increasing bacterial resistance towards antibiotics and, consequently, accumulation of antibacterial residues in chicken products and increased consumer's demand for products without antibacterial residues, alternative solutions that could substitute antibiotics without affecting productivity or product quality should be attempted. Recently, natural replacements such as ginger, etheric oils, organic acids, garlic prebiotics, immune stimulants and plant extracts were used to improve productiveness, and body performance, prevent pathogenic microorganisms, and reduce antibacterial activity usage in poultry manufacturing. The utilization of a single alternative or a combination of variable replacements and perfect surveillance and flock health might improve the profits and sustain the productivity of poultry. This chapter aimed at summarizing the recent knowledge and information regarding the utilization

* **Corresponding author Mohamed E. Abd El-Hack:** Department of Poultry, Faculty of Agriculture, Zagazig University, Zagazig 44511, Egypt; E-mail: m.ezzat@zu.edu.eg

of ginger and its derivatives as natural alternatives or supplements in poultry feed and their impacts on poultry productivity, meat and egg traits in addition to economic efficacy.

Keywords: Antibacterial, Ginger, Meat quality, Natural alternatives, Poultry, Productivity.

INTRODUCTION

Nowadays, the uncontrolled usage of antibacterial antibiotics in poultry feed provokes serious complications. Many crucial reasons are found that forbid the utilization of antibiotics like drug residues in poultry meat and develop resistance of bacterial species against many drugs. Therefore, their usage has been controlled in different nations due to the spread of antibiotic-tolerant human pathogens [1 - 5]. The European Uniting Community firstly applied prohibition rules on using antibiotics in poultry feed in January 2006. However, the antibacterial administrated in sub-treating doses as antibacterial growth promoters had been separated [6]. To overcome the undesirable productivity and susceptibility to infections after removing antibacterial from poultry diets, attempts have been made to find other applicable replacements. Food additives might be nutritional or non-nutritional substances that control the nutritional substance availability in the diet. Natural growth promoters such as prebiotics, probiotics, enzymes, and botanical substances could be used as food supplements for broilers instead of synthetic types [7 - 9]. Recently, growth promoters like probiotics, yeast cultivations, organic acids, enzymes, prebiotics, plant-derived essential oils and some herbs' extracts have been utilized [10 - 16]. Ginger rhizome (*Zingiber officinale*) has been used as a remedy or tenderness seasoning. The ginger can be used as a substitute for antibacterial growth promoters due to increased poultry productivity, improved feed palatability and accessibility, and improved secretion of digestive enzymes [17]. Using antibiotics or synthetic supplements as food supplements in livestock feed is hazardous because of their potential toxicological properties. Besides, the tendency for using natural harvests has been increased to minimize the usage of synthetic components [18]. Therefore, therapeutic herbs or derivatives are believed to substitute synthetic compounds and achieve consumer demands and common market competition [19, 20].

Zingiber officinale Roscoe, relevant to the *Zingiberaceae*, generally known as ginger, is a monocotyledonous herbal plant and one of the chief common food-flavoring supplements globally [21]. Recently, several therapeutic effects of ginger have been recognized, for example, anti-inflammatory, analgesic, digestive moderating agent, antioxidant and antimicrobial properties [22]. Ginger might be a substitute for common synthetic growth promoters such as antibacterial agents

[23]. The characteristic ginger feel is derived from the unsteady materials that consist of zingerone, shogaols and gingerols [24]. This chapter highlighted the effects of ginger and its byproducts as natural supplements and feed additives on the poultry carcass characters, growth performance, egg and meat features, productivity, food digestibility, immunogenic reaction, some serum biochemical parameters, and bacteriological infection of poultry.

IMPACT OF GINGER OIL

Ginger oil has been derived from roots of *Z. officinale*, whose constituents are affected by the geographical zone, method of extraction and roots' status. Ginger oil's antibacterial, antifungal, analgesic, immunological, and anti-inflammatory effects have been preclinically studied [25]. Ginger oils have been well-studied and they are considered safe without detrimental effects. Owing to their broad pharmacological actions, their effects on digestive and respiratory diseases are of prodigious interest.

BIOCHEMICAL COMPONENTS OF GINGER EXTRACT

The biochemical constituents of ginger oil and its prospective antioxidant and anti-inflammatory effects have been studied [26]. By Gas Chromatography-Mass Spectrometry (GC–MS), ginger oil consists of zingiberene, which accounts for 31% of the total composition, curcumin represents 15.4% and sesquiphellandrene represents 14.02%. Ginger oil contains hydroxyl and α, α-diphenyl-β-picrylhydrazyl (DPPH) radicals and superoxide, which reduce fat peroxidation. Intraperitoneal injection of ginger oil causing the suppression of phorbol12-myristate-13-acetate commenced superoxide chemicals liberated by the macrophages. On the other hand, oral therapy of ginger oil for one month as a minimum definitively enhanced superoxide dismutase (SOD), glutathione and glutathione reductase levels [26].

According to the source of rhizomes, the product of ginger oil is changed from 1.0 to 3% [27]. Moreover, the origin of rhizome, freshness or dehydration, and extraction techniques affect ginger oils' chemical composition.

The basic ginger oil was extracted from fresh roots of *Zingiber officinale* by GC–MS [28]. The latter authors reported the existence of 69 elements, accounting for 96.93% of the total oil. The essential element was zingiberene representing 28.62%, followed by camphene representing 9.32%, curcumin representing 9.09% and lastly, β-phellandrene representing 7.97%. Examination of the oleoresin showed the presence of 34 constituents, calculating 88.63% of the total oleoresin. The main constituents were trans-6-shogaol representing 26.23%, trans-10-shogaol representing 13.0%, α -zingiberene representing 9.66% and 10-

gingerdione representing 6.80%. Moreover, the basic oil was found to have a 100% antifungal effect against *Fusarium oxysporum*, whereas the oleoresin displayed a 100% antifungal effect against *Aspergillus niger*. The authors mentioned above demonstrated the increased antioxidant effect of sunflower oil compared with the basic oil or synthetic antioxidants (BHT and BHA). Additionally, the basic oil was derived from *Z. officinale* roots by exposure of the fresh roots to hydro-distillation. The oil was analyzed either by GC or GC-MS techniques to evaluate its antimicrobial effect using the discdiffusion technique [29]. The results revealed that the basic oil had been characterized by a great percentage of sesquiterpenes representing 66.66%, monoterpenes representing 17.28%, and aliphatic elements representing 13.58%. The main sesquiterpene constituents were zingiberene representing 46.71%, valencene representing 7.61%, β-funebrene representing 3.09%, and selina-4(14),7(11)-diene representing 1.03%. Moreover, monoterpenes were classified as citronellyl n-butyrate representing 19.34%, followed by β-phellandrene representing 3.70%, camphene representing 2.59% and α-pinene representing 1.09%. Additionally, unsteady oils derived from Sikkim's two main prevalent cultivars, known as Bhaisa and Majulay, were separated and differentiated by GC-MS or GC techniques [30]. Sixty constituents calculating about 94.9% and 92.6% of the Bhaisa and Majulay oils, respectively, had been identified. The considerable elements of Bhaisa oil were geranyl acetate representing 18.8%, followed by zingiberene representing 16.3%, and lastly, geranial representing 8.2%. In comparison, the constituents of Majulay oil were zingiberene representing 19.8%, followed by geranial representing 16.5%. The Bhaisa oil had major constituents of oxygenated ingredients representing 43.1% compared to various ginger cultivar oils.

The basic oil was extracted from the raw ethanolic extract of ginger from Tanahu, Gorkha and the Sindhupalchhowk region of Nepal [31]. Essential oils of ginger were derived using hydro-distillation yielding, Gorkha representing 1.9%, Sindhupalchowk representing 1.8% and Tanahu representing 1.1%. GC-MS method revealed the presence of 105 elements from Sindhupalchowk ginger, 85 ingredients from Tanahu ginger and 88 elements from Gorkha ginger, and the main constituents, sesquiterpenes and monoterpenes derivatives. After essential oil extraction, soxhlet extraction was carried out for the 95% ethanolic derivation from fresh ginger and the deposit ginger.

The basic oils derived from the water of ginger or steam distillation gathered from Vietnam were 2.1% and 2.05%, respectively [32]. The essential constituents of the basic oil separated by the aqueous distillation method were curcumin representing 11.7% followed by β-bisabolene representing 4.1%, while those of the basic oil prepared by vapor distillation method was curcumin, accounting

12.6% followed by α-zingiberene representing 10.3%, β-bisabolene representing 8.1% and β-sesquiphellandrene representing 7.4%.

The basic oils derived from the water of ginger or steam distillation gathered from Vietnam were 2.1% and 2.05%, respectively [32]. The essential constituents of the basic oil separated by the aqueous distillation method were curcumin representing 11.7% followed by β-bisabolene representing 4.1%, while those of the basic oil prepared by vapor distillation method was curcumin, accounting 12.6% followed by α-zingiberene representing 10.3%, β-bisabolene representing 8.1% and β-sesquiphellandrene representing 7.4%.

The basic oils derived from *Zingiber officinale* roots were analyzed by GC–MS or TLC techniques. The TLC-supported bioautography procedure demonstrated their antimicrobial and antioxidant components [33]. The ginger oil was characterized by the high content of sesquiterpene hydrocarbons, comprising 27.16% β-sesquiphellandrene, 15.29% caryophyllene, 13.97% zingiberene, 10.52% α-farnesene and 6.62% curcumin (Fig. 1).

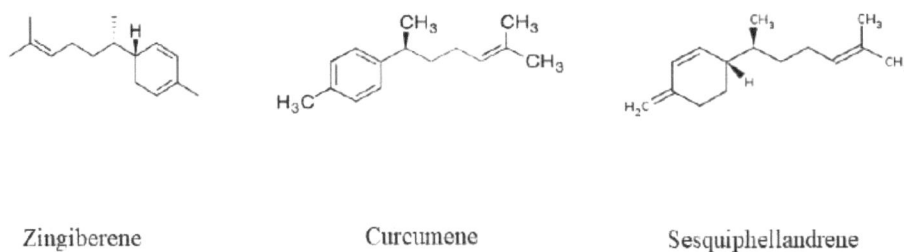

Zingiberene   Curcumene   Sesquiphellandrene

Fig. (1). Chemical composition of ginger oil.

VALUABLE EFFECTS OF GINGER SUPPLEMENTATION IN POULTRY

Impact of Ginger Supplementation on Carcass Weight

Supplementation of ginger oil for at least fifty days without changes in body weight gain (BWG) and feed conversion ratio (FCR) among the control birds fed a basal diet or 10 mg/kg/day of ginger oil was demonstrated [34]. The effect of ginger oil on growth performance and bacterial count of broiler chicken was assessed [35]. 200 male and female one-day-old chicks of Ross 308 were divided into four treated groups for seven weeks at the following doses; 0 (Control), 10, 20 and 40 mg/kg/day. The results reported no obvious changes in feed intake, FCR and BWG between the examined birds. Another study estimated four variable foods that included 100 µl/kg BW of H_2O as control, 50, 100, and 150 µl/kg BW of basic oil of ginger to evaluate the effects of basic ginger oil on the oxidative or antioxidant biochemical characters, growth parameters as well as the histological architecture of testes and consequently fertility in Japanese quails [36]. The results showed that basic oil's growth charactereristics were not altered, irrespective of its dosage. However, the left testis was increased in weight at a dosage of 100 or 150 µl/kg BW, comparable with the other sets.

The effects of basic ginger oil on growth and laying characters and antioxidants of egg yolk and cholesterol value in Japanese quails have been studied [37]. This study was conducted on 80 birds evenly divided into four categories orally treated daily, 100 µl/kg BW H_2O as a control and 50, 100, and 150 µl/kg BW of essential oil originating from ginger roots. The results revealed that feed conversion ratio and body weight gain were not changed by oral application of ginger essential oil. Oral administration of essential oil of ginger rhizomes showed no recognizable effects on intestine, gizzard, liver, and heart weights comparable with the control group. However, oral application of essential oil derived from ginger roots improved the eggs' weight in Japanese quails compared with the control group.

The effects of ginger supplementation on the growth performance, carcass features, antioxidant status and blood parameters in broiler chicks subjected to heat stress ($32 \pm 2°C$ for 8 h per day) were evaluated [38]. Male day-old broiler chicks were randomly and equally divided into six examined groups; plain diet without additive as a control, essential diet combined with vitamin E (100 mg/kg) as a positive control, essential diet containing either 7.5 or 15 g/kg of ginger root powder, and essential diet enclosing 75 or 150 mg/kg BW of ginger essential oil. The results revealed that at 22 days of age, the group fed 7.5 g/kg of ginger root powder showed considerably improved body weight and weight gain comparable to the control group. There were no clear variations between the treated categories concerning body weight gain, feed intake characters, feed conversion ratio, and blood parameters after 42 and 49 days of age. In all treated groups, total antioxidant capacity was increased and Malondialdehyde (MDA) level was decreased in serum compared to the control group.

The supplementation of ginger and garlic essential oils mixture on growth parameters of broiler chicks during 42 days was investigated [39]. Eighty-four seven days-old Arbor Acres were randomly and evenly divided into four groups. The first category was given a direct feed with no supplement. In contrast, the other categories were nourished plain diets admixed with variable levels of combinations of ginger and garlic essential oils at a concentration of (100 g garlic + 100 g ginger)/ton, (200 g garlic + 200 g ginger)/ton and (300 g garlic + 300 g ginger)/ton. The results showed no obvious changes in body weight gain and feed conversion ratio among all examined categories.

The effects of black seed oil (*Nigella sativa* L.), ginger oil (*Zingiber officinale* Roscoe), oregano oil (*Origanum vulgare* L.) and thyme oil (*Thymus vulgaris* L.) as therapeutic herbs on the productivity and serum parameters of Gimmizah chicks were evaluated [40]. A total of 405 one-month-old Gimmizah chicks were haphazardly allocated to 9 groups of three replicates for each. All birds were maintained until 16 weeks of age. The 1st, 2nd, 3rd and 4th chicks' groups received a plain feed mingled with 1 ml of the black seed, ginger, oregano, or thyme /kg diet. The 5th, 6th, 7th and 8th chicks' groups were nourished with a basic food with supplementation of 0.5, 1.0, 1.5 and 2.0 ml/kg diet of an equivalent combination of the aforementioned herbal plants together, respectively. The 9th chicks' group was utilized as a control. The findings revealed that the maximum BWG was noted for the chicks fed a basic regimen containing a 1 ml oregano/kg diet. Moreover, the impacts of the basic diet containing essential oil of ginger (50, 100, or 200 mg/kg diet) in addition to mannan-oligosaccharide (2 g/kg diet) on growth characteristics of broiler chicks were investigated [41]. 375, one-day-old male broiler chicks were randomly and evenly divided into five groups. The results showed that the birds fed on ginger essential oil and mannan-oligosaccharide mixture (200 mg/kg diet) exhibited an increased body weight gain from the first day to 42 days of age compared to control birds.

To evaluate the effect of essential oil of *Zingiber officinale* on body weight gain in broiler chicks, 105 one-day-old male broiler chicks (Ross 308) were raised for 42 days to evaluate variable foods as growth promoters [42]. Broiler chicks were divided into seven categories; plain diet, plain diet + flavomycin (10 mg/kg diet), plain diet + extract of *Yucca schidigera* (120 mg/kg diet), plain diet + essential oil of *Oreganum vulgare* (120 mg/kg diet), plain diet + essential oil of *Thymus vulgaris* (120 mg/kg diet), plain diet + essential oil of *Syzygium aromaticum* (120 mg/kg diet) and plain diet + essential oil of *Zingiber officinale* (120 mg/kg diet). The results indicated that *Zingiber officinale* improved the body weight gain compared with the other examined categories. Fig. (**2**) shows the shape of ginger roots.

Fig. (2). The shape of ginger roots.

Impact of Ginger and its Products on Carcass Characters

The dietary supplementation of *Zingiber officinale* promoted the growth character and beneficial microflora [42]. Nutritional supplementation of *Z. officinale, S. aromaticum,* or *O. vulgare* reduced the entire measurement of the gastrointestinal tracts of administrated chickens but increased the jejunum mass. The weight of the liver was diminished in quails fed with the basic oil of ginger [36].

It has been found that most of the organs' size and carcass characteristics were not deteriorated during the treatment, except for a decrease in the liver mass of chickens exposed to garlic oil compared with control or ginger oil-treated birds. Similarly, a decrease was observed in the head mass ratio of chickens fed with essential oils compared with control birds. The relative mass of organs might be decreased in response to different doses except for the gizzard and head compared with control birds. Male broiler chickens had been designated less than female chickens [34].

Oral administration of essential oil derived from ginger roots (0, 0.25%, 0.5% and 0.75%) had no considerable effects on gizzard, intestine mass, liver and heart comparable to the control quails; however, the quantity of the fat of abdomen was considerably decreased in all treated birds [43]. Moreover, carcass quality and meat characteristics were not influenced by supplementation of a mixture of ginger and garlic essential oils in the broiler feed [39].

Impact of Ginger and its Products on Egg Quality and Production

The impact of dietary supplementation of ginger oil on egg quality and production in Japanese quail has been investigated [43]. Eight weeks-aged quails were randomly assigned into four groups, containing 4 different concentrations of ginger oil (0, 0.25%, 0.5% and 0.75%). There were no significant changes in egg production and the cumulative egg production rate in the first period. Still, there was significant superiority after the addition of ginger oil compared with the control birds in the subsequent periods and a considerable increase in egg mass in all treated groups during the 4 stages of the experiment. Albumin weight was increased during the fourth period, while there were no major alterations in the first three stages. There was a considerable change in the shell mass in the second stage but no considerable variations in other stages for this feature. In addition, there were no considerable changes in the yolk diameter, high yolk, yolk index, yolk weight, shell thickness, and high albumin. Besides, supplementation of ginger oil resulted in a considerable increase in egg productivity with no considerable variations in egg quality characteristics in Japanese quail. Compared with control birds, egg mass was increased in quails subjected to essential oil derived from ginger roots regardless of dosage [43].

In poultry, the essential oil extracted from ginger roots might be used to reduce lipid peroxidation in germinal tissues and, consequently, improve fertility. Additionally, oral administration of 100–150 µl/kg BW of basic oil extracted from ginger roots in laying Japanese quails maximized the effects on egg mass without any detrimental influences on the feed uptake and body weight gain [36]. Instead, oral administration of basic oil derived from ginger roots did not affect egg productivity and the weekly weight of eggs [37]. The impact of ginger extracted oil on laying hens' productivity, antioxidant, and immunological characteristics wasinvestigated [44]. Control hens were given the plain diet while the treated birds were fed basic feeds containing 0.1% ginger extracted oil. It was found that ginger extract improved the rate of laying with daily egg mass.

Impact of Ginger on Reproductive Function

Administration of essential oil of ginger 100 and 150 µl/kg BW improves the fertility ratio compared with the control birds [36]. On the other hand, the effects of essential oil extracted from ginger roots on some reproductive characteristics of layers were assessed [45]. Eighty female Japanese quails, three weeks of age, were randomly allocated into four dietary groups. Japanese quails of the 1st group were orally administered with clean H_2O (100 µl/kg BW) from the 3rd till the 13th week. In contrast, the 2nd, 3rd and 4th groups were orally administered with essential oil derived from ginger roots 50, 100 and 150 µl/kg BW, respectively, at the same time. No clear improvements were noticed for carcass and individual ovary relative weights. At the same time, the individual uterine weight was increased in a dose-dependent manner. Moreover, hatchability rate, fertility and chick weight were noticeably improved in the birds exposed to 100 and 150 µl/kg BW compared with those of the control group. In addition, fetal mortalities were decreased with any dose amount of the essential oil of ginger rhizomes [45].

Impact of Ginger on Blood Parameters

The effect of ginger and garlic essential oils on various serum parameters of broilers was evaluated [34, 35]. Forty two-day-old Arbor acres chicks were assigned into four categories. The four groups were administered with the four doses; 0 (Control), 10, 20 and 40 mg/kg/day accordingly *via* a stomach tube. There were no apparent differences reported in serum aspartate transaminases (AST), alanine transaminase (ALT), and serum creatinine owing to ginger treatment, indicating that none of the administered three doses of ginger the oil was detrimental. On the other hand, malondialdehyde (MDA), transaminases, total cholesterol and triglycerides were reduced in exposed quails [36]. The characteristics of antioxidant enzymes, globulin and total protein in serum were increased in treated birds compared with the control birds. Besides, oral administration of 100 or 150 µl/kg BW of essential oil extracted from ginger roots decreased the serum characteristics of transaminases (AST or ALT), low-density lipoprotein (LDL), and total cholesterol compared to the control birds and significantly decreased egg or serum values of cholesterols with no any detrimental effects on feed uptake and BWG in laying Japanese quails [37].

The effects of feed supplementation of ginger or thyme basic oils on broiler chickens' immunological, biochemical and hematological parameters were evaluated [46]. 105, one-day-old Ross 208 broiler chicks were allocated into 7 categories. The control group was fed with only a plain diet. Thyme oil was administrated at a dosage of 100 mg/kg BW (T100), 200 mg/kg BW (T200) and 300 mg/kg BW (T300). The ginger oil was given at a dosage of 100 mg/kg BW (G100), 200 mg/kg BW (G200) and 300 mg/kg BW (G300). Hematological examination showed a noticeable increase in hemoglobin or packed cell volume in the T200 group and the total leukocytic count, particularly heterophils percent in G100 and T200 categories. Globulin or total protein was decreased in G200 or G300 classes, while the class G100 showed a superior effect on lipid characteristics.

Administration of 150 mg/kg BW essential oil derived from ginger led to increased total SOD function in the liver tissue compared with the control birds [38]. There was a reduction in the hepatic malondialdehyde level in the groups given basic oil or ginger powder comparable with the control group. Catalase enzymes, glutathione peroxidase (GSH-PX), and total SOD in the erythrocytes showed insignificant variations among the treated groups. The whole exposed groups revealed increased total antioxidant capacity (TAOC) and decreased serum MDA values compared with the control group. Besides, insignificant changes were observed in the serum creatinine values and transaminases (AST or ALT) after the administration of ginger oil at 10 mg, 20 mg and 40 mg/kg BW per day, indicating that neither of these doses was toxic to birds [35].

Supplementation of variable therapeutic herbs, either alone or in a combination, in whatever the dose, considerably reduced the ALT activity and the total serum cholesterol level compared with the control birds [40]. Moreover, LH, FSH, estradiol and serum total proteins values were significantly increased in a dose-dependent manner [45].

The effects of supplementation of ginger extract on atherosclerosis caused by high serum cholesterol were examined because of its probable antioxidant activity [47]. There was a noticeable reduction in triglycerides, phospholipids and VLDL or LDL cholesterol in both aortic tissue homogenate and serum related to a high dosage of ginger extract. Additionally, the antioxidant activity of the ginger extract was assessed *via* its radical scavenging action (RSA), using the rapid 1,1-diphenyl-2- picrylhydrazyl (DPPH) technique which exhibited the greatest antioxidant activity accompanying a high dosage of ginger extract.

The ginger extract didn't alter the role of TAOC and GSH-PX but only reduced MDA value and altered the serum SOD character of the poultry [44]. Besides, the ginger extract didn't change the globulin, albumin, and serum total protein concentrations but improved lysozyme activity. Additionally, ginger extract decreased prostaglandin E2 (PGE2) value. Moreover, blood cholesterol levels were reduced by ingestion of the essential oils of ginger (100 mg/kg) and mannan-oligosaccharide-supplemented feed [41]. Both S. aromaticum and Z. *officinale* dietary supplementation [42].

Supplementation of 100 or 150 µl/kg BW of essential oil extracted from ginger roots significantly reduced the serum ALT, AST, total cholesterol and LDL-cholesterol compared with the control results [37]. Oral administration of 100–150 µl/kg BW of basic oil derived from ginger roots decreased the blood cholesterol values without any detrimental effects on feed uptake and BWG in laying Japanese quails [36].

Impact of Ginger and its Products on Microbial Infections

Increased dosage of ginger essential oils quantitatively reduced *Escherichia coli*, like most Enterobacteriaceae populations and *Shigella* and *Salmonella* species in the ileocecal fillings compared with the control results [34]. Colony-forming units (CFU) of *Staphylococci spp* were almost similar among the birds exposed to ginger oil but were decreased compared with the control data.

Administration of ginger essential oil showed explicit antibacterial action against *Pseudomonas aeruginosa*, *Escherichia coli* and *Staphylococcus aureus*, and antifungal action against *Candida albicans* and *Aspergillus niger* [29]. Essential oil of ginger mainly comprised great monoterpenes as sesquiterpenes and revealed significant antimicrobial activity against most pathogens.

The antibacterial activity of 8% ginger extract from residue ginger and fresh ginger after extraction of essential oil against *Klebsiella spp, Staphylococcus aureus* and *Escherichia coli* was evaluated *via* the inhibition zone and compared with chloramphenicol findings [31]. Moreover, the DPPH radical scavenging technique has achieved antioxidant activity in an extensive range of dosages. Meanwhile, residue ginger or fresh ginger after essential oil extraction exhibited the supreme degree of inhibition comparable to ascorbic acid, while the essential oil didn't possess antioxidative action.

The ginger essential oil extracted by hydrodistillation had an antibacterial effect with a minimal concentration of 6.25 mg/ml against three gram-positive bacterial strains and suppressing both *L. monocytogenes* and B. cereus [48]. The antibacterial activity of ginger essential oils extracted *via* the agar diffusion method has been evaluated [32]. The studied microorganisms were some gram-negative bacteria such as *Pseudomonas aeruginosa, Escherichia coli* and *Salmonella abony*; some gram-positive strains such as *Bacillus pumilus, Bacillus subtilis, Staph. epidermidis* and *Staph. aureus*; some fungal strains such as *Botrytis cinerea, Aspergillus niger, Rhizopus nigricans, Penicillium spp.* and some yeast strains such as *Saccharomyces cerevisiae* and *Candida albicans*. The essential oil displayed a little activity against the studied gram-negative and gram-positive strains while exhibiting a possible fungicidal effect against the studied fungal species. Supplementation of several medicinal herbs, either separately or in combination, whatever their concentration, reduced the gut total aerobic and anaerobic bacterial populations in addition to coliforms count compared with the control data [40].

The antibacterial effect of basic oils extracted from roots of *Z. officinale* or *Curcuma longa* L. was examined [49]. The extraction procedures of essential oils were achieved by GC–MS technique. It had been approved that ginger essential oil was more efficient than turmeric oil, both of which were thought to have a bacteriostatic effect (minimum inhibitory concentration "MIC" 2500–5000 µg/mL) or bactericidal effect (MIC 5000–10000 µg/mL). Therefore, the essential oil of ginger was believed to be an alternative supplement for controlling enteric Salmonella infection.

Dietary supplementation of *Z. officinale* improved beneficial bacterial count such as lactic acid bacteria in the small intestine, particularly jejunum [42]. Ginger-extracted oil exhibited a potent suppressing effect against some bacterial species and some pathogenic fungi, with MIC ranging between 20 - 120 µg/mL depending on the bacterial species [33]. *E. coli* or other *Enterobacteria* such as *Shigella* and *Salmonella* counts in the fillings of intestines were significantly decreased by increasing oil doses comparable with the control cases [35].

Impact of Ginger on Egg and Meat Quality

The effects of ginger and its products on the quality of poultry meat and the organic structure of meat are not well studied in previous literature. Ginger is considered an essential source of the plant proteolytic enzyme. The extracts of ginger possess proteolytic action causing an elevation in the collagen solvability and proteolysis in exposed chicken musculature [58]. Moreover, the effect of ginger supplementation on the performance and meat quality was studied where the application of 0.5% ginger in the diet reduced the plasma glucose and fats of meat comparable to the control birds [59]. It had been reported that the application of ginger for chickens at the level of (0.5, 1.0, 1.5 and 2.0%) resulted in a considerable reduction in the weight of fat compared with chickens fed ginger-free feed [60]. Moreover, the addition of ginger in broiler diets as an alternative for antibiotics resulted in a slight improvement in the tenderness and pH of broiler chicken meat while resulting in the loss of cooking and water holding capacity compared to the control chicken.

Ginger significantly improved the egg quality. The addition of ginger at 100 g/ton in the diet for 8 weeks in Hyline Brown laying hens improved the egg quality by intensifying the albumin content and Haugh unit of eggs compared to the control hens [61]. Besides, the addition of ginger in the diets of layers improved the T-SOD activity and decreased the cholesterol and MDA content in the yolk compared with control birds [57, 61]. Moreover, it had been declared that blood antioxidant enzymes were increased, accompanying improved Haugh unit in ginger root-fed birds [62]. This might be attributed to the radical-scavenging effect of antioxidant constituents and phenolic compounds in the ginger extract, which decreases phospholipid oxidative status [63] and improves the carcass organs' function, which is revealed by lowered serum concentrations of AST and ALT, consequently stimulating the synthesis and production of antioxidative enzymes [64].

Additionally, ginger root powder reduced the cholesterol level of yolk [65]. Besides, the activity of ginger in lowering the cholesterol level was also noted in broilers [18]. Moreover, the reduction in cholesterol level in the yolk might be attributed to the changes of HDL-metabolism, implicated in hindering cholesterol passage. The ginger extract might be utilized as a supplement to improve production of eggs with low-cholesterol levels, which is highly preferable because high cholesterol is a predisposing factor for atherosclerosis [66].

The Economic Impact of Ginger

In poultry, the ginger powder might reduce the cost to benefit percentage and increase the economic feasibility. The outcome of ginger supplementation on the carcass quality, growth performance and economic value of broiler chickens has been studied [67]. The administration of ground ginger at 50 ml/liter of drinking H2O or at 14 g/kg diet increased the profits and net gain and provided the minimum cost-benefit ratio compared with the control birds [67]. Moreover, garlic, ginger, and their mixture has been assessed on the growth features and economic efficiency in broiler chickens [68]. The birds' marketing profit was more remarkable in 1% of the ginger-fed birds than in the control birds.

Throughout the total time of the experiment, the cost of feeding was greater in the ginger-fed birds and combined garlic and ginger-fed birds by 1% than in the control birds. Return over the cost of feeding was lesser in the ginger group and combined garlic and ginger group by 1% than in the control birds. Lastly, it might be summarized that the ginger didn't positively or negatively affect the return over feed cost. Similarly, dietary supplementation of ginger showed no variations in feed cost per kg BWG for broiler chickens [69].

CONCLUSION

It could be concluded that ginger and its derivatives are safe as feed additives because they have no acute toxicological effects. According to several pieces of literature, ginger feed supplementation reduced abdominal fat and stomach mass, but with no effects on broilers' body profiles and growth characteristics. It could also be indicated that diet supplementation with ginger improved broiler chickens' immune response and antioxidant activity. It might be attributed to ginger having strong antioxidative activity and improved regular antibodies formation. Moreover, ginger has been used as an alternative for antibacterial therapeutics in fowl ventures. Ginger extract has been used as a food supplement to produce hypocholesterolemic eggs, which is favorable as cholesterol predisposes to atherosclerosis and heart illnesses. Several studies revealed that ginger and its products exhibited identical impacts to antibiotics in fowls. Furthermore, the dosage and administration method for those replacements for antibiotics remain essential to be efficient in poultry manufacturing.

CONSENT FOR PUBLICATION

Not applicable.

CONFLICT OF INTEREST

The author declares no conflict of interest, financial or otherwise.

ACKNOWLEDGEMENTS

Declared none.

REFERENCES

[1] El-Hack MEA, Mahgoub SA, Alagawany M, Dhama K. Influences of dietary supplementation of antimicrobial cold pressed oils mixture on growth performance and intestinal microflora of growing Japanese quails. Int J Pharmacol 2015; 11(7): 689-96. [Google Scholar]. [http://dx.doi.org/10.3923/ijp.2015.689.696]

[2] Abd El-Hack ME, Mahgoub SA, Hussein MMA, Saadeldin IM. Improving growth performance and health status of meat-type quail by supplementing the diet with black cumin cold-pressed oil as a natural alternative for antibiotics. Environ Sci Pollut Res Int 2018; 25(2): 1157-67.

[http://dx.doi.org/10.1007/s11356-017-0514-0] [PMID: 29079983]

[3]   Hussein MMA, Abd El-Hack ME, Mahgoub SA, Saadeldin IM, Swelum AA. Effects of clove (*Syzygium aromaticum*) oil on quail growth, carcass traits, blood components, meat quality, and intestinal microbiota. Poult Sci 2019; 98(1): 319-29.
[http://dx.doi.org/10.3382/ps/pey348] [PMID: 30165540]

[4]   Kishawy ATY, Amer SA, Abd El-Hack ME, Saadeldin IM, Swelum AA. The impact of dietary linseed oil and pomegranate peel extract on broiler growth, carcass traits, serum lipid profile, and meat fatty acid, phenol, and flavonoid contents. Asian-Australas J Anim Sci 2019; 32(8): 1161-71.
[http://dx.doi.org/10.5713/ajas.18.0522] [PMID: 30744351]

[5]   Mahgoub SAM, El-Hack MEA, Saadeldin IM, Hussein MA, Swelum AA, Alagawany M. Impact of *Rosmarinus officinalis* cold-pressed oil on health, growth performance, intestinal bacterial populations, and immunocompetence of Japanese quail. Poult Sci 2019; 98(5): 2139-49.
[http://dx.doi.org/10.3382/ps/pey568] [PMID: 30590789]

[6]   Cervantes H. Banning antibiotic growth promoters: Learning from the European experience. Poult Int 2006; 10-2.

[7]   Elgeddawy SA, Shaheen HM, El-Sayed YS, *et al.* Effects of the dietary inclusion of a probiotic or prebiotic on florfenicol pharmacokinetic profile in broiler chicken. J Anim Physiol Anim Nutr (Berl) 2020; 104(2): 549-57.
[http://dx.doi.org/10.1111/jpn.13317] [PMID: 32017274]

[8]   Mohamed LA, El-Hindawy MM, Alagawany M, Salah AS, El-Sayed SAA. Effect of low- or high-CP diet with cold-pressed oil supplementation on growth, immunity and antioxidant indices of growing quail. J Anim Physiol Anim Nutr (Berl) 2019; 103(5): 1380-7.
[http://dx.doi.org/10.1111/jpn.13121] [PMID: 31141220]

[9]   Reda FM, Alagawany M, Mahmoud HK, Mahgoub SA, Elnesr SS. Use of red pepper oil in quail diets and its effect on performance, carcass measurements, intestinal microbiota, antioxidant indices, immunity and blood constituents. Animal 2019; 1-9. [Google Scholar]. [CrossRef].
[PMID: 31826776]

[10]   Alagawany M, Elnesr SS, Farag MR. The role of exogenous enzymes in promoting growth and improving nutrient digestibility in poultry. Majallah-i Tahqiqat-i Dampizishki-i Iran 2018; 19(3): 157-64. [Google Scholar].
[PMID: 30349560]

[11]   Ezzat Abd El-Hack M, Alagawany M, Ragab Farag M, *et al.* Beneficial impacts of thymol essential oil on health and production of animals, fish and poultry: a review. J Essent Oil Res 2016; 28(5): 365-82. [Google Scholar]. [CrossRef].
[http://dx.doi.org/10.1080/10412905.2016.1153002]

[12]   Abd El-Hack ME, Abdelnour SA, Taha AE, *et al.* Herbs as thermoregulatory agents in poultry: An overview. Sci Total Environ 2020; 703: 134399.
[http://dx.doi.org/10.1016/j.scitotenv.2019.134399] [PMID: 31757531]

[13]   Alagawany M, Abd El-Hack ME, Farag MR, Sachan S, Karthik K, Dhama K. The use of probiotics as eco-friendly alternatives for antibiotics in poultry nutrition. Environ Sci Pollut Res Int 2018; 25(11): 10611-8.
[http://dx.doi.org/10.1007/s11356-018-1687-x] [PMID: 29532377]

[14]   Alagawany M, Elnesr SS, Farag MR, *et al.* Licorice (Glycyrrhiza glabra) herb as an eco-friendly additive to promote poultry health—Current Knowledge and Prospects. Animals (Basel) 2019; 9: 536. [Google Scholar]. [CrossRef].
[http://dx.doi.org/10.3390/ani9080536]

[15]   Gado AR, Ellakany HF, Elbestawy AR, *et al.* Herbal medicine additives as powerful agents to control and prevent avian influenza virus in poultry–a review. Ann Anim Sci 2019; 19(4): 905-35. [Google Scholar]. [CrossRef].

[http://dx.doi.org/10.2478/aoas-2019-0043]

[16]   Elnesr SS, Ropy A, Abdel-Razik AH. Effect of dietary sodium butyrate supplementation on growth, blood biochemistry, haematology and histomorphometry of intestine and immune organs of Japanese quail. Animal 2019; 13(6): 1234-44.
[http://dx.doi.org/10.1017/S1751731118002732] [PMID: 30333074]

[17]   Kothari D, Lee WD, Niu KM, Kim SK. The Genus *Allium* as Poultry Feed Additive: A Review. Animals (Basel) 2019; 9(12): 1032.
[http://dx.doi.org/10.3390/ani9121032] [PMID: 31779230]

[18]   Zhang GF, Yang ZB, Wang Y, Yang WR, Jiang SZ, Gai GS. Effects of ginger root (*Zingiber officinale*) processed to different particle sizes on growth performance, antioxidant status, and serum metabolites of broiler chickens. Poult Sci 2009; 88(10): 2159-66.
[http://dx.doi.org/10.3382/ps.2009-00165] [PMID: 19762870]

[19]   Ali BH, Blunden G, Tanira MO, Nemmar A. Some phytochemical, pharmacological and toxicological properties of ginger (*Zingiber officinale* Roscoe): A review of recent research. Food Chem Toxicol 2008; 46(2): 409-20.
[http://dx.doi.org/10.1016/j.fct.2007.09.085] [PMID: 17950516]

[20]   Onu PN. Evaluation of two herbal spices as feed additives for finisher broilers. Biotechnol Anim Husb 2010; 26(5-6): 383-92. [Google Scholar]. [CrossRef].
[http://dx.doi.org/10.2298/BAH1006383O]

[21]   Wang WH, Wang ZM. [Studies of commonly used traditional medicine-ginger]. Zhongguo Zhongyao Zazhi 2005; 30(20): 1569-73. [Google Scholar]. [PubMed].
[PMID: 16422532]

[22]   Khan RU, Naz S, Nikousefat Z, *et al.* Potential applications of ginger (*Zingiber officinale*) in poultry diets. Worlds Poult Sci J 2012; 68(2): 245-52. [Google Scholar]. [CrossRef].
[http://dx.doi.org/10.1017/S004393391200030X]

[23]   Asghar A, Farooq M, Mian M, Khurshid A. Economics of broiler production in Mardan division. J Rural Dev Adm 2000; 32: 56-65. [Google Scholar].

[24]   An K, Zhao D, Wang Z, Wu J, Xu Y, Xiao G. Comparison of different drying methods on Chinese ginger (*Zingiber officinale* Roscoe): Changes in volatiles, chemical profile, antioxidant properties, and microstructure. Food Chem 2016; 197(Pt B): 1292-300.
[http://dx.doi.org/10.1016/j.foodchem.2015.11.033] [PMID: 26675871]

[25]   Mahboubi M. *Zingiber officinale* Rosc. essential oil, a review on its composition and bioactivity. Clinical Phytoscience 2019; 5(1): 6. [Google Scholar]. [CrossRef].
[http://dx.doi.org/10.1186/s40816-018-0097-4]

[26]   Jeena K, Liju VB, Kuttan R. Antioxidant, anti-inflammatory and antinociceptive activities of essential oil from ginger. Indian J Physiol Pharmacol 2013; 57(1): 51-62. [Google Scholar]. [PubMed].
[PMID: 24020099]

[27]   Govindarajan V. Ginger-Chemistry, Technology and Quality Evaluation: Part I-CRC Critical Reviews in Food Science and Nutrition. Quensland 1982; 17: 98. [Google Scholar].

[28]   Singh G, Maurya S, Catalan C, de Lampasona MP. Studies on essential oils, Part 42: chemical, antifungal, antioxidant and sprout suppressant studies on ginger essential oil and its oleoresin. Flavour Fragrance J 2005; 20(1): 1-6. [Google Scholar]. [CrossRef].
[http://dx.doi.org/10.1002/ffj.1373]

[29]   Kumar Sharma P, Singh V, Ali M. Chemical composition and antimicrobial activity of fresh rhizome essential oil of *Zingiber officinale* Roscoe. Pharmacogn J 2016; 8(3): 185-90. [Google Scholar]. [CrossRef].
[http://dx.doi.org/10.5530/pj.2016.3.3]

[30]   Sasidharan I, Venugopal VV, Menon AN. Essential oil composition of two unique ginger (*Zingiber*

officinale Roscoe) cultivars from Sikkim. Nat Prod Res 2012; 26(19): 1759-64.
[http://dx.doi.org/10.1080/14786419.2011.571215] [PMID: 21985708]

[31] Bhattarai K, Pokharel B, Maharjan S, Adhikari S. Chemical Constituents and Biological Activities of Ginger Rhizomes from Three Different Regions of Nepal. J. Nutri. Diet Probiotics 2018; 1: 180005. [Google Scholar].

[32] Stoyanova A, Konakchiev A, Damyanova S, Stoilova I, Suu PT. Composition and antimicrobial activity of ginger essential oil from Vietnam. J Essent Oil-Bear Plants 2006; 9(1): 93-8. [Google Scholar]. [CrossRef].
[http://dx.doi.org/10.1080/0972060X.2006.10643478]

[33] El-Baroty GS, El-Baky HA, Farag RS, Saleh MA. Characterization of antioxidant and antimicrobial compounds of cinnamon and ginger essential oils. Afr J Biochem Res 2010; 4: 167-74. [Google Scholar].

[34] Dieumou FE, Teguia A, Kuite JR, Tamokou JD, Fonge BN, Dongmo MC. Effects of ginger (*Z. officinale*) and garlic (*Allium sativum*) essential oils on growth performance and gut microbial population of broiler chickens. Livest Res Rural Dev 2009; 21: 23-32. [Google Scholar].

[35] Shanoon AK, Jassim MS, Amin QH, Ezaddin IN. Effects of Ginger (*Zingiber officinale*) Oil on Growth Performance and Microbial Population of Broiler Ross 308. Int J Poult Sci 2012; 11(9): 589-93. [Google Scholar]. [CrossRef].
[http://dx.doi.org/10.3923/ijps.2012.589.593]

[36] Herve T, Raphaël KJ, Ferdinand N, *et al.* 2018 Growth performance, serum biochemical profile, oxidative status, and fertility traits in male Japanese quail fed on ginger (*Zingiber officinale*, roscoe) essential oil. Vet Med Int 2018; 2018: 1-8.
[http://dx.doi.org/10.1155/2018/7682060] [PMID: 30050674]

[37] Herve T, Raphaël KJ, Ferdinand N, *et al.* Effects of Ginger (*Zingiber officinale,* Roscoe) Essential Oil on Growth and Laying Performances, Serum Metabolites, and Egg Yolk Antioxidant and Cholesterol Status in Laying Japanese Quail. J Vet Med 2019; 2019: 1-8.
[http://dx.doi.org/10.1155/2019/7857504] [PMID: 31001562]

[38] Habibi R, Sadeghi GH, Karimi A. Effect of different concentrations of ginger root powder and its essential oil on growth performance, serum metabolites and antioxidant status in broiler chicks under heat stress. Br Poult Sci 2014; 55(2): 228-37.
[http://dx.doi.org/10.1080/00071668.2014.887830] [PMID: 24697550]

[39] Mohamed NES. Response of Broiler Chicks to Diets Containing Mixture Garlic and Ginger Essential Oils as Natural Growth Promoter. Khartoum, Sudan: Sudan University for Science and Technology 2015.

[40] E.; ELnaggar, A.S. Growth and physiological response of gimmizah chicks to dietary supplementation with ginger, black seeds, thyme and oregano oil as natural feed additives. Egypt Poult Sci J 2016; 36: 1163-82. [Google Scholar].

[41] Ghasemi HA, Taherpour K. Comparison of broiler performance, blood biochemistry, hematology and immune response when feed diets were supplemented with ginger essential oils or mannan-oligosaccharide. Iran J Vet Med 2015; 9: 195-205. [Google Scholar].

[42] Tekeli A, Celik L, Kutlu HR, Gorgulu M. Effect of dietary supplemental plant extracts on performance, carcass characteristics, digestive system development, intestinal microflora and some blood parameters of broiler chicks. Proceedings of the 12th European Poultry Conference. Verona, Italy. 2006; pp. 10-4.

[43] Osman A, El-Araby GM, Taha H. Potential use as a bio-preservative from lupin protein hydrolysate generated by alcalase in food system. J Appl Biol Biotechnol 2016; 4: 076-81.

[44] An S, Liu G, Guo X, An Y, Wang R. Ginger extract enhances antioxidant ability and immunity of layers. Anim Nutr 2019; 5(4): 407-9.

[http://dx.doi.org/10.1016/j.aninu.2019.05.003] [PMID: 31890918]

[45]   Tchoffo H, Ngoula F, Kana JR, Kenfack A, Ngoumtsop VH, Vemo NB. Effects of ginger (*Zingiber officinale*) rhizomes essential oil on some reproductive parameters in laying Japanese quail (*Coturnix coturnix japonica*). Advances in Reproductive Sciences 2017; 5(4): 64-74. [Google Scholar]. [CrossRef].
       [http://dx.doi.org/10.4236/arsci.2017.54008]

[46]   Saleh N, Allam T, El-Latif AA, Ghazy E. The effects of dietary supplementation of different levels of thyme (*Thymus vulgaris*) and ginger (*Zingiber officinale*) essential oils on performance, hematological, biochemical and immunological parameters of broiler chickens. Glob Vet 2014; 12: 736-44. [Google Scholar].

[47]   AlTahtawy R, ElBastawesy A, Monem M, Zekry Z, AlMehdar H, ElMerzabani M. Antioxidant activity of the volatile oils of Zingiber officinale (ginger). Spatula DD - Peer Reviewed Journal on Complementary Medicine and Drug Discovery 2011; 1(1): 1-8. [Google Scholar]. [CrossRef].
       [http://dx.doi.org/10.5455/spatula.201012091111419]

[48]   Natta L, Orapin K, Krittika N, Pantip B. Essential oil from five Zingiberaceae for anti food-borne bacteria. Int Food Res J 2008; 15: 337-46. [Google Scholar].

[49]   Majolo C, Nascimento VP, Chagas EC, Chaves FCM. Antimicrobial activity of essential oil from *Curcuma longa* and *Zingiber officinale* rhizomes against enteric Salmonella isolated from chicken. Rev Bras Plantas Med 2014; 16: 505-12. [Google Scholar]. [CrossRef].
       [http://dx.doi.org/10.1590/1983-084X/13_109]

[50]   Qorbanpour M, Fahim T, Javandel F, *et al.* Effect of Dietary Ginger (*Zingiber officinale* Roscoe) and Multi-Strain Probiotic on Growth and Carcass Traits, Blood Biochemistry, Immune Responses and Intestinal Microflora in Broiler Chickens. Animals (Basel) 2018; 8(7): 117.
       [http://dx.doi.org/10.3390/ani8070117] [PMID: 30011890]

[51]   Alizadeh-Navaei R, Roozbeh F, Saravi M, Pouramir M, Jalali F, Moghadamnia AA. Investigation of the effect of ginger on the lipid levels. A double blind controlled clinical trial. Saudi Med J 2008; 29(9): 1280-4. [Google Scholar].
       [PMID: 18813412]

[52]   Hong JC, Steiner T, Aufy A, Lien TF. Effects of supplemental essential oil on growth performance, lipid metabolites and immunity, intestinal characteristics, microbiota and carcass traits in broilers. Livest Sci 2012; 144(3): 253-62. [Google Scholar]. [CrossRef].
       [http://dx.doi.org/10.1016/j.livsci.2011.12.008]

[53]   Jagetia GC, Baliga MS, Venkatesh P, Ulloor JN. Influence of ginger rhizome (*Zingiber officinale* Rosc) on survival, glutathione and lipid peroxidation in mice after whole-body exposure to gamma radiation. Radiat Res 2003; 160(5): 584-92.
       [http://dx.doi.org/10.1667/RR3057] [PMID: 14565823]

[54]   Shewita RS, Taha AE. Influence of dietary supplementation of ginger powder at different levels on growth performance, haematological profiles, slaughter traits and gut morphometry of broiler chickens. S Afr J Anim Sci 2018; 48: 997-1008. [Google Scholar]. [CrossRef].

[55]   Azhir D, Zakeri A, Rezapour AK. Effect of ginger powder rhizome on humeral immunity of broiler chickens. Eur J Exp Biol 2012; 2: 2090-2. [Google Scholar].

[56]   Arkan BM, Mohammed AM, Ali J. Effect of ginger (*Zingiber officinale*) on performance and blood serum parameters of broilers. Int J Poult Sci 2012; 91: 143-6. [Google Scholar].

[57]   Zhao X, Yang ZB, Yang WR, Wang Y, Jiang SZ, Zhang GG. Effects of ginger root (*Zingiber officinale*) on laying performance and antioxidant status of laying hens and on dietary oxidation stability. Poult Sci 2011; 90(8): 1720-7.
       [http://dx.doi.org/10.3382/ps.2010-01280] [PMID: 21753209]

[58]   Naveena BM, Mendiratta SK. Tenderisation of spent hen meat using ginger extract. Br Poult Sci 2001;

42(3): 344-9.
[http://dx.doi.org/10.1080/00071660120055313] [PMID: 11469554]

[59]   Attia YA, Saber SH, Abd-El-Hamid EA, Radwan MW. Response of broiler chickens to dietary supplementation of ginger (*Zingiber officinale*) continuously or intermittently in comparison with prebiotics. Egypt Poult Sci 2017; 37: 523-43. [Google Scholar].

[60]   Herawati , Marjuki . Herawati; Marjuki. The effect of feeding red ginger (*Zingiber officinale* Rosc.) as phytobiotic on broiler slaughter weight and meat quality. Int J Poult Sci 2011; 10(12): 983-6. [Google Scholar]. [CrossRef].
[http://dx.doi.org/10.3923/ijps.2011.983.986]

[61]   Wen C, Gu Y, Tao Z, Cheng Z, Wang T, Zhou Y. Effects of Ginger Extract on Laying Performance, Egg Quality, and Antioxidant Status of Laying Hens. Animals (Basel) 2019; 9(11): 857.
[http://dx.doi.org/10.3390/ani9110857] [PMID: 31652863]

[62]   Yang CW, Ding X, Zhao X, Guo YX, Mu AL, Yang ZB. Effects of star anise (Illicium verum Hook. f.), Salvia miltiorrhiza (Salvia miltiorrhiza Bge) and ginger root (*Zingiber officinale* Roscoe) on laying performance, antioxidant status and egg quality of laying hens. Europ Poult Sci 2017.

[63]   Si W, Chen YP, Zhang J, Chen ZY, Chung HY. Antioxidant activities of ginger extract and its constituents toward lipids. Food Chem 2018; 239: 1117-25.
[http://dx.doi.org/10.1016/j.foodchem.2017.07.055] [PMID: 28873530]

[64]   Li J, Wang S, Yao L, *et al.* 6-gingerol ameliorates age-related hepatic steatosis: Association with regulating lipogenesis, fatty acid oxidation, oxidative stress and mitochondrial dysfunction. Toxicol Appl Pharmacol 2019; 362: 125-35.
[http://dx.doi.org/10.1016/j.taap.2018.11.001] [PMID: 30408433]

[65]   Gurbuz Y, Salih YG. Influence of sumac (*Rhus Coriaria* L.) and ginger (*Zingiber officinale*) on egg yolk fatty acid, cholesterol and blood parameters in laying hens. J Anim Physiol Anim Nutr (Berl) 2017; 101(6): 1316-23.
[http://dx.doi.org/10.1111/jpn.12652] [PMID: 28160334]

[66]   Abdollahi AM, Virtanen HEK, Voutilainen S, *et al.* Egg consumption, cholesterol intake, and risk of incident stroke in men: the Kuopio Ischaemic Heart Disease Risk Factor Study. Am J Clin Nutr 2019; 110(1): 169-76.
[http://dx.doi.org/10.1093/ajcn/nqz066] [PMID: 31095282]

[67]   Oleforuh-Okoleh VU, Chukwu GC, Adeolu AI. Effect of ground ginger and garlic on the growth performance, carcass quality and economics of production of broiler chickens. Glob. J Bio-Dcience Biotechnol 2014; 3: 225-9. [Google Scholar].

[68]   Karangiya VK, Savsani HH, Patil SS, *et al.* Effect of dietary supplementation of garlic, ginger and their combination on feed intake, growth performance and economics in commercial broilers. Vet World 2016; 9(3): 245-50.
[http://dx.doi.org/10.14202/vetworld.2016.245-250] [PMID: 27057106]

[69]   Mohammed AA, Yusuf M. Evaluation of ginger (*Zingiber officinale*) as a feed additive in broiler diets. Livest Res Rural Dev 2011; 23: 202.http://www.lrrd.org/lrrd23/9/moha23202.htm

CHAPTER 4

Use of Cinnamon and its Derivatives in Poultry Nutrition

Rana M. Bilal[1], Faiz ul Hassan[2], Majed Rafeeq[3], Mayada R. Farag[4], Mohamed E. Abd El-Hack[5,*], Mahmoud Madkour[6] and Mahmoud Alagawany[5,*]

[1] *College of Veterinary and Animal Sciences, The Islamia University of Bahawalpur, 63100, Pakistan*

[2] *Institute of Animal & Dairy Sciences, Faculty of Animal Husbandry, University of Agriculture, Faisalabad, 38040, Pakistan*

[3] *University of Balochista Quetta, Pakistan*

[4] *Forensic Medicine and Toxicology Department, Veterinary Medicine Faculty, Zagazig University, Zagazig 44519, Egypt*

[5] *Poultry Department, Faculty of Agriculture, Zagazig University, Zagazig 44511, Egypt*

[6] *Animal Production Department, National Research Centre, Dokki, 12622 Giza, Egypt*

Abstract: The recent trend toward banning the use of antibiotics in poultry feed as a growth promoter directs the scientific community to look for natural alternatives with potential growth-promoting and immunomodulating properties. Phytogenic feed additives have attracted significant attention as alternatives to antibiotics to improve growth performance and enhance immune responses. They have anti-inflammatory, antioxidant, antiviral, and antifungal properties, depending on their chemical structure and composition. Scientists are using these non-conventional ingredients as feed additives in the form of oil or powder. Essential oils (EO) are volatile liquids produced from aromatic plants. Their application has gained momentum in controlling cholesterol as free radical scavengers, anti-microbials, antifungals, and stimulants of digestive enzymes. EO's possible antimicrobial features against harmful pathogens are primarily associated with the high content of volatile components in oils. The current chapter highlights the beneficial impact of cinnamon oil as a feed additive on poultry growth performance, meat quality, carcass traits, and its hypo-cholesterolaemic impact, antioxidant act, microbiological aspects, and immunomodulatory effects.

Keywords: Cinnamon, Essential oil, Hypo-cholesterolaemic property, an antibiotic alternative, Poultry.

---

* **Corresponding author Mahmoud Alagawany and Mohamed E. Abd El-Hack:** Poultry Department, Faculty of Agriculture, Zagazig University, Zagazig 44511, Egypt; E-mails: dr.mahmoud.alagwany@gmail.com and dr.mohamed.e.abdalhaq@gmail.com

INTRODUCTION

The poultry industry is the largest food-providing sector globally, and it is considered one of the primary livelihoods of rural communities [1]. According to many research data, it is proved that the dietary combination of natural products containing bioactive components and poultry products may beneficially impact human health [1, 2]. Essential oils derived from different plant parts are frequently used in cosmetics as a lush fragrance, skincare, and herbal perfumery. They are also used for medicinal purposes as potent antifungals that can possess inhibitory effects against harmful pathogens [3, 4].

Moreover, these natural herbs have gained much attention as growth promoters in today's poultry production because of their pronounced effects on birds' health and performance. These herbs and their derived products are replacing antibiotic growth promoters (AGP) to some extent. Besides, their local and economic availability makes them more precious as natural feed additives for the poultry industry. Apart from these beneficial effects, they positively impact digestion and absorption processes [5]. Recently, various herbs and extracts have been used to increase birds' overall productivity. Cinnamon essential oils (CEOs) and their major compounds have strong antibacterial effects against *Escherichia coli, Parahemolyticus, Staphylococcus epidermis, Enterococcus faecalis, Pseudomonas aeruginosa, Staphylococcus aureus,* and *Salmonella sp* [6].

Furthermore, diets fortified with cinnamon and its extracted oils have inhibitory effects against pathogenic bacteria and improve poultry performance by enhancing their digestive capacity [7]. The observation recorded by another study [8] indicated that cinnamon and its oil also have potent analgesic and hypocholesterolemic properties and may serve as strong free radical scavengers. This chapter describes the beneficial applications and new aspects of cinnamon and its derivatives, which will be valuable for physiologists, scientists, nutritionists, veterinarians, pharmacists, pharmaceutical industries, and poultry breeders.

CHEMICAL COMPOSITION

Various scientific reports have been published on quantifying and identifying compounds present in cinnamon. The essential volatile oil of cinnamon (CEO) is mainly obtained from its bark and leaf. The concentration of some cinnamon oil constituents is presented in Table **1**. The chemical constituents of CEO are immensely varied among the different plant parts, such as (roots, leaves, bark, and fruit). Moreover, the oil extraction method also influences its production. The essential oil of the bark of *Cinnamomum altissimum Kosterm* contains important compounds like α-terpineol (7.8%), methyl eugenol (12.8%), linalool (36.0%), and limonene (8.3%). The whole phenolic constituents represent about 50 μg GAE/mg oil. Also, the extract's antioxidant action was 345.2 μM Fe^{+2}/g dry mass using ferric reducing antioxidant power assay (FRAP) and with a concentration of IC50 of 38.5 μg/mL by the method DPPH (1,1-diphenyl-2-picrylhydrazyl) assay. It was observed that *Cinnamomum verum* leaf oil has an optimum volatile component (3.23%) [8].

Table 1. Some chemical constituents are identified in the cinnamon oil (leaf and bark).

Compound	Concentration (%) in Cinnamon Leafoil	Concentration (%) in Cinnamon Bark Oil
Eugenol	74.9	0.39-2.37
Benzaldehyde	0.1	0.23-0.31
Benzyl benzoate	3.0	0.01-0.37
Benzyl alcohol	0.2	0.14
Caryophyllene oxide	0.5	0.35
Cinnamaldehyde	1.1	62.09-89.31
Cinnamyl acetate	1.8	1.48-2.44
1,8-Cineole	0.6	1.02
Camphene	0.3	0.08-0.12
Linalool	2.5	1.6-4.08
α-Pinene	1.2	0.37-0.50
β-Phellandrene	0.2	0.23-0.25
α-Cubebene	0.9	0.12-0.21
Limonene	0.5	0.19-0.33
Cymene	0.8	0.02-1.31
α-Humulene	0.6	0.01-0.28
Myrcene	0.1	0.05-0.40
β-Pinene	0.3	0.07-0.15
Delta-3-Carene	0.6	0.37
β-Caryophyllene	4.1	0.89-2.05
Phenylethyl alcohol	0.1	0.15
α-Terpinene	0.1	0.03
α-Phellandrene	0.9	0.01
α-Terpineol	0.3	0.01
α-Thujene	0.2	-
Safrole	1.3	-

(Table 1) cont.....

Compound	Concentration (%) in Cinnamon Leafoil	Concentration (%) in Cinnamon Bark Oil
Styrene	0.1	-
Elemene	-	0.08-0.33
Borneol	-	0.01-0.12
Coumarin	-	0.41-0.47
Benzenepropanal	-	0.41
Hinesol	-	0.36
T-cadinol	-	2.47
α-Muurolene	-	4.32
α-Amorphene	-	1.98

Studies have shown that the essential oil of cinnamon species encompasses approximately fifteen vital chemical constituents. It is found that Cinnamaldehyde cinnamaldehyde (67.57%) is abundantly present in *Cinnamomum Verum* as compared to other wild cinnamon species. *Cinnamomum sinharajense* possesses comparatively higher oil (3.53%) content, while *Cinnamomum rivulorum* was found to contain the lowest stem bark oil (0.51%) and leaf oil (0.41%). However, the leaf portions of *Cinnamomum sinharajense* have a high quantity of the aromatic oily liquid, eugenol (87.53%). The composition of CEO essential oil extracted from its leaf, bark, fruit, and root indicates that it has y-cadinene (36.0%), 8-caryophyllene (5.6%), and T-cadinol (7.7%) compounds [9].

It is found that 84% of CEO is present in three isoprene units *i.e.*, sesquiterpenes; however, the other parts of cinnamon contain< 9% of this group of components.

Furthermore, the critical constituents of cinnamon leaf and bark oils were organic compounds such as phenylpropanoids. Still, a class of isoprenoids monoterpenes was the most decadent part almost (95%) of the root oil. Kasim *et al.* [10] noticed that the percentage of cinnamon oil production and the obtained chemical constituents depend on the extraction methods. The most significant amount of production was exhibited by soxhlet extraction with dichloromethane fraction (DCM-F) (5.22%), hexane (3.84%), and petroleum ether fractions (3.71%). But, the hydro-distillation procedure showed the lower CEO extraction (only 1.82%).

Results revealed nine major volatile compounds in the CEO extracted by hydro-distillation *viz*, ether, aldehydes, alcohols, ester, ketone, alkenes, and carboxylic acids. The trans-cinnamaldehyde is a flavored compound, mainly in the bark of cinnamon. Its concentration was 86.67% by the soxhlet extraction method using hexane. Atiphasaworn *et al.* [11] measured the volatile components of EO *from*

Cinnamomum bejolghota bark by using gas chromatography-mass spectrometry GC-MS and identified about 36 volatile compounds, among them the terpinen-4-ol and 1, 8-cineole, borneol, γ-terpineol were the most prominent compounds.

The oils derived from the flowers of *Cinnamomum bejolghota* possess decisive antifungal and antibacterial actions. An insignificant inhibitory concentration (MIC) of *Cinnamomum bejolghota bark* oil against bacteria extended from 31.25 - 62.50 μg/mL, while the inhibition actions against the fungal pathogens (125–500 μg/mL =MIC) were moderate. The quickly exerted antimicrobial function of *Cinnamomum bejolghota* bark oil was associated with 1, 8-cineole and borneol linalool, γ-terpineolnatural terpinen-4-ol terpene alcohol. Simsek *et al*. [12], found that the main constituents in the essential oil of bark of *Cinnamomum zeylanicum Lauraceae* were 88.2% cinnamaldehyde, 1.0% eugenol, and 8.0% benzyl alcohol with Gas chromatography (GC) and Gas Chromatography-Mass Spectrometry (GC-MS) systems [12]. The extracted oil from the bark of *Cinnamomum zeylanicum* has an exclusive, fragrant, mono-terpene that is a natural sourceof trans-cinnamaldehyde(45.62%). The main constituent liable for the cinnamon barks' aroma and therapeutic properties was 3-phenyl, 2-propenal (87.013%) [13].

EFFECTS ON GROWTH PERFORMANCE

Body Weight and Body Weight Gain

It has been observed that powder and oil of cinnamon (*Cinnamomum izeylanicum*) enhance birds' overall growth performance. Broilers fed with 2 g/kg cinnamon gained significantly higher body weight [14]. Likewise, in a 49-day study period, body weight gain (BWG) was higher in groups fed with ajwain essential oil and CEO [15]. Further, advocating the positive effect of CEO on quail performance, it was observed that quails fed with a concentration of (200mg/kg) cinnamon oil gained more weight at 21–35 days in comparison to cinnamon powder, (virginiamycin) antibiotic, and symbiotic [16].

Commercially available phytogenic growth promoters include alkaloids sanguinarine, capsicum oleoresin carvacrol, cinnamaldehyde, and chelerythrine to evaluate growth performance. It was concluded that additives did not exhibit any beneficial impact on growth [17]. The observation recorded by Shirzadegan [18] showed that 0.5% cinnamon powder resulted in significantly higher growth performance indices of broiler chicks. Furthermore, broiler chicks gained more weight with a diet fortified with CEO than control group chicks [19]. Apart from the beneficial effect of CEO compounds as growth promoters in poultry feed, Lee *et al*. [20] reported that female broiler chicks failed to produce a remarkable change in weight gain after the cinnamaldehyde supplementation.

Similarly, supplementation of broiler diets with 0.5 to 2 g/kg cinnamon did not positively affect their growth [21]. Additionally, supplementation of the quail diet with 250 and 500 mg/kg cinnamon oil (*Cinnamomum zeylanicum* L.) failed to impact the birds' weight gain [22] positively. Moreover, adding 0.5 or 1.0 mL/ kg of cinnamon oil to the broiler diet had no significant impact on body weight [23].

In today's scenario, the world is much concerned with implementing feasible approaches that may limit pathogenic bacterial growth and promote overall animal productivity by competitive exclusion, modulation of the intestinal system, and more nutrient absorption. In this view, scientists also used EO to promote animal productivity, particularly avian species. They concluded that EO could beneficially modify intestinal microarchitecture, enhance the immune status, and trigger nutrient digestion and absorption. Further, it was also observed that EO improves feed efficiency and growth performance by promoting immunogenic activity, modifying gut microbiota, elevating secretion of endogenous digestive enzymes, and eliciting antioxidant, antibacterial, and antiviral effects [24 - 26]. The outcome of cinnamon oil on body weight gain is presented in Fig. (**1**).

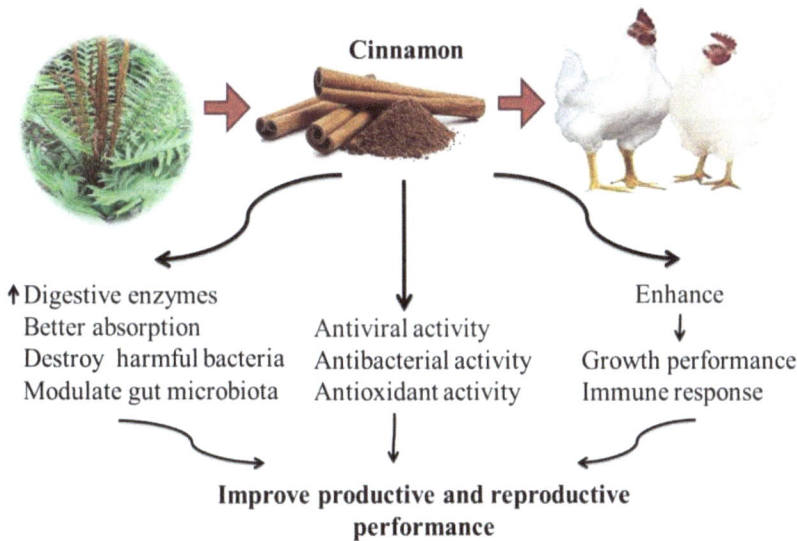

Fig. (1). The outcome of cinnamon oil on body weight gain.

Feed Utilization

The inclusion of cinnamon-derived products such as its oil and powder has become debatable. Still, most researchers agree that it may positively impact avian growth *via* improving its feed efficiency and body weight gain [12 - 28]. Birds fed with cinnamon oil at 500 ppm exhibited better growth [29].

Likewise, Pathak *et al.* [30] recorded that enramycin at a dose of 125 mg/kg, or a mixture of 500 mg/kg calcium formate and cinnamaldehyde significantly promoted the feed efficiency of *E.coli* (10^8 bacteria/bird) challenged broilers during the 2nd week of production and suggested that EO and organic acid could replace the antibiotic. A mixture of thyme and cinnamon essential oils at 200 ppm significantly improved feed efficiency, feed intake (FI), and final weight relative to the control group [19]. Similarly, it was recorded that dietary (200 mg/kg) cinnamon oil significantly improved the feed conversion ratio (FCR) of quails relative to control birds (0–35 days) without affecting the FI [27].

On the other side, the trial conducted by Simsek *et al.* [12] reported that the inclusion of cinnamon oil failed to produce a positive impact on feed utilization. Similarly, laying hens subjected to cold stress conditions (8.8 ± 3 °C) and fed a diet containing Zn and CEO (combined or single) did not show any positive impact on feed efficiency [28]. Interestingly [16], it was noted that the addition of 200 mg/kg virginiamycin or cinnamon oil in the quail diet positively impacted feed efficiency. Quails fortified with a dose of 250 and 500 mg/kg cinnamon oil (*Cinnamomum zeylanicum L.*) showed no improvement in feed efficiency [22]. Hernandez *et al.* [31] showed that broiler chicks fed a diet containing 200 ppm essential oil extract (EOE) from cinnamon, pepper, and oregano had no remarkable change in parameters of FI or FCR during the rearing stage of production. The study conducted by Lee *et al.* [20] indicated that a group of broilers fed with cinnamaldehyde failed to significantly influence the FI and FCR traits. Moreover, the author further added that the additive negatively impacted water intake.

CARCASS TRAITS

Cinnamon has been reported to promote the unsaturated fatty acids of carcass and alleviate its content of saturated fatty acids (SFA) [32]. Dietary supplementation with *Cinnamomum zeylanicum L.* essential oil at 250 and 500 mg/kg in the quail diet failed to affect quail carcass characteristics [22] significantly. Similarly, Hernandez *et al.* [31] observed that 200 ppm EOE failed to exhibit any change in the weights of proventriculus, pancreas, gizzard, liver, and intestine contents in broiler chickens. Further, CEO and ajwain essential oils at 3 and 4 g/kg levels posed no significant change in broiler carcass traits [15]. Furthermore, a broiler diet fortified with 0.5 or 1.0 mL per kg CEO showed no substantial change in internal organs weights and carcass characteristics [23].

BLOOD PARAMETERS

Most studies were carried out to evaluate the impact of CEO on poultry birds' blood biochemical profiles. Reports suggested that CEO supplementation has beneficially modulated immunity, lipid profile and antioxidant activity. A mixture of CEO and Zn significantly decreased serum triglycerides levels in a study on laying hens reared under cold stress conditions [28]. Also, Al-Kassie [19] and Ciftci *et al.* [29] reported that the CEO could promote serum glutathione peroxidase and catalase enzyme activities, omega-6 fatty acids, total unsaturated fatty acid ratio and blood phagocytic activity. Abudabos *et al.* [33] pointed out that the concentrations of plasma thiobarbituric acid reactive substances(TBARS), total globulin and serum protein were improved in birds infected when the birds were subjected to a diet supplemented with anise, yucca extract, cinnamaldehyde, carvacrol, oregano and thyme essential oils. Broilers were supplemented with a diet containing bamboo leaf flavonoid (BLF) and CEO; they showed a significant change in the liver malondialdehyde (MDA) contents at the 3rd week and serum IgM contents at the 6th week of age [34].

Moreover, Quails showed a reduction in plasma cholesterol and triglycerides when fed with a diet having oregano essential oil (OEO) plus CEO or *mannanoligosaccharide* [35]. The hypo-cholesterol function of EO and its derived compounds might be associated with the discouraging effects on 3-hydroxy-3-methylglutaryl coenzyme A (HMG-CoA reductase) [36], which is a principal enzyme in cholesterol synthesis [37]. Furthermore, spices such as linalool and eugenol cuminaldehyde were exhibited to prevent lipid peroxidation *via* lowering the production of free radicals and MDA formation and consequently limiting the cell membrane destruction [38].

INTESTINAL MICROBIOTA

It has been reported that modern nutritional approaches show a significant role in boosting poultry birds' health and performance [38]. Birds possessing balance gut microbiota may have good feed efficiency [39]. According to Mehdipour *et al.* [16] and Yang *et al.*, [34] CEOs may act as potential antimicrobial agents in birds and behave as potential food bio-preservative agents. The inclusion of 100 mg/kg CEO may promote the growth of *Lactobacillus* and *Bifidobacterium* and inhibit cecal *E. coli* growth in broilers [34]. Similarly, Hammer [40] and Zhao [41] added that folk herbal essential oils viz: cinnamaldehyde, thymol, and carvacrol controlled the growth and population of harmful bacteria, including *Salmonella* and *E. coli.*

Volatile compounds such as cinnamaldehyde, eugenol, and carvacrol found in cinnamon essential oil exhibited noticeable antimicrobial and antifungal activity

[42]. In an *in vitro* study, it was recorded that CEO successfully inhibited the growth of *E.coli, Klebsiella sp. Listeria monocytogenes, Rhizomucor sp* and *Bacillus sp* [43].

Further, in an *in vitro* study on the cecal contents, the inclusion of 20 mM of cinnamaldehyde, eugenol, carvacrol, or thymol successfully reduced *Campylobacter* growth after an incubation period of 15s. However, at the eight h of incubation, ten mM concentrations of these ingredients were found to be much more effective in decreasing the *C. jejuni* count by at least 5-log CFUs/mL [44].

Same as above, a mixture of cinnamaldehyde and thymol inhibited the growth and proliferation of pathogens and promoted overall performance by beneficially modulating the gut microarchitecture in meat birds [45, 46]. In another study, broiler chick's given a dose of 300 mg/kg of cinnamon bark oil decreased *E. coli* growth of ceca [47].

Cinnamon and its derived compound cinnamaldehyde positively impact birds' gut health when combined with capsaicin and carvacrol [48].

Quails fed a diet containing 200 ppm/kg diet CEO promoted the growth of ileal Lactobacillus and reduced ileal coliforms count relative to the control group and other group containing antibiotic (virginiamycin) [27].

Moreover, volatile components such as carvacrol influence the pH equilibrium of inorganic ions by breaking the membrane integrity [49].

A study by Pathak *et al.* [30] suggested that broiler fed a diet containing a mixture of calcium formate and cinnamaldehyde (500 mg/kg feed) exhibited no significant difference in ileal and cecal total bacterial counts, *Lactobacillus* and *E.coli.* However, the additive inhibited the growth of *Salmonella* counts significantly. This may be attributed to the ability of the herbal extract to break the organism's cell structure. They may enhance digestion and absorption hence promoting the overall productivity [50]. While the CEO application may positively impact health and growth performance, immunogenic responses, lipid profile, antioxidant and antibacterial functions of poultry birds [51].

These beneficial impacts may be associated with the phenolic compounds in cinnamon oil that may break the cell membranes of bacteria such as *E. coli.*

Furthermore, essential oil enhances mucus production in the gut, consequently lowering pathogenic bacterial attachment to the epithelial cells of the intestinal lumen [48].

CONCLUSION

These key results suggest that the inclusion of cinnamon essential oil extracts in poultry diets as feed additives may exhibit positive growth-promoting, free radical scavenger activity, as well as hypo-cholesterolaemic, immunomodulatory, and microbiological activities. Cinnamon and its derived compounds may be used as an effective growth promoter replacing antibiotics for better health protection, and economical, environmental, and performance features of the poultry production sector.

CONSENT FOR PUBLICATION

Not applicable.

CONFLICT OF INTEREST

The author declares no conflict of interest, financial or otherwise.

ACKNOWLEDGEMENTS

Declared none.

REFERENCES

[1]   Alagawany M, Elnesr SS, Farag MR, *et al.* Use of licorice (*Glycyrrhiza glabra*) herb as a feed additive in poultry: Current knowledge and prospects. Animals (Basel) 2019; 9(8): 536.
        [http://dx.doi.org/10.3390/ani9080536] [PMID: 31394812]

[2]   Abd El-Hack ME, Alagawany M, Shaheen H, *et al.* Ginger and Its Derivatives as Promising Alternatives to Antibiotics in Poultry Feed. Animals (Basel) 2020; 10(3): 452.
        [http://dx.doi.org/10.3390/ani10030452] [PMID: 32182754]

[3]   Alagawany M, Elnesr SS, Farag MR. Use of liquorice (*Glycyrrhiza glabra*) in poultry nutrition: Global impacts on performance, carcass and meat quality. Worlds Poult Sci J 2019; 75(2): 293-304.
        [https://doi.org/10.1017/S0043933919000059].
        [http://dx.doi.org/10.1017/S0043933919000059]

[4]   Gutiérrez RMP, Mitchell S, Solis RV. *Psidium guajava*: A review of its traditional uses, phytochemistry and pharmacology. J Ethnopharmacol 2008; 117(1): 1-27.
        [http://dx.doi.org/10.1016/j.jep.2008.01.025] [PMID: 18353572]

[5]   Talat Guler , Mehmet Ciftci , Dalkilic B, Bestami D, Simsek G. The effect of an essential oil mix derived from oregano, clove and aniseed on broiler performance. Int J Poult Sci 2005; 4(11): 879-84.
        [http://dx.doi.org/10.3923/ijps.2005.879.884]

[6]   El Atki Y, Aouam I, El Kamari F, *et al.* Antibacterial activity of cinnamon essential oils and their synergistic potential with antibiotics. J Adv Pharm Technol Res 2019; 10(2): 63. [https://doi.org /10.4103/japtr.JAPTR_366_18].
        [http://dx.doi.org/10.4103/japtr.JAPTR_366_18]

[7]   Wenk C. Why all the discussion about herbs? Lyons, TP. Nicholasville, KY, USA: Nottingham University Press 2000; pp. 79-96.

[8]   Baig MW, Nasir B, Waseem D, Majid M, Khan MZI, Haq I. *Withametelin*: a biologically active

withanolide in cancer, inflammation, pain and depression. Saudi Pharm J 2020; 28(12): 1526-37. [http://dx.doi.org/10.1016/j.jsps.2020.09.021] [PMID: 33424246]

[9]   Paranagama PA, Wimalasena S, Jayatilake GS, Jayawardena AL, Senanayake UM, Mubarak AMA. Comparison of essential oil constituents of bark, leaf, root and fruit of Cinnamon (Cinnamomum zeylanicum Blum) grown in Sri Lanka. J Natn Sci Foundation Sri Lanka 2001; 29: 147-53.

[10]   Kasim NN, Ismail SNAS, Masdar ND, Hamid FA, Nawawi WI. Extraction and potential of cinnamon essential oil towards repellency and insecticidal activity. Int J Sci Res 2014; 4(7): 2250-3153.

[11]   Atiphasaworn P, Monggoot S, Pripdeevech P. Chemical composition, antibacterial and antifungal activities of *Cinamomum bejolghota* bark oil from Thailand J Appl. Pharm Sci 2017; 7: 69-73. [https://doi.org/10.7324/JAPS.2017.70409].

[12]   Simsek ÜG, Ciftci M, Doğan G, Özçelik M. Antioxidant activity of cinnamon bark oil (*Cinnamomum zeylanicum L.*) in Japanese quails under thermo neutral and heat stressed conditions. Kafkas Univ Ve. Fak Derg 2013; 19: 889-94. [https://doi.org/10.9775/kvfd.2013.9049].

[13]   El-Baroty GS, El-Baky HA, Farag RS, Saleh MA. Characterization of antioxidant and antimicrobial compounds of cinnamon and ginger essential oils. Afr J Biochem Res 2010; 4: 167-74. [https://doi.org/10.5897/AJBR.9000056].

[14]   Toghyani M, Toghyani M, Gheisari A, Ghalamkari G, Eghbalsaied S. Evaluation of cinnamon and garlic as antibiotic growth promoter substitutions on performance, immune responses, serum biochemical and haematological parameters in broiler chicks. Livest Sci 2011; 138(1-3): 167-73. [https://doi.org/10.1016/j.livsci.2010.12.018]. [http://dx.doi.org/10.1016/j.livsci.2010.12.018]

[15]   Devi PC, Samanta AK, Das B, Kalita G, Behera PS, Barman S. Effect of plant extracts and essential oil blend as alternatives to antibiotic growth promoters on growth performance, nutrient utilization and carcass characteristics of broiler chicken. Indian J Anim Nutr 2018; 35(4): 421-7. [https://doi.org/10.5958/2231-6744.2018.00064.6]. [http://dx.doi.org/10.5958/2231-6744.2018.00064.6]

[16]   Mehdipour Z, Afsharmanesh M, Sami M. Effects of dietary synbiotic and cinnamon (*Cinnamomum verum*) supplementation on growth performance and meat quality in Japanese quail. Livest Sci 2013; 154(1-3): 152-7. [https://doi.org/10.1016/j.livsci.2013.03.014]. [http://dx.doi.org/10.1016/j.livsci.2013.03.014]

[17]   Muhl A, Liebert F. Growth, nutrient utilization and threonine requirement of growing chicken fed threonine limiting diets with commercial blends of phytogenic feed additives. J Poult Sci 2007; 44(3): 297-304. [https://doi.org/10.2141/jpsa.44.297]. [http://dx.doi.org/10.2141/jpsa.44.297]

[18]   Shirzadegan K. Reactions of modern broiler chickens to administration of cinnamon powder in the diet. Iran J Appl Anim Sci 2014; 4: 367-71.

[19]   Al-Kassie GA. Influence of two plant extracts derived from thyme and Cinnamon on broiler performance. Pak Vet J 2009; 29: 169-73.

[20]   Lee KW, Everts H, Kappert HJ, Frehner M, Losa R, Beynen AC. Effects of dietary essential oil components on growth performance, digestive enzymes and lipid metabolism in female broiler chickens. Br Poult Sci 2003; 44(3): 450-7. [http://dx.doi.org/10.1080/0007166031000085508] [PMID: 12964629]

[21]   Koochaksaraie RR, Irani M, Gharavysi S. The effects of cinnamon powder feeding on some blood metabolites in broiler chicks. Rev Bras Cienc Avic 2011; 13(3): 197-202. [https://doi.org/10.1590/S1516-635X2011000300006]. [http://dx.doi.org/10.1590/S1516-635X2011000300006]

[22]   Tonbak F, Çiftçi M. Effects of cinnamon oil (*Cinnamomum zeylanicum L.*) supplemented to ration on growth performance and carcass characteristics in heat-stressed Japanese quails. Firat Univ Saglik

Bilim Vet Derg 2012; 26: 157-64.

[23]   Symeon GK, Athanasiou A, Lykos N, *et al.* The effects of dietary Cinnamon (*Cinnamomum zeylanicum*) oil supplementation on broiler feeding behaviour, growth performance, carcass traits and meat quality characteristics. Ann Anim Sci 2014; 14(4): 883-95. [https://doi.org/10.2478/aoas-201--0047].
[http://dx.doi.org/10.2478/aoas-2014-0047]

[24]   Saeed M, Kamboh AA, Syed SF, *et al.* Phytochemistry and beneficial impacts of cinnamon (*Cinnamomum zeylanicum*) as a dietary supplement in poultry diets. Worlds Poult Sci J 2018; 74(2): 331-46. [https://doi.org/10.1017/S0043933918000235].
[http://dx.doi.org/10.1017/S0043933918000235]

[25]   Mahgoub SAM, El-Hack MEA, Saadeldin IM, Hussein MA, Swelum AA, Alagawany M. Impact of *Rosmarinus officinalis* cold-pressed oil on health, growth performance, intestinal bacterial populations, and immunocompetence of Japanese quail. Poult Sci 2019; 98(5): 2139-49.
[http://dx.doi.org/10.3382/ps/pey568] [PMID: 30590789]

[26]   Abo Ghanima MM, Elsadek MF, Taha AE, *et al.* Effect of housing system and rosemary and cinnamon essential oils on layers performance, egg quality, haematological traits, blood chemistry, immunity, and antioxidant. Animals (Basel) 2020; 10(2): 245.
[http://dx.doi.org/10.3390/ani10020245] [PMID: 32033082]

[27]   Mehdipour Z, Afsharmanesh M. Evaluation of synbiotic and Cinnamon (*Cinnamomum verum*) as antibiotic growth promoter substitutions on growth performance, intestinal microbial populations and blood parameters in Japanese quail. J Livest Sci Technol 2018; 6: 1-8. [https://doi.org/10.22103/jlst.2018.10558.1200].

[28]   Torki M, Akbari M, Kaviani K. Single and combined effects of zinc and cinnamon essential oil in diet on productive performance, egg quality traits, and blood parameters of laying hens reared under cold stress condition. Int J Biometeorol 2015; 59(9): 1169-77.
[http://dx.doi.org/10.1007/s00484-014-0928-z] [PMID: 25376631]

[29]   Ciftci M, Dalkilic B, Cerci IH, Guler T, Ertas ON, Arslan O. Influence of dietary cinnamon oil supplementation on performance and carcass characteristics in broilers. J Appl Anim Res 2009; 36(1): 125-8. [https://doi.org /10.1080/09712119.2009.9707045].
[http://dx.doi.org/10.1080/09712119.2009.9707045]

[30]   Pathak M, Mandal GP, Patra AK, Samanta I, Pradhan S, Haldar S. Effects of dietary supplementation of *cinnamaldehyde* and formic acid on growth performance, intestinal microbiota and immune response in broiler chickens. Anim Prod Sci 2017; 57(5): 821-7. [https://doi.org /10.1071/AN15816].
[http://dx.doi.org/10.1071/AN15816]

[31]   Hernández F, Madrid J, García V, Orengo J, Megías MD. Influence of two plant extracts on broilers performance, digestibility, and digestive organ size. Poult Sci 2004; 83(2): 169-74.
[http://dx.doi.org/10.1093/ps/83.2.169] [PMID: 14979566]

[32]   Dalkilic B, Ciftci M, Guler T, Cerci IH, Ertas ON, Guvenc M. Influence of dietary cinnamon oil supplementation on fatty acid composition of liver and abdominal fat in broiler chicken. J Appl Anim Res 2009; 35(2): 173-6. [https://doi.org /10.1080/09712119.2009.9707011].
[http://dx.doi.org/10.1080/09712119.2009.9707011]

[33]   Abudabos AM, Alyemni AH, Dafalla YM, Khan RU. The effect of phytogenics on growth traits, blood biochemical and intestinal histology in broiler chickens exposed to *Clostridium perfringens* challenge. J Appl Anim Res 2018; 46(1): 691-5. [https://doi.org /10.1080/09712119.2017.1383258].
[http://dx.doi.org/10.1080/09712119.2017.1383258]

[34]   Yang Y, Zhao L, Shao Y, *et al.* Effects of dietary graded levels of cinnamon essential oil and its combination with bamboo leaf flavonoid on immune function, antioxidative ability and intestinal microbiota of broilers. J Integr Agric 2019; 18(9): 2123-32. [https://doi.org /10.1016/S2095-3119(19)62566-9].

[http://dx.doi.org/10.1016/S2095-3119(19)62566-9]

[35]   Sarica S, Corduk M, Yarim GF, Yenisehirli G, Karatas U. Effects of novel feed additives in wheat based diets on performance, carcass and intestinal tract characteristics of quail. S Afr J Anim Sci 2009; 39(2): 39.
[http://dx.doi.org/10.4314/sajas.v39i2.44388]

[36]   Elson CE, Underbakke GL, Hanson P, Shrago E, Wainberg RH, Qureshi AA. Impact of lemongrass oil, an essential oil, on serum cholesterol. Lipids 1989; 24(8): 677-9.
[http://dx.doi.org/10.1007/BF02535203] [PMID: 2586227]

[37]   Goldstein JL, Brown MS. Regulation of the mevalonate pathway. Nature 1990; 343(6257): 425-30.
[http://dx.doi.org/10.1038/343425a0] [PMID: 1967820]

[38]   Naidu KA. Eugenol — an inhibitor of lipoxygenase-dependent lipid peroxidation. Prostaglandins Leukot Essent Fatty Acids 1995; 53(5): 381-3.
[http://dx.doi.org/10.1016/0952-3278(95)90060-8] [PMID: 8596779]

[39]   Elnesr SS, Ropy A, Abdel-Razik AH. Effect of dietary sodium butyrate supplementation on growth, blood biochemistry, haematology and histomorphometry of intestine and immune organs of Japanese quail. Animal 2019; 13(6): 1234-44.
[http://dx.doi.org/10.1017/S1751731118002732] [PMID: 30333074]

[40]   Hammer KA, Carson CF, Riley TV. Antimicrobial activity of essential oils and other plant extracts. J Appl Microbiol 1999; 86(6): 985-90.
[http://dx.doi.org/10.1046/j.1365-2672.1999.00780.x] [PMID: 10438227]

[41]   Zhao LL, Liao XD, Zhang LY, Luo XG, Lu L. Bacteriostatic effects of plant extracts and their compounds on chicken pathogenic bacteria *in vitro*. Chinese J Anim Nutr 2017; 29: 3277-86.

[42]   Basílico MZ, Basílico JC. Inhibitory effects of some spice essential oils on Aspergillus ochraceus NRRL 3174 growth and ochratoxin A production. Lett Appl Microbiol 1999; 29(4): 238-41.
[http://dx.doi.org/10.1046/j.1365-2672.1999.00621.x] [PMID: 10583751]

[43]   Gupta C, Garg AP, Uniyal RC, Kumari A. Comparative analysis of the antimicrobial activity of cinnamon oil and cinnamon extract on some food-borne microbes. Afr J Microbiol Res 2008; 2: 247-51. [https://doi.org/10.5897/AJMR.9000180].

[44]   Kollanoor Johny A, Darre MJ, Donoghue AM, Donoghue DJ, Venkitanarayanan K. Antibacterial effect of trans-cinnamaldehyde, eugenol, carvacrol, and thymol on *Salmonella Enteritidis* and *Campylobacter jejuni* in chicken cecal contents *in vitro*. J Appl Poult Res 2010; 19(3): 237-44.
[http://dx.doi.org/10.3382/japr.2010-00181]

[45]   Ouwehand AC, Tiihonen K, Kettunen H, Peuranen S, Schulze H, Rautonen N. *In vitro* effects of essential oils on potential pathogens and beneficial members of the normal microbiota. Vet Med (Praha) 2010; 55(2): 71-8.
[http://dx.doi.org/10.17221/152/2009-VETMED]

[46]   Tiihonen K, Kettunen H, Bento MHL, *et al.* The effect of feeding essential oils on broiler performance and gut microbiota. Br Poult Sci 2010; 51(3): 381-92.
[http://dx.doi.org/10.1080/00071668.2010.496446] [PMID: 20680873]

[47]   Chowdhury S, Mandal GP, Patra AK, *et al.* Different essential oils in diets of broiler chickens: 2. Gut microbes and morphology, immune response, and some blood profile and antioxidant enzymes. Anim Feed Sci Technol 2018; 236: 39-47. [https://doi.org/10.1016/j.anifeedsci.2017.12.003].
[http://dx.doi.org/10.1016/j.anifeedsci.2017.12.003]

[48]   Jamroz D, Wiliczkiewicz A, Wertelecki T, Orda J, Skorupińska J. Use of active substances of plant origin in chicken diets based on maize and locally grown cereals. Br Poult Sci 2005; 46(4): 485-93.
[http://dx.doi.org/10.1080/00071660500191056] [PMID: 16268107]

[49]   Lambert RJW, Skandamis PN, Coote PJ, Nychas GJE. A study of the minimum inhibitory concentration and mode of action of oregano essential oil, thymol and carvacrol. J Appl Microbiol

2001; 91(3): 453-62.
[http://dx.doi.org/10.1046/j.1365-2672.2001.01428.x] [PMID: 11556910]

[50] Faix Š, Faixová Z, Plachá I, Koppel J. Effect of *Cinnamomum zeylanicum* essential oil on antioxidative status in broiler chickens. Acta Vet Brno 2009; 78(3): 411-7.
[http://dx.doi.org/10.2754/avb200978030411]

[51] Altop A, Erener G, Duru ME, Isik K. Effects of essential oils from *Liquidambar orientalis* Mill. leaves on growth performance, carcass and some organ traits, some blood metabolites and intestinal microbiota in broilers. Br Poult Sci 2018; 59(1): 121-7.
[http://dx.doi.org/10.1080/00071668.2017.1400657] [PMID: 29094608]

CHAPTER 5

Clove (*Syzygium aromaticum*) and its Derivatives in Poultry Feed

Mahmoud Alagawany[1,*], Mohamed E. Abd El-Hack[1,*], Muhammad Saeed[2], Shaaban S. Elnesr[3] and Mayada R. Farag[4]

[1] *Department of Poultry, Faculty of Agriculture, Zagazig University, Zagazig 44511, Egypt*

[2] *Cholistan Veterinary University, Bahawalpur 63100, Pakistan*

[3] *Poultry Production Department, Faculty of Agriculture, Fayoum University, Fayoum 63514, Egypt*

[4] *Forensic Medicine and Toxicology Department, Veterinary Medicine Faculty, Zagazig University, Zagazig 44519, Egypt*

Abstract: Production of safe and healthy poultry diets of high profitability is the central aim of poultry men. This safety is achieved by using natural products as growth stimulants. Natural feed additives such as medicinal products derived from herbs and spices are mainly used in the poultry feed industry as appetite and enzyme secretion stimulants. The use of clove (*Syzygium aromaticum*) and its derivatives has lately received much greater attention as an alternative to traditional antibiotics. The clove exhibited strong antibacterial, antioxidant, anti-septic and anti-inflammatory properties and appetite and digestion stimulants. The clove and its derivatives contain bioactive components, including eugenol, eugenyl acetate, β-caryophyllene, salicylic acid, ferulic acid, caffeic acid, ellagic acid, kaempferol, methyl amyl ketone, humulene, gallotannic acid, and crategolic acid that have beneficial effects. Eugenol is the main bioactive component present in the clove. The potential advantages of utilizing clove extracts in poultry diets include improved growth performance, egg production and feed conversion ratio, enhanced digestion, and down-regulated disease incidence. From the available literature, clove and its essential oil is one of the beneficial plant extracts to increase growth performance in poultry by improving the intestinal microbiota population. Clove extract contains various molecules (principally eugenol) that have self-biological activities in poultry physiology and metabolism. This chapter includes information on clove and its derivatives in poultry nutrition.

Keywords: Antibacterial, Antioxidant, Clove, Eugenol, Poultry nutrition.

---

[*] **Corresponding author Mahmoud Alagawany and Mohamed E. Abd El-Hack:** Department of Poultry, Faculty of Agriculture, Zagazig University, Zagazig 44511, Egypt; E-mails: mmalagwany@zu.edu.eg and dr.mohamed.e.abdalhaq@gmail.com

INTRODUCTION

The clove (*Syzygium aromaticum,* Fig. **1**) is a precious, valuable, and important spice of the family *Myrtaceae* used as a food preservative and its medicinal benefits are known for centuries. Clove has a historical background and is believed to be originated in the 1ˢᵗ century BC [1]. Its unique odor and sweet taste are commonly used as a spice worldwide [2]. For the growth of clove, well-drained, organic matter-rich loam soil is required. The optimum temperature required for clove growth is 20-30°C, whereas above 10°C constant temperature is vital for it. This specie barely tolerates the waterlogged soggy conditions. The geographical area with 150-300 cm rainfall is considered for its growth [2]. It has a good concentration of fiber and minerals like magnesium, potassium and calcium. The clove also has a high concentration of manganese [3]. Besides, the clove is also a good source of omega-6 fatty acids.

Fig. (1). Clove (*Syzygium aromaticum*).

Clove is an important source of various phenolic compounds, including hydroxybenzoic acid, hydroxycinamic acids and hydroxyphenyl propenes. Eugenol is the main bioactive component in the clove, present in a concentration of 9381.7 to 14650 mg/100 g of fresh plant weight. Gallic acid is also found in a 783.5 mg/100 g fresh weight [4]. Around 25 to 20% volatile oil is present in good quality oil containing 70-85% eugenol, 10-15% eugenol acetate and 5-12% beta-caryophyllene. Some other constituents are responsible for the characteristic and pleasant fragrance. However, these constituents are present in minor quantities and include kaempferol, methyl amyl ketone, α-humulene, β- humulene, gallotannic acid, methyl salicylate, benzaldehyde and crategolic acid [5]. Some other phenolic acids present in the clove are salicylic acid, ferulic acid, caffeic acid and ellagic acid, whereas flavonoids are present in trace amounts, including kaempferol and quercetin. Appreciable amounts of essential oil are present in aerial parts of clove [5 - 10].

One of the main phytochemicals present in the clove essential oil (CEO) is eugenol and shows bactericidal action by increasing the permeability of the cell membrane and burst of the cytoplasmic membrane. The CEO and eugenol have antibacterial potential against gram-negative bacteria like *salmonella Typhimurium, salmonella enteritidis, E. coli* and *Pseudomonas aeruginosa,* and gram-positive bacteria like *Listeria monocytogenes, streptococcus pyogenes, streptococcus pneumonia, staphylococcus aureus, staphylococcus epidermis, Bacillus subtilis* and *Bacillus cereus* [11 - 13]. Researchers have reported the antiviral activity of clove, where Eugenin that is isolated from CEO exhibited an inhibitory response (dose: 10 μg/ml) against the herpes simplex virus [14]. Similarly, researchers found that administering eugenol (intravenous and intragastric route) in rabbits compared to paracetamol showed greater fever-reducing potential [1]. The clove and its derivatives are found to be active against fungi. They act as fungicide agents by increasing the cell permeability and altering the cell morphology of the different fungi, namely, *Aspergillus species, Candida albicans, Dermatophyte species, Epidermophyton floccosum, Fusarium moniliforme, Fusarium oxysporum, Mucor species, Microsporum canis, Microsporum gypseum, Trichophyton mentagrophytes, Trichophyton rubrum* [15 - 17]. The main objectives of the following sections are to provide an overview of the use of clove and its derivatives in poultry feeding. Lastly, the nutritional value of the clove plants and their effect on immunity, performance, and other health aspects will be very valuable for physiologists, scientists, nutritionists, veterinarians and pharmacists.

NUTRITIVE VALUE OF CLOVE

The clove (*Syzygium aromaticum*) is a precious, valuable, and important spice of *Myrtaceae.* Due to its unique odor and sweet taste, the clove is commonly used as a spice worldwide [2]. Besides its taste and odor, clove is rich in nutrient content. The nutrient composition may vary in different parts of the world due to the agro-climatic conditions under which cloves are cultivated and the various methods used to process and prepare these dried clove buds [18]. The clove contains good fiber, magnesium, potassium and calcium. The clove also has a high concentration of manganese [3].

Additionally, clove is also a good source of omega-6 fatty acids. According to the US Department of Agriculture, the clove contains 9.87% moisture, 5.97% protein, 65.53% carbohydrates, 33.9% dietary fiber, 13% lipids, and 5.67% ash contents. The detailed nutrient composition of the clove is given in Table **1**.

Table 1. Nutrient value of clove.

Nutrients	Value per 100g	Units
Water	9.87	g
Energy	274	Kcal
Protein	5.97	g
Carbohydrates	65.53	g
Fiber	33.9	g
Sugar	2.38	g
Total lipids	13	g
Saturated Fatty Acids	3.95	g
Monounsaturated Fatty Acids	1.39	g
Polyunsaturated Fatty Acids	3.61	g
Ash	5.63	g
Minerals	-	-
Calcium	632	mg
Copper	0.37	mg
Iron	11.83	mg
Magnesium	259	mg
Manganese	60.13	mg
Phosphorus	104	mg
Potassium	1020	mg
Sodium	277	mg
Zinc	2.32	mg
Selenium	7.2	mcg
Vitamins	-	-
Vitamin A	160	IU
Vitamin E	8.82	mg
Vitamin K	141.8	mcg
Vitamin C	0.2	mg
Thiamin	0.16	mg
Riboflavin	0.22	mg
Pantothenic acid	0.51	mg
Vitamin B6	0.39	mg
Niacin	1.56	mg
Folate	25	mcg

BENEFICIAL APPLICATION OF CLOVE AND ITS DERIVATIVES

Antimicrobial Properties

Clove and its derivatives have good antimicrobial potential. Eugenol, as the main phytochemicals present in the CEO is active against gram-positive and gram-negative bacteria. The eugenol shows bactericidal action by increasing the permeability of the cell membrane and burst of the cytoplasmic membrane. The CEO and eugenol have antibacterial potential against gram-negative bacteria like *Salmonella typhimurium, Salmonella enteritidis, E. coli,* and *Pseudomonas aeruginosa,* and gram-positive bacteria like *Listeria monocytogenes, Streptococcus pyogenes, Streptococcus pneumonia, Staphylococcus aureus, Staphylococcus epidermis, Bacillus subtilis* and *Bacillus cereus* [11 - 13].

Clove and its derivatives are found to be active against fungi. They act as fungicide agents by increasing the cell permeability and altering the cell morphology of the different fungi, namely, *Aspergillus species, Candida albicans, Dermatophyte species, Epidermophyton floccosum, Fusarium moniliforme, Fusarium oxysporum, Mucor species, Microsporum canis, Microsporum gypseum, Trichophyton mentagrophytes* and *Trichophyton rubrum* [15 - 17]. The aqueous extract of clove, containing eugenol combined with the acyclovir, has exhibited antiviral efficacy against the *herpes simplex virus type 1* and *influenza A virus.* Eugenol is effective against *herpes simplex virus type 2* [19]. Moreover, other phytochemicals found in the clove have the potential against different viruses, like *β*-caryophyllene against the *herpes virus* and Eugeniin have the antiviral activity against *herpes virus* strains and hepatitis C by blocking the DNA polymerase enzyme in the virus [20 - 22]. In anti-parasitic activity, *Leishmania donovani, Leishmania major, Leishmania tropica, Leishmania amazonensis* and *Trypanosoma cruzi* are sensitive to eugenol [23, 24]. Besides, piroplasm parasites and chloroquine-resistant *Plasmodium falciparum* are sensitive to the clove's methanolic extract [25]. Moreover, eugenol has been reported to be fatal for the growth and multiplication of many parasites, namely, *Haemonchus contortus, Fasciola gigantica, Giardia lamblia* and *Schistosoma mansoni* [26, 27].

Antibacterial Activity of Clove

Clove and its derivatives are natural antibacterial agents. The clove has antibacterial activity against many gram-positive and gram-negative bacteria. This antibacterial activity is due to the biochemical compounds found in the clove. These biochemical compounds are mainly eugenol, eugenyl acetate, β-caryophyllene, 2-heptanone, α-humulene, methyl salicylate, iso-eugenol, methyl-eugenol, phenylpropanoids, dehydrodieugenol, trans-confireryl aldehyde, biflorin, kaempferol, rhamnetin, myricetin, gallic acid, ellagic acid and oleanolic acid. The mechanism of antibacterial activity of these compounds is considered by challenging the cell membrane permeability. These compounds can denature proteins and react with the cell membrane phospholipids, increasing the permeability of the cell membrane, which disrupts the cytoplasmic membrane [28]. The clove showed the best results against the food-borne gram-positive bacteria like *Staphylococcus aureus, Bacillus cereus, Enterococcus Faecalis and Listeria monocytogenes*, and against gram-negative bacteria like non-toxigenic strains of *Escherichia Coli* [29], *Yersinia enterocolitica, Salmonella choleraeseus, Penicillium auregenosa* and *Salmonella typhimurium* [28, 30]. In another study, the antibacterial activity of clove oil against gram-positive bacteria (*Carnobacterium divergens and Staphylococcus aureus*) and gram-negative bacteria (*Salmonella typhimurium, Escherichia coli, and Serratia liquefaciens*) was reported by broth microdilution method [31, 32].

A study was conducted on the antibacterial activity by using different spices like mint, cinnamon, mustard, ginger, garlic, and clove. The 3% aqueous extract of clove showed comprehensive bactericidal activity against all tested food-borne pathogens such as *Escherichia coli, Staphylococcus aureus,* and *Bacillus cereus*, and showed good inhibitory action at 1% concentration of aqueous extract [22]. The clove and its derivatives are also effective against antibiotic-resistant bacteria. The aqueous and ethanolic extracts of clove buds showed the inhibitory action against the clinical isolates of bacteria, which were multidrug-resistant to methicillin, beta-lactams, aminoglycosides, tetracyclines, fluoroquinolones, and macrolide antibiotics [28]. Moreover, the clove combined with different antibiotics showed a synergistic effect against *S. aureus* isolate [33].

Antioxidant Activity of Clove

Free radicals are atoms or molecules which are highly reactive with other cellular structures because they contain unpaired electrons. Free radicals are natural by-products of continuing biochemical reactions in the body. Antioxidant activity is a characteristic of a certain agent to reduce the load of these free radicals [34] and it is measured by the oxygen radical absorption capacity (ORAC) value, a standard developed by the US Department of Agriculture for the antioxidant activity comparison [2]. Due to having high levels of phenolic compounds and other compounds, the clove is considered the best antioxidant present in nature. The clove has an ORAC value of over 10 million, much higher than the other antioxidants present in nature [3]. As a result of metabolism, the reactive oxygen species (ROS) produced in living organisms cause tissue damage and cell death, which impede the normal cell function and lead to different chronic diseases [35]. The main antioxidant agents found in the clove are phenolic and flavonoid compounds [1] and there is a positive correlation between polyphenols and antioxidant capability. The antioxidant activity of clove is comparable to other antioxidants like alpha-tocopherol, butylated hydroxytoluene, and butylated hydroxyl anisole [10]. The clove can encounter ROS and their higher polyphenol content derivatives like eugenol, β-caryophyllene, kaempferol, rhamnetin, gallic acid biflorin, and myricetin [28]. Therefore, among spices, clove has a higher concentration of polyphenols and higher antioxidant capability [22, 36]. One drop of clove oil has 400 times higher antioxidant activity than blueberries [1]. Moreover, the clove has higher antioxidative activity than onion, garlic, pepper, cinnamon, mint and ginger.

BENEFICIAL APPLICATION OF CLOVE AND ITS DERIVATIVES IN POULTRY

Effect of Clove and its Derivatives on Poultry Performance

The use of clove and its derivatives in poultry nutrition has shown affirmative results in improving the growth performance, as illustrated in Fig. (**2**). The inclusion of clove powder at a level of 0.5% can be used as a growth promoter in the quail [37]. This was confirmed by Al-Mufarrej *et al.* [38], who showed that clove powder (10 g/kg diet) could be used to increase the growth rate without affecting the health of the intestine and liver of broilers. Boyraz and Özcan [39] stated that the diet supplemented with clove extract (400 mg/kg) could be useful as a natural growth promoter for poultry instead of antibiotics and boost the performance parameters. Also, Dalkiliç and Güler [40] clarified that the clove extract (400 mg/kg) as a natural and safe feed additive has affirmative impacts on the performance and digestion process and can be considered a substitute natural growth promoter for poultry instead of antibiotics. Mukhtar [41] pointed out that the chicken diet supplemented with clove oil level of 600 mg/kg diet improved body weight and feed conversion ratio, in addition to decreasing the mortality rate. The supplementation of clove essential oil (450 mg/kg) in broiler diets during the whole period of the experiment (0-42 d) significantly improved body weight gain and feed conversion ratio [42]. Mohammadi *et al.* [43] indicated that broiler chicken diets fortified with clove oil (100 and 300 mg/kg) improved the performance and feed conversion ratio *via* augmented gastrointestinal villus height. The inclusion of 450 mg clove oil/kg in broiler diets from 23 to 42 days of age significantly improved body weight gain and feed conversion ratio compared with the control group [44]. The diet supplemented with a combination of 0.8% clove flower oil plus 0.8% clove powder and the drinking water treated with 0.4% aqueous clove extract improved the performance of broiler chickens compared with the control [45]. The broiler diet increased the live body weight with clove essential oil (300 to 500 mg/kg) [46].

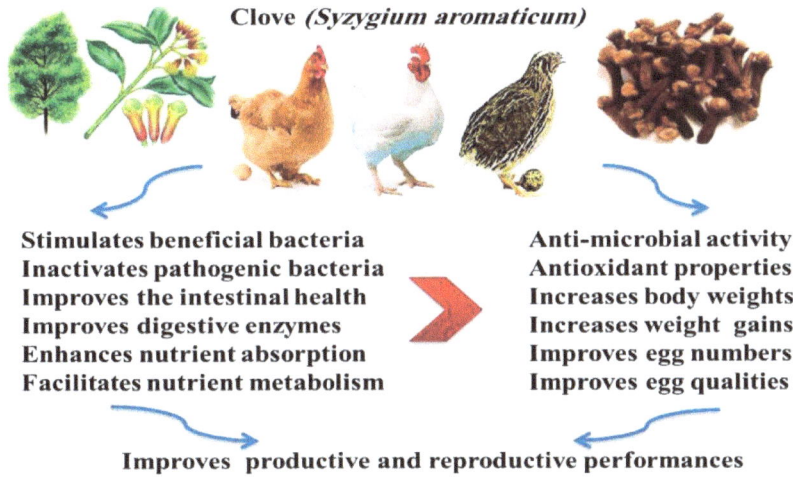

Fig. (2). Beneficial activities of clove in poultry.

The application of clove as a growth promoter in the poultry diet has been successfully reported. Adu *et al*. [47] stated that the clove leaf meal possessed antibacterial properties, capable of stimulating the growth of beneficial bacteria, inactivating pathogenic bacteria, and facilitating nutrient metabolism and absorption in the gastrointestinal tract to enhance the growth performance of broiler chickens. Tahir *et al*. [48] illustrated that adding 1% eugenol of clove leaves essential oil in the diet provided optimum body weight gain and final body weight and the best FCR for broilers at six weeks of age. The use of eugenol from clove leaves essential oil in the diet is expected to increase broiler performance to yield high meat products with the best quality. It is possible because eugenol has antioxidant and antimicrobial properties and functions as a growth booster. Besides, the eugenol compound can improve digestive enzymes to enhance livestock growth, in which a blend of essential oil components stimulates the secretion of digestive enzymes in chicken [49]. The supplementation of aqueous extract of clove flower at levels of 1 and 5% in drinking water significantly enhanced the production efficiency of birds represented by improvement of body weight, body weight gain, feed conversion ratio, water consumption rate, and reduction of mortality rate [50].

The improvement in the productive performance of poultry after feeding on diets containing clove and its derivatives is due to its distinctive properties. The clove possesses growth-promoting properties, such as appetite and digestion stimulating properties [51] as well as antimicrobial properties [52]. It has been reported that the clove stimulates bile salt secretion and digestive enzyme activities in the intestinal mucosa and pancreas [53]. It has been stated that clove oil is super-rich in manganese, and trace minerals are necessary for protein and carbohydrate metabolism to improve broiler performance [41]. Clove has inhibitory impacts on a large number of food-borne pathogens. The inhibitory impacts of clove on microbes are attributed to the fact that this herb contains essential oils, which in turn are composed of components such as eugenol [54]. The better growth rate of birds fed clove could be due to the presence of eugenol. Thus, the use of this herb is given much attention in the poultry industry.

Effect of Clove and its Derivatives on Egg Production and Quality

Dietary supplementation with clove and its oil positively affected egg production. Clove oils contain some volatile oils and saturated and unsaturated fatty acids, which positively impact egg production, egg weight, FCR, sexual maturity age, and ileum morphology of laying quail [55]. Kaya *et al.* [56] observed an improvement in egg production of laying hens with the addition of clove. Gandomani *et al.* [57] presented that the addition of clove buds in diets of laying hens significantly affected egg weight. Arpášová *et al.* [58] showed that adding 1 ml clove oil/kg had significantly influenced the albumen index and the addition of 0.6 and 1 ml clove oil/kg had significantly influenced Haugh units. Dietary clove essential oil prevented this contamination in eggs in an experimental infection [59] may be due to the antibacterial properties of clove oil. Eugenol in clove essential oil can prevent the growth of pathogens. The dietary inclusion of clove bud powder reduced egg yolk n-6 to n-3 ratio and improved its oxidative stability by decreasing malondialdehyde [60]. This may be due to the antioxidant properties of clove. The addition of clove oil to the quail diet significantly increased the concentrations of antioxidant enzymes in blood plasma [61].

CONCLUSION

This chapter demonstrated that clove and its derivatives as feed supplements could be an effective growth promoter. The clove and its derivatives have antioxidant, growth-promoting and antimicrobial properties. The major findings of this work pointed out that the supplementation of clove or its derivatives in poultry diets had beneficial effects on the growth performance, egg production and antimicrobial and antioxidant activities. Clove can be used as a potential alternative to antibiotics for more safety in the health of the poultry industry. The authors hope this chapter is comprehensive enough for the inquiring scholars seeking collective topics relevant to clove and its derivatives as promising feed additives. Several investigations are needed to shed more light on the impact of cloves or their derivatives on poultry production.

CONSENT FOR PUBLICATION

Not applicable.

CONFLICT OF INTEREST

The author declares no conflict of interest, financial or otherwise.

ACKNOWLEDGEMENTS

Declared none.

REFERENCES

[1]   Hussain S, Rahman R, Mushtaq A, El Zerey-Belaskri A. Clove: A review of a precious species with multiple uses. Int J Chem Biochem Sci 2017; 11: 129-33.

[2]   Milind P, Deepa K. Clove: a champion spice. Int J Res Ayurveda Pharm 2011; 2: 47-54.

[3]   Bhowmik D, Kumar KS, Yadav A, Srivastava S, Paswan S, Dutta AS. Recent trends in Indian traditional herbs Syzygium aromaticum and its health benefits. J Pharmacogn Phytochem 2012; 1: 13-22.

[4]   Neveu V, Perez-Jiménez J, Vos F, *et al.* Phenol-Explorer: an online comprehensive database on polyphenol contents in foods. Database (Oxford) 2010; 2010(0): bap024.
[http://dx.doi.org/10.1093/database/bap024] [PMID: 20428313]

[5]   Gulcin W, Şatb İG, Beydemira Ş, Elmastaşc M, Küfrevioglu Öİ. Comparison of antioxidant activity of clove (Eugenia caryophylata Thunb) buds and lavender (*Lavandula stoechas* L.). Food Chem 2004; 87(3): 393-400.
[http://dx.doi.org/10.1016/j.foodchem.2003.12.008]

[6]   Bamdad F, Kadivar M, Keramat J. Evaluation of phenolic content and antioxidant activity of Iranian caraway in comparison with clove and BHT using model systems and vegetable oil. Int J Food Sci Technol 2006; 41(s1): 20-7.
[http://dx.doi.org/10.1111/j.1365-2621.2006.01238.x]

[7]   Jirovetz L, Buchbauer G, Stoilova I, Stoyanova A, Krastanov A, Schmidt E. Chemical composition and antioxidant properties of clove leaf essential oil. J Agric Food Chem 2006; 54(17): 6303-7.
[http://dx.doi.org/10.1021/jf060608c] [PMID: 16910723]

[8]   Dudonné S, Vitrac X, Coutière P, Woillez M, Mérillon JM. Comparative study of antioxidant properties and total phenolic content of 30 plant extracts of industrial interest using DPPH, ABTS, FRAP, SOD, and ORAC assays. J Agric Food Chem 2009; 57(5): 1768-74.
[http://dx.doi.org/10.1021/jf803011r] [PMID: 19199445]

[9]   Pérez-Jiménez J, Neveu V, Vos F, Scalbert A. Identification of the 100 richest dietary sources of polyphenols: an application of the Phenol-Explorer database. Eur J Clin Nutr 2010; 64(S3) (Suppl. 3): S112-20.
[http://dx.doi.org/10.1038/ejcn.2010.221] [PMID: 21045839]

[10]   Gülçin İ, Elmastaş M, Aboul-Enein HY. Antioxidant activity of clove oil – A powerful antioxidant source. Arab J Chem 2012; 5(4): 489-99.
[http://dx.doi.org/10.1016/j.arabjc.2010.09.016]

[11]   Smith-Palmer A, Stewart J, Fyfe L. Antimicrobial properties of plant essential oils and essences against five important food-borne pathogens. Lett Appl Microbiol 1998; 26(2): 118-22.
[http://dx.doi.org/10.1046/j.1472-765X.1998.00303.x] [PMID: 9569693]

[12] Du W-X, Olsen CW, Avena-Bustillos RJ, McHugh TH, Levin CE, Friedman M. Effect of allspice, cinnamon, and clove bud oils in edible apple fi lms on physical properties and antimicrobial activities. J Food Sci 2009; 74(7): M372-8.
[http://dx.doi.org/10.1111/j.1750-3841.2009.01282.x]

[13] Pichika MR, Mak K-K, Kamal MB, *et al.* A comprehensive review on eugenol's antimicrobial properties and industry applications: A transformation from ethnomedicine to industry. Pharmacogn Rev 2019; 13(25): 1-9.
[http://dx.doi.org/10.4103/phrev.phrev_46_18]

[14] Chaieb K, Hajlaoui H, Zmantar T, *et al.* The chemical composition and biological activity of clove essential oil,Eugenia caryophyllata (*Syzigium aromaticum* L. Myrtaceae): a short review. Phytother Res 2007; 21(6): 501-6.
[http://dx.doi.org/10.1002/ptr.2124] [PMID: 17380552]

[15] Park MJ, Gwak KS, Yang I, *et al.* Antifungal activities of the essential oils in *Syzygium aromaticum* (L.) Merr. Et Perry and *Leptospermum petersonii* Bailey and their constituents against various dermatophytes. J Microbiol 2007; 45(5): 460-5.
[PMID: 17978807]

[16] Pinto E, Vale-Silva L, Cavaleiro C, Salgueiro L. Antifungal activity of the clove essential oil from *Syzygium aromaticum* on Candida, Aspergillus and dermatophyte species. J Med Microbiol 2009; 58(11): 1454-62.
[http://dx.doi.org/10.1099/jmm.0.010538-0] [PMID: 19589904]

[17] Rana IS, Rana AS, Rajak RC. Evaluation of antifungal activity in essential oil of the *Syzygium aromaticum* (L.) by extraction, purification and analysis of its main component eugenol. Braz J Microbiol 2011; 42(4): 1269-77.
[http://dx.doi.org/10.1590/S1517-83822011000400004] [PMID: 24031751]

[18] Al-Jasass FM, Al-Jasser MS. Chemical composition and fatty acid content of some spices and herbs under Saudi Arabia conditions. Sci World J 2012; 2012: 1-5.
[http://dx.doi.org/10.1100/2012/859892] [PMID: 23319888]

[19] El-Saber Batiha G, Alkazmi LM, Wasef LG, Beshbishy AM, Nadwa EH, Rashwan EK. *Syzygium aromaticum* L. (Myrtaceae): Traditional Uses, Bioactive Chemical Constituents, Pharmacological and Toxicological Activities. Biomolecules 2020; 10(2): 202.
[http://dx.doi.org/10.3390/biom10020202] [PMID: 32019140]

[20] Hussein G, Miyashiro H, Nakamura N, Hattori M, Kakiuchi N, Shimotohno K. Inhibitory effects of Sudanese medicinal plant extracts on hepatitis C virus (HCV) protease. Phytother Res 2000; 14(7): 510-6.
[http://dx.doi.org/10.1002/1099-1573(200011)14:7<510::AID-PTR646>3.0.CO;2-B] [PMID: 11054840]

[21] Astani A, Reichling J, Schnitzler P. Screening for antiviral activities of isolated compounds from essential oils. Evid Based Complement Alternat Med 2011; 2011: 1-8.
[http://dx.doi.org/10.1093/ecam/nep187] [PMID: 20008902]

[22] Cortés-Rojas DF, de Souza CRF, Oliveira WP. Clove (*Syzygium aromaticum*): a precious spice. Asian Pac J Trop Biomed 2014; 4(2): 90-6.
[http://dx.doi.org/10.1016/S2221-1691(14)60215-X] [PMID: 25182278]

[23] Ueda-Nakamura T, Mendonça-Filho RR, Morgado-Díaz JA, *et al.* Antileishmanial activity of Eugenol-rich essential oil from *Ocimum gratissimum*. Parasitol Int 2006; 55(2): 99-105.
[http://dx.doi.org/10.1016/j.parint.2005.10.006] [PMID: 16343984]

[24] Santoro GF, Cardoso MG, Guimarães LGL, Mendonça LZ, Soares MJ. Trypanosoma cruzi: Activity of essential oils from *Achillea millefolium* L., *Syzygium aromaticum* L. and *Ocimum basilicum* L. on epimastigotes and trypomastigotes. Exp Parasitol 2007; 116(3): 283-90.
[http://dx.doi.org/10.1016/j.exppara.2007.01.018] [PMID: 17349626]

[25] Batiha GES, Beshbishy AM, Tayebwa DS, Shaheen HM, Yokoyama N, Igarashi I. Inhibitory effects of *Syzygium aromaticum* and *Camellia sinensis* methanolic extracts on the growth of Babesia and Theileria parasites. Ticks Tick Borne Dis 2019; 10(5): 949-58.
[http://dx.doi.org/10.1016/j.ttbdis.2019.04.016] [PMID: 31101552]

[26] Machado M, Dinis AM, Salgueiro L, Custódio JBA, Cavaleiro C, Sousa MC. Anti-Giardia activity of *Syzygium aromaticum* essential oil and eugenol: Effects on growth, viability, adherence and ultrastructure. Exp Parasitol 2011; 127(4): 732-9.
[http://dx.doi.org/10.1016/j.exppara.2011.01.011] [PMID: 21272580]

[27] El-kady AM, Ahmad AA, Hassan TM, El-Deek HEM, Fouad SS, Al-Thaqfan SS. Eugenol, a potential schistosomicidal agent with anti-inflammatory and antifibrotic effects against *Schistosoma mansoni,* induced liver pathology. Infect Drug Resist 2019; 12: 709-19.
[http://dx.doi.org/10.2147/IDR.S196544] [PMID: 30992676]

[28] Mittal M, Gupta N, Parashar P, Mehra V, Khatri M. Phytochemical evaluation and pharmacological activity of *Syzygium aromaticum*: a comprehensive review. Int J Pharm Pharm Sci 2014; 6: 67-72.

[29] Burt SA, Reinders RD. Antibacterial activity of selected plant essential oils against *Escherichia coli* O157:H7. Lett Appl Microbiol 2003; 36(3): 162-7.
[http://dx.doi.org/10.1046/j.1472-765X.2003.01285.x] [PMID: 12581376]

[30] Hu Q, Zhou M, wei S. Progress on the antimicrobial activity research of clove oil and eugenol in the food antisepsis field. J Food Sci 2018; 83(6): 1476-83.
[http://dx.doi.org/10.1111/1750-3841.14180] [PMID: 29802735]

[31] Zengin H, Baysal A. Antibacterial and antioxidant activity of essential oil terpenes against pathogenic and spoilage-forming bacteria and cell structure-activity relationships evaluated by SEM microscopy. Molecules 2014; 19(11): 17773-98.
[http://dx.doi.org/10.3390/molecules191117773] [PMID: 25372394]

[32] Kaur K, Kaushal S. Phytochemistry and pharmacological aspects of *Syzygium aromaticum*: A review. J Pharmacogn Phytochem 2019; 8: 398-406.

[33] Haroun MF, Al-Kayali RS. Synergistic effect of *Thymbra spicata* L. extracts with antibiotics against multidrug- resistant *Staphylococcus aureus* and *Klebsiella pneumoniae* strains. Iran J Basic Med Sci 2016; 19(11): 1193-200.
[PMID: 27917275]

[34] Alfikri FN, Pujiarti R, Wibisono MG, Hardiyanto EB. Yield, quality, and antioxidant activity of Clove (*Syzygium aromaticum* L.) bud oil at the different phenological stages in young and mature trees. Scientifica (Cairo) 2020; 2020: 1-8.
[http://dx.doi.org/10.1155/2020/9701701] [PMID: 32566363]

[35] Pulikottil SJ, Nath S. Potential of clove of *Syzygium aromaticum* in development of a therapeutic agent for periodontal disease: A review. S Afr Dent J 2015; 70: 108-15.

[36] Shan B, Cai YZ, Sun M, Corke H. Antioxidant capacity of 26 spice extracts and characterization of their phenolic constituents. J Agric Food Chem 2005; 53(20): 7749-59.
[http://dx.doi.org/10.1021/jf051513y] [PMID: 16190627]

[37] Tariq H, Raman Rao PV, Raghuvanshi RS, Mondal BC, Singh SK. Effect of Aloe vera and clove powder supplementation on carcass characteristics, composition and serum enzymes of Japanese quails. Vet World 2015; 8(5): 664-8.
[http://dx.doi.org/10.14202/vetworld.2015.664-668] [PMID: 27047153]

[38] Al-Mufarrej SI, Al-Baadani HH, Fazea EH. Effect of level of inclusion of clove (<i>Syzygium aromaticum</i>) powder in the diet on growth and histological changes in the intestines and livers of broiler chickens. S Afr J Anim Sci 2019; 49(1): 166-75.
[http://dx.doi.org/10.4314/sajas.v49i1.19]

[39] Boyraz N, Özcan M. Inhibition of phytopathogenic fungi by essential oil, hydrosol, ground material

and extract of summer savory (*Satureja hortensis* L.) growing wild in Turkey. Int J Food Microbiol 2006; 107(3): 238-42.
[http://dx.doi.org/10.1016/j.ijfoodmicro.2005.10.002] [PMID: 16330123]

[40] Dalkiliç B, Güler T. The effects of clove extract supplementation on performance and digestibility of nutrients in broilers. FU Sag Bil Vet Derg 2009; 3: 161-6.

[41] Mukhtar MA. The effect of dietary clove oil on broiler performance. Aust J Basic Appl Sci 2011; 5(7): 49-51.

[42] Mehr MA, Hassanabadi A, Moghaddam HN, Kermanshahi H. Supplementation of clove essential oils and probiotic to the broiler's diet on performance, carcass traits and blood components. Iran J Appl Anim Sci 2014; 4(1): 117-22.

[43] Mohammadi Z, Ghazanfari S, Moradi MA. Effect of supplementing clove essential oil to the diet on microflora population, intestinal morphology, blood parameters and performance of broilers. Eur Polit Sci 2014; 78: 1-11.

[44] Azadegan MM, Hassanabadi A, Nasiri MH, Kermanshahi H. Supplementation of clove essential oils and probiotic to the broiler's diet on performance, carcass traits and blood components. Res Opin Anim Vet Sci 2014; 4: 218-23.

[45] Salman KAAA, Ibrahim DK. Test the activity of supplementation clove (*Eugenia caryophyllus*) powder, oil and aqueous extract to diet and drinking water on performance of broiler chickens exposed to heat stress. Int J Poult Sci 2012; 11(10): 635-40.
[http://dx.doi.org/10.3923/ijps.2012.635.640]

[46] Valenzuela-Grijalva NV, Pinelli-Saavedra A, Muhlia-Almazan A, Domínguez-Díaz D, González-Ríos H. Dietary inclusion effects of phytochemicals as growth promoters in animal production. J Anim Sci Technol 2017; 59(1): 8.
[http://dx.doi.org/10.1186/s40781-017-0133-9] [PMID: 28428891]

[47] Adu OA, Gbore FA, Oloruntola OD, Falowo AB, Olarotimi OJ. The effects of Myristica fragrans seed meal and *Syzygium aromaticum* leaf meal dietary supplementation on growth performance and oxidative status of broiler chicken. Bull Natl Res Cent 2020; 44(1): 149.
[http://dx.doi.org/10.1186/s42269-020-00396-8]

[48] Tahir M, Chuzaem S, Widodo E, Hafsah . The performance of broilers given eugenol of clove leaf essential oil as a feed additive. Russ J Agric Soc-Econ Sci 2019; 95(11): 200-5.
[http://dx.doi.org/10.18551/rjoas.2019-11.28]

[49] Williams P, Losa R. The use of essential oils and their compounds in poultry nutrition. World Poult 2001; 17(4): 14-5.

[50] AL-Zuhairi ZA, Abdulrazzaq M. 15-study the effect of adding aqueous extract of clove (*Eugenia caryophyllus*) to drinking water in productivity and physiological efficiency of broiler chicken. Bas J Vet Res 2018; 17(1): 165-75.
[http://dx.doi.org/10.33762/bvetr.2018.144949]

[51] Çabuk M, Alcicek A, Bozkurt M, İmre N. Antimicrobial properties of the essential oils isolated from aromatic plants and using possibility as alternative feed additives. Proceedings of the 2nd National Animal Nutrition Congress. 2003 Sep 8-20; Konya. 184-7.

[52] Dorman HJD, Deans SG. Antimicrobial agents from plants: antibacterial activity of plant volatile oils. J Appl Microbiol 2000; 88(2): 308-16.
[http://dx.doi.org/10.1046/j.1365-2672.2000.00969.x] [PMID: 10736000]

[53] Lee KW, Everts H, Kappert HJ, Wouterse H, Frehner M, Beynen AC. Cinnamaldehyde, but not thymol, counteracts the carboxymethyl cellulose-induced growth depression in female broiler chickens. Int J Poult Sci 2004; 3(9): 608-12.
[http://dx.doi.org/10.3923/ijps.2004.608.612]

[54] Asha K, Sunil B, George GP. Effect of clove on the bacterial quality and shelf life of chicken meat. J

Mater Sci Technol 2014; 2(2): 37-9.

[55]   Wasman PH, Mustafa MAG. The dietary impact of clove and cinnamon powders and oil supplementations on the performance, ileum morphology, and intestine bacterial population of quails. Plant Arch 2020; 20(1): 1503-9.

[56]   Kaya H, Kaya A, Celebi S, Macit M. Effects of dietary supplementation of essential oils and vitamin e on performance, egg quality and *escherichia coli* count in excreta. Agric Res Commun Cent 2013; 47(6): 515-20.

[57]   Taheri Gandomani V, Mahdavi AH, Rahmani HR, Riasi A, Jahanian E. Effects of different levels of clove bud (*Syzygium aromaticum*) on performance, intestinal microbial colonization, jejunal morphology, and immunocompetence of laying hens fed different n-6 to n-3 ratios. Livest Sci 2014; 167: 236-48.
       [http://dx.doi.org/10.1016/j.livsci.2014.05.006]

[58]   Arpášová H, Gálik B, Fik M, Pistová V. The effect of the clove essential oil to the production and quality of laying hens eggs. Lucr Stiint Zooteh Biotehnol 2017; 50(1): 1-5.

[59]   Ordóñez G, Llopis N, Peñalver P. Efficacy of eugenol against a Salmonella enterica serovar enteritidis experimental infection in comercial layers in production. J Appl Poult Res 2008; 17(3): 376-82.
       [http://dx.doi.org/10.3382/japr.2007-00109]

[60]   Alizadeh MR, Mahdavi AH, Rahmani HR, Jahanian E. Effects of different levels of clove bud (*Syzygium aromaticum*) on yolk biochemical parameters and fatty acids profile, yolk oxidative stability, and ovarian follicle numbers of laying hens receiving different n-6 to n-3 ratios. Anim Feed Sci Technol 2015; 206: 67-75.
       [http://dx.doi.org/10.1016/j.anifeedsci.2015.05.007]

[61]   Mustafa MA, Wasman PH. The impact of powders and oil additives of cinnamon and clove in quails diet as antistressor and antioxidant during hot months. Iraqi J Agric Sci 2020; 51(3): 760-6.
       [http://dx.doi.org/10.36103/ijas.v51i3.1031]

Pomegranate (*Punica Granatum* L): Beneficial Impacts, Health Benefits and Uses in Poultry Nutrition

Youssef A. Attia[1,2], Ayman E. Taha[3], Mohamed E. Abd El-Hack[4,*], Mohamed Abdo[5,6], Ahmed I. Abo-Ahmed[7], Mahmoud A. Emam[8], Karima El Naggar[9], Mervat A. Abdel-Latif[10], Nader R. Abdelsalam[11] and Mahmoud Alagawany[4]

[1] *Animal and Poultry Production Department, Faculty of Agriculture Damanhour University, Damanhour, Egypt*

[2] *Department of Agriculture, Faculty of Environmental Sciences, King Abdulaziz University, 21589, Jeddah, Kingdom of Saudi Arabia*

[3] *Department of Animal Husbandry and Animal Wealth Development, Faculty of Veterinary Medicine, Alexandria University, Rasheed, 22758 Edfina, Egypt*

[4] *Department of Poultry, Faculty of Agriculture, Zagazig University, Zagazig 44511, Egypt*

[5] *Department of Animal Histology and Anatomy, School of Veterinary Medicine, Badr University in Cairo (BUC), Egypt*

[6] *Department of Anatomy and Embryology- Faculty of Veterinary Medicine- University of Sadat City, Egypt*

[7] *Department of Anatomy and Embryology, Faculty of Veterinary Medicine, Benha University, Toukh 13736, Egypt*

[8] *Department of Histology, Faculty of Veterinary Medicine, Benha University, Toukh 13736, Egypt*

[9] *Department of Nutrition and Clinical Nutrition, Faculty of Veterinary Medicine, Alexandria University, Egypt*

[10] *Department of Nutrition and Veterinary Clinical Nutrition, Faculty of Veterinary Medicine, Damanhour University, Damanhour 22511, Egypt*

[11] *Agricultural Botany Department, Faculty of Agriculture (Saba Basha), Alexandria University, 21531 Alexandria, Egypt*

Abstract: *Punica Grantum* L is an ancient, magical and distinctive fruit. It is local to the Mediterranean basin and has been broadly utilized in traditional pharmaceuticals in numerous nations. The extracts collected from various parts (peels, seeds, juice and flowers) of this natural fruit can be used as multiple additives for practice because of its polyphenolic contents. Polyphenols found in *P. Grantum* have been shown to have

* **Corresponding author Mohamed E. Abd El-Hack:** Department of Poultry, Faculty of Agriculture, Zagazig University, Zagazig 44511, Egypt; E-mail: m.ezzat@zu.edu.eg

various pharmacological activities such as anti-inflammatory, antioxidant, antimicrobial, anti-diarrheal, immunomodulatory, anti-carcinogenic, and wound healing promotors. Moreover, they are reported to have anti-cestodial, anti-nematodal and anti-protozoan activities. *P. Grantum* L or its by-products supplementation can play a major role in poultry nutrition by enhancing immunity, scavenging free radicals, and inhibiting antimicrobial activity, leading to improved poultry performance. Owing to its functions above, it can be a potential substitute for modulating immune functions and gut microbiota to relieve diarrhea and enteritis, preventing colibacillosis and coccidiosis in chickens. Moreover, it is reported that polyphenols and tannins of *P. Grantum* act as an antioxidant by scavenging reactive oxygen species and preventing lipid oxidation and inflammatory molecule production. This chapter highlights the work done in the recent past on *P. Grantum*. Despite the voluminous pharmacological properties of *P. Grantum,* its usage in the chicken ration is limited. So, this chapter aims to broaden the information of researchers, veterinary advisors, and poultry nutritionists to recommend *P. Grantum* as a safe, natural added substance in poultry feed to substitute the synthetic additives for nourishment purposes.

Keywords: Chicken, Growth performance, Immunity, *Punica granatum* L. by-products.

INTRODUCTION

Currently, traditional medicinal plants for health care have gained great worldwide attention. Phytobiotics, natural bioactive compounds derived from plants, are mainly flavonoids, alkaloids, tannins, terpenes, steroids, and essential oils [1, 2]. Recently, phytobiotics have been considered antimicrobial growth promoters due to their antimicrobial, immunomodulatory, and antioxidant activities [3, 4]. At present, the European Union has restricted antibiotics supplementation in poultry production due to concerns about bacterial resistance in humans. Consequently, poultry nutritionists and researchers are pursuing alternative feed additives for commercial poultry species to improve their productivity *via* maintaining gut health and improving the feed utilization and quality of products obtained from these birds [5, 6]. Recently, herbal plants, powders, and extracts have been used as alternative feed additives in many poultry studies [7 - 10]. Pomegranate, a fruit from the Punicaceae family, has been used for various purposes, commonly eaten raw, served as a drink, used in the cocktail and ground after drying and used in traditional recipes. Besides, it is used in the food, dyes and cosmetics industry [11, 12].

Additionally, to eliminate environmental pollution, discarded parts (peels, rinds and seeds) of the pomegranate can be a nutritious feed additive for animals [13] and poultry. Pomegranate is enriched with several phytochemicals, mainly polyphenols [14]. Its peels accounted for about half of its weight. They were reported to contain several phenolics, tannins (ellagitannin, pedunculagin, punicalagin and punicalin), flavonoids (kaempferol, quercetin and luteolin glycosides) and phenolic acids (caffeic acid, coumaric acid, chlorogenic acid, phlorodizin, rutin and ferulic acid) [15, 16]. Arils also contain many polyphenols, such as gallic acid, chlorogenic acid, catechin, ferulic acid, coumaric acid, and caffeic acid [17]. In *in vitro* investigations reported an antimicrobial capacity of pomegranate peel and seed owing to their phenolic contents [18, 19]. The health benefits of pomegranate are innumerable. Oral administration of pomegranate peel powder enhances humoral and cellular immunity [20].

Additionally, pomegranate extract has anti-inflammatory, antioxidant, anti-cancer, cardio- and neuroprotective effects [21] and enhances wound healing [22, 23]. Pomegranate peel extract is effective against different types of bacteria [24, 25]. This effect is owed to its polyphenol content [15, 26]. Also, the oil content of pomegranate seed is rich in an immunomodulatory compound, punicic acid [27]. Several properties of pomegranate are beneficial and further studies concerning its use in poultry nutrition are needed. The use of pomegranate peel has no growth-promoting effect in broiler chickens [28]. After an extensive review of literature, it is established that various parts of pomegranate (juice, bark, seeds and leaves) have gained concern and studied in different animal species. Despite the exciting advantageous capacities of pomegranate, many investigations of its applications in poultry nourishment are needed. So, this reappraisal aims to provide a modern understanding to veterinary analysts and poultry nutritionists to conduct more inquiries about pomegranate to assess its utilization as a safe, natural feed additive in poultry feed.

PHYTOCHEMICALS IN POMEGRANATE

In the last few years, substantial advancement has been made to understand pomegranate's mechanisms and pharmacological mode of action. Individual parts (roots, bark peels, rind, seeds and leaves) have their own medicinal and therapeutic benefits. The most therapeutically valuable phytochemical constituents of pomegranate are anthocyanins, anthocyanidins, ellagic acid, ellagitannins, punicic acid, flavonoids, estrogenic flavonols and flavones. Important phytochemical constituents of pomegranate and their structures are presented in Table **1** and Fig. (**1**).

Table 1. Principal phytochemical constituents of pomegranate.

Plant Component	Phytochemical Constituents
Pomegranate Juice	Anthocyanins, caffeic acid, glucose, ellagic acid, ascorbic acid, gallic acid, catechins, rutin, quercetin [29 - 32]
Pomegranate seed oil	Punicic acid, ellagic acid, sterols, fatty acids [33, 34],

(Table 1) cont.....

Plant Component	Phytochemical Constituents
Pomegranate pericarp	Phenolic punicalagins, anthocyanidins, flavonoids, gallic acid fatty acids, quercetin, catechin, rutin [35]
Pomegranate leaves	Ellagic acid, fatty acids [36, 37]

Plant Component	Phytochemical Constituents
Pomegranate flowers	Gallic acid, triterpenoids, oleanolic acid, ursolic acid, [16, 38, 39]
Pomegranate bark and roots	Punicalin, punicalagin, piperidine alkaloids [21]

Benefiial bioactive properties

- Radicakl scavenging activity
- A-amylase inhibitory activity
- Modulatory activity on hepatocyte glucose uptake
- Protective effect against oxidative stress-induced hepatocyte injury

Polyphenolic composition

- Gallic acid
- Caffeic acid
- Epicatechin
- Ferulic acid
- Querceetin
- Vanillic acid

Punica granatum L. (pomegrante) fruit peel acetone extract

Fig. (1). Phenolic composition and beneficial bioactive properties of pomegranate fruit.

CONVENTIONAL USES OF POMEGRANATE FRUIT

Pomegranate is considered an effective pharmaceutical plant in ancient nations like China, Egypt and Greece. Moreover, all its parts have numerous therapeutic benefits. It is extensively used to treat helminths infestations, fungal infections, diarrhea, and stomatitis. Moreover, the root extracts are suggested to be useful for gynecological problems. Furthermore, it has great effectiveness in treating snake bites, burns, leprosy and diabetes [21]. In addition, the old Chinese civilization paid attention to red pomegranate juice, known as 'soul concentrate', which acts as an antiaging factor [21].

CHEMICAL ANALYSIS OF POMEGRANATE

The chemical components of pomegranate are shown in Table **2** and Fig. (**2**).

Table 2. Chemical constituents of pomegranate.

Item	Concentration
Crude Protein	60.80%
Moisture	90.41%
Crude Ash	20.82%
Total polyphenols	145.91 mg/g
Hydrolysable tannins	14.26 mg/g
Total flavonoids,	57.59 mg/g

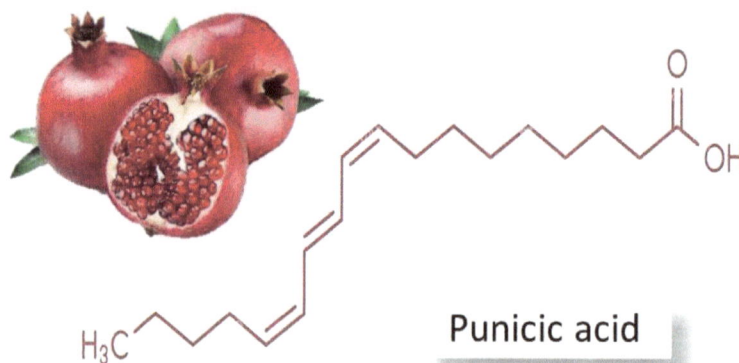

Fig. (2). Chemical structure of Punicic acid.

BIOLOGICAL PROPERTIES AND THERAPEUTIC APPLICATIONS

Several studies have reported pomegranate's antioxidant, anti-inflammatory, antimicrobial, anti-cancer, and immuno-stimulant properties. They are summarized in Table **3** and Fig. (**3**).

Table 3. Summary of different therapeutic activity of Pomegranate.

Activity	References
Antioxidant activity	[41 - 43]
Anti-inflammatory activity	[44 - 46]

(Table 3) cont.....

Activity	References
Hypoglycemic and cholesterol-lowering activities	[16, 38, 47, 48]
Antimicrobial activity	[24, 49, 50, 52]

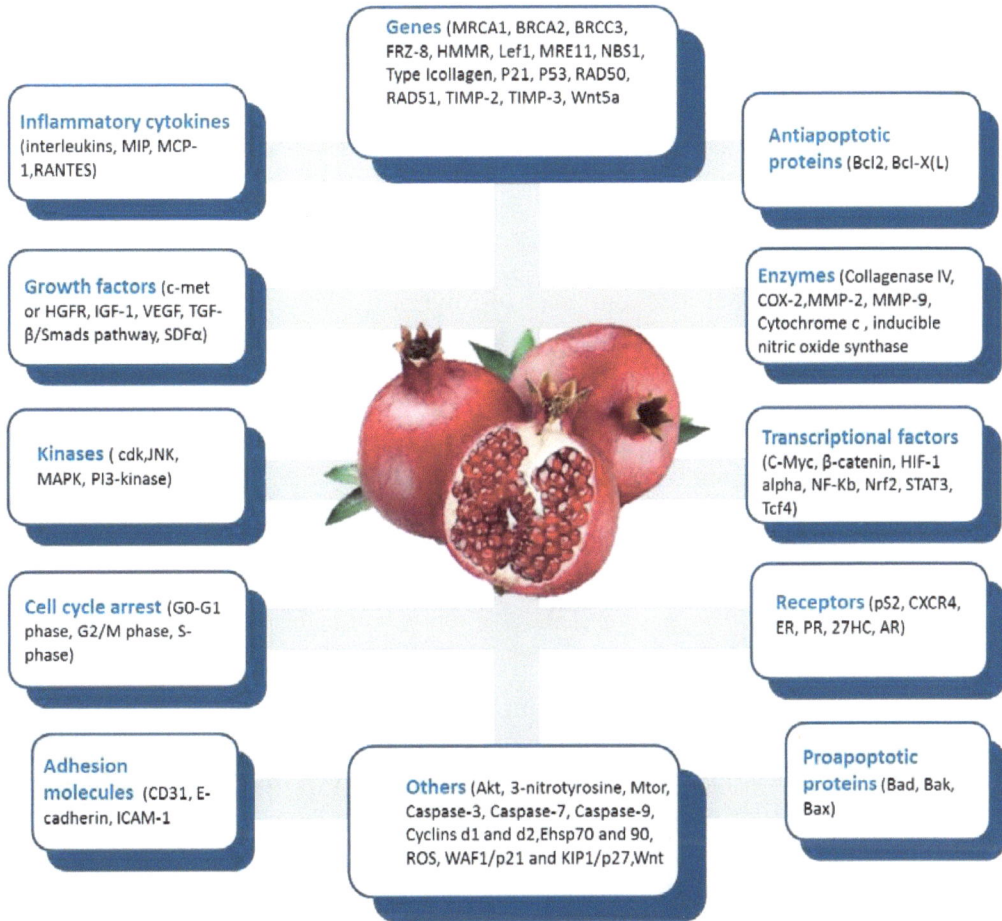

Genes (MRCA1, BRCA2, BRCC3, FRZ-8, HMMR, Lef1, MRE11, NBS1, Type Icollagen, P21, P53, RAD50, RAD51, TIMP-2, TIMP-3, Wnt5a

Inflammatory cytokines (interleukins, MIP, MCP-1,RANTES)

Antiapoptotic proteins (Bcl2, Bcl-X(L))

Growth factors (c-met or HGFR, IGF-1, VEGF, TGF-β/Smads pathway, SDFα)

Enzymes (Collagenase IV, COX-2,MMP-2, MMP-9, Cytochrome c , inducible nitric oxide synthase

Kinases (cdk,JNK, MAPK, PI3-kinase)

Transcriptional factors (C-Myc, β-catenin, HIF-1 alpha, NF-Kb, Nrf2, STAT3, Tcf4)

Cell cycle arrest (G0-G1 phase, G2/M phase, S-phase)

Receptors (pS2, CXCR4, ER, PR, 27HC, AR)

Adhesion molecules (CD31, E-cadherin, ICAM-1

Others (Akt, 3-nitrotyrosine, Mtor, Caspase-3, Caspase-7, Caspase-9, Cyclins d1 and d2,Ehsp70 and 90, ROS, WAF1/p21 and KIP1/p27,Wnt

Proapoptotic proteins (Bad, Bak, Bax)

Fig. (3). Some biological properties and therapeutic applications of pomegranate.

Antioxidant Activity

Polyphenols, including flavones, flavonols, condensed and hydrolyzable tannins and alkaloids such as pelletitierine are the dominant compounds in pomegranate. Hydrolyzable tannins constitute about 90% of the antioxidant activity of polyphenols found in pomegranate juice [40]. The antioxidant activity of pomegranate juice is more potent three times than green tea due to its high polyphenol content. Pomegranate extract is enriched in flavonoid that possesses antioxidant activity *via* decreasing hydrogen peroxide concentrations [41]. Moreover, the antioxidant capacity of anthocyanins present in pomegranate extract is higher than vitamins E, C and A [42, 43].

Anti-inflammatory Activity

In the last decades, the research exploring the anti-inflammatory property of pomegranate is significantly increased. Oral administration of pomegranate juice or Pomegranate seed oil can decrease the expressions of IL-1β, IL-6, IL-8, TNF-α and NF-kB in many body organs [44, 45]. Granatin B is a marker for pomegranate's anti-inflammatory effect [46].

Glucose and Lipid Metabolism Activities

Pomegranate is considered a cholesterol-lowering plant and protector against cardiovascular diseases [47]. Pomegranate can decrease cholesterol absorption and increase its excretion in feces by influencing the enzymes involved in cholesterol metabolism. Additionally, pomegranate flowers are used in Unani and Ayurvedic medicines for treating diabetic disorders [16] where the extract of pomegranate flower enhances insulin sensitivity and decreases glucose levels by inhibiting alpha-glucosidase, thus reducing the conversion of sucrose to glucose [38]. Moreover, Bagri *et al.* [48] suggested the utilization of pomegranate as a food supplement to protect against glucose metabolic disorders.

Antimicrobial Activity

Pomegranate sources are considered an alternative therapy against multidrug-resistant pathogenic bacteria. Pomegranate and its extracts have antimicrobial activity [24, 49]. Tannins, one of the pomegranate constituents, are lethal for microorganisms [50]. Moreover, phenols can decrease the substrate availability to microorganisms and interfere with bacterial protein secretions [51]. Chief phenolics in pomegranate are ellagic acid, ellagitannin, gallic acid, and punicalagin, which have antimicrobial activities. Both ellagitannins and punicalagins modulate gut microbiota and inhibit clostridial growth without any harmful effect on gut microbiota [52]. Additionally, pomegranate fruit and peels extracts have antimicrobial activity against *Staphylococcus aureus*, *Listeria monocytogenes*, *Escherichia coli* and *Yersinia enterocolitica* [24].

STUDIES ON POULTRY LIVESTOCK

Researchers have shown great interest in improving the feed utilization and growth and reducing mortalities using supplementation of pomegranate pomegranate-derived additives in poultry feed. Ahmed and Yang [53] reported that pomegranate by-product at 0.5-1.0% supplement could improve the chicken immune response dose-dependently. Moreover, Ahmed *et al*. [54] reported that pomegranate at a 2% supplemental level could improve the composition of the thigh and breast muscles of broilers. Also, Ahmed *et al*. [55] revealed that dietary inclusion of fermented pomegranate by-products in the broiler diet by up to 2% improved the nutritional quality, fatty acid profile and shelf life of broiler meat. Additionally, Szymczyk *et al*. [56] reported a significant improvement in breast meat's fatty acid composition. Also, Saleh *et al*. [57] recorded improved oxidative status and quality of breast meat in broiler chickens fed pomegranate peel extract (200 and 300 mg/kg feed). In addition, the study above reported that the antioxidant potential of pomegranate peel extract is equal to that of α-tocopherol in refrigerated meat. Another recent trial of the same lab concluded that supplementing broiler diets with 0.2 and 0.3 g kg^{-1} PPE resulted in the enrichment of broiler thigh meat with long-chain PUFA n-3 and enhanced their antioxidant and quality characteristics [58]. Recently, the antioxidant status and quality of broiler meat were investigated by Akuru *et al*. [59] as birds were supplemented with different levels of pomegranate peel powder meal (PPPM). They reported that 2 g/kg PPPM significantly improved the water-binding capacity of broiler breast meat by lowering the cooking loss of meat, and this meat showed the highest ABTS scavenging ability. In another recent report, Ghosh *et al*. [60] indicated that the administration of pomegranate peel at 50 ml/L in the drinking water of broiler chickens enhanced their growth performance with a hypolipidemic effect and stimulated the antioxidant status as it reduced lipid peroxidation and enhanced reduced glutathione and catalase enzyme activity.

On the other hand, Bostami *et al*. [61] demonstrated that fermented pomegranate by-products' supplementation in a broiler diet at different levels (0.5, 1 and 2%) enhanced their body weight gain, lowered fecal pH and gas emission (NH$_3$, H$_2$S and SO$_2$) and considered as economical feed ingredients that could be used in poultry diet. Moreover, Yaseen *et al*. [62] reported that pomegranate peel extract inclusion (0.05 and 0.1% /kg diet) reduced the abdominal fat deposition in broiler chickens, the hypolipidemic effect of which was confirmed by lowered serum concentrations of total cholesterol, TG, VLDL and LDL-cholesterol, as well as immunostimulatory effect occurring through increasing serum lysozyme content. Abbas *et al*. [63] investigated the partial replacement of yellow corn with untraditional feedstuff as pomegranate peel powder (PPP) in the Japanese quail diet. Authors found that quails can effectively utilize the 7.5% substitution level of PPP, which was reflected by the improved product performance through enhanced FCR, egg production and jejunal morphology. In addition, Rezvani *et al*. [64] reported the beneficial effect of pomegranate seed extract supplementation (2%) in fat-containing diets of broiler chicken as it improved the fat digestibility and increased the *Lactobacillus* count in the caecum. In the same direction, Sarıca, Ürkmez [65] summarized that PPE inclusion in the broiler diet enhanced their performance and could be used as a natural antimicrobial compound instead of probiotics as it was effective in lowering the concentrations of total coliform bacteria and *E. coli*, while increased the *Lactobacillus* bacteria concentration in the ileum. Likewise, Hamady *et al.* [66] described that PPE dietary inclusion significantly improved productive broiler performance and dressing percentage while reducing the total aerobic bacteria in chickens' intestines, thereby boosting health. Rama Rao *et al.* [67] concluded that the addition of PPM (250 and 500 mg/kg) in broiler diets enhanced their humoral immune response against the ND vaccine and the activity of glutathione peroxidase enzyme in the liver, as well as reduced lipid peroxidation in their hepatic tissue without any adverse effects on performance and carcass traits. Kishawy *et al.* [68] reported that pomegranate peel extract (PPE) and sage oil could be used to ameliorate the adverse effects of oxidized oils as they enhanced broiler performance (improved gain and FCR), immune response (stimulated humoral and cellular immunity), and lowered the serum concentration of lipid profile parameters. Recently, Kishawy *et al.* [69], found that the addition of 0.05% PPE to soybean oil or linseed oil in broiler rations resulted in improved broiler growth performance, while the higher level (0.1% PPE) reduced serum lipid and abdominal fat levels and increased flavonoids in broiler muscles. On the other hand, Manterys *et al.* [70] found that pomegranate seed oil administration in broiler diet showed immunomodulatory effect, altered the lipid profile, increased c9,t11 conjugated linoleic acid (CLA) concentration deposition in the adipose tissue while, had no effect on the liver-related parameters as ALT and AST.

As a way of combating the heat stress (HS) problem in poultry farms, Hosseini-Vashan, and Raei-Moghadam [71] found that supplementation of pomegranate pulp (higher than 7% of diet), alleviated the adverse effects of HS as it enhanced their performance, reduced serum lipid profiles, especially cholesterol and LDL and improved antioxidant status of HS-subjected birds. Oxidative stress was applied to broiler chicken supplemented with 1% pomegranate peel. The results showed that pomegranate peel alleviated this stress by improving its production performance, serving as an antioxidant in blood plasma in SOD, catalase and GPx and reduced MDA values [72]. Prashanth Kumar *et al.* [73] evaluated the efficiency of PPE in broiler chicken reared in the summer season. They realized that dietary supplementation at 100 mg/kg diet significantly lowered cholesterol levels without any negative impact on their growth or immune status. In addition, Abdel Baset *et al.* [74] reported that the addition of PPP to broiler diets (2 and 4 g/kg diet) positively enhanced birds' growth rate, blood serum metabolites, increased the level of IgM and lysozyme improving their immunological response and the quality of meat.

Moreover, Nagla *et al.* [75] confirmed the antibacterial activity of PPE against

Salmonella Enteritidis infection in broilers which exerted similar activity as a commercial antibiotic and antioxidant effects and significantly (P<0.05) improved both blood parameters and meat quality. In another study, the combined effects of *Punica granatum* and *Terminalia chebula* (PgTc) extract were observed in chickens challenged with a*vian* pathogenic *Escherichia coli*. The results stated an improved survival rate in a dose-dependent manner. Moreover, PgTc significantly enhances chicken immunity *via* modulating the expressions of TLR2, TLR4 and TLR5 and inhibits inflammation *via* reducing the levels of the inflammatory cytokine; IL-8, IL-1β, TNF-α, VCAM1 and ICAM1 [76]. The investigation of pomegranate utilization in the poultry diet was extended to the layer diet. The inclusion of pomegranate seed pulp (5, 10 and 15%) in the laying hen diet is reported to improve the egg production without any adverse effects on egg quality [77]. Another trial was conducted by Rezvani *et al*. [64] in the laying hens using pomegranate molasses added in the drinking water with different levels up to 1%. They indicated a positive effect of this additive on the egg quality in terms of yolk color, albumin index and Haugh unit and it could be used as a preservative for egg texture and quality.

Direct Additive Effects on Meat

Many studies reported the direct application of pomegranate extract and juice in chicken production and meat products, improving shelf-life and consumers' acceptability. Lytou *et al*. [78] reported a decreased spoilage microbes and an increased shelf-life of chicken breast fillets immersed in pomegranate juice. Another study reported that the marination of chicken breast meat in pomegranate juice could potentially reduce the bacterial growth and sensorial deterioration in the refrigerator for up to 12 days [79]. Additionally, pomegranate peel extract has a synergistic antioxidant effect and extends the shelf life of refrigerated meat [80]. Moreover, pomegranate juice is one of the best inhibitory additives for microbial growth in chicken breast meat [81]. Another study detected that extracts of pomegranate and kinnow by-products are strong natural antioxidant sources, so these fruits by-products can be used to reduce the oxidative damages in poultry meat products [81].

Additionally, Kanatt *et al*. [82] indicated that agro-industrial by-products such as PPE could be used as a natural preservative in the food industry. It showed antioxidative effect and effective antimicrobial action against *Staphylococcus aureus* and *Bacillus cereus* and enhanced chicken meat products' shelf life by 2–3 weeks during chilled storage. Some general properties of pomegranate on poultry livestock and their performance are summarized in Fig. (**4**).

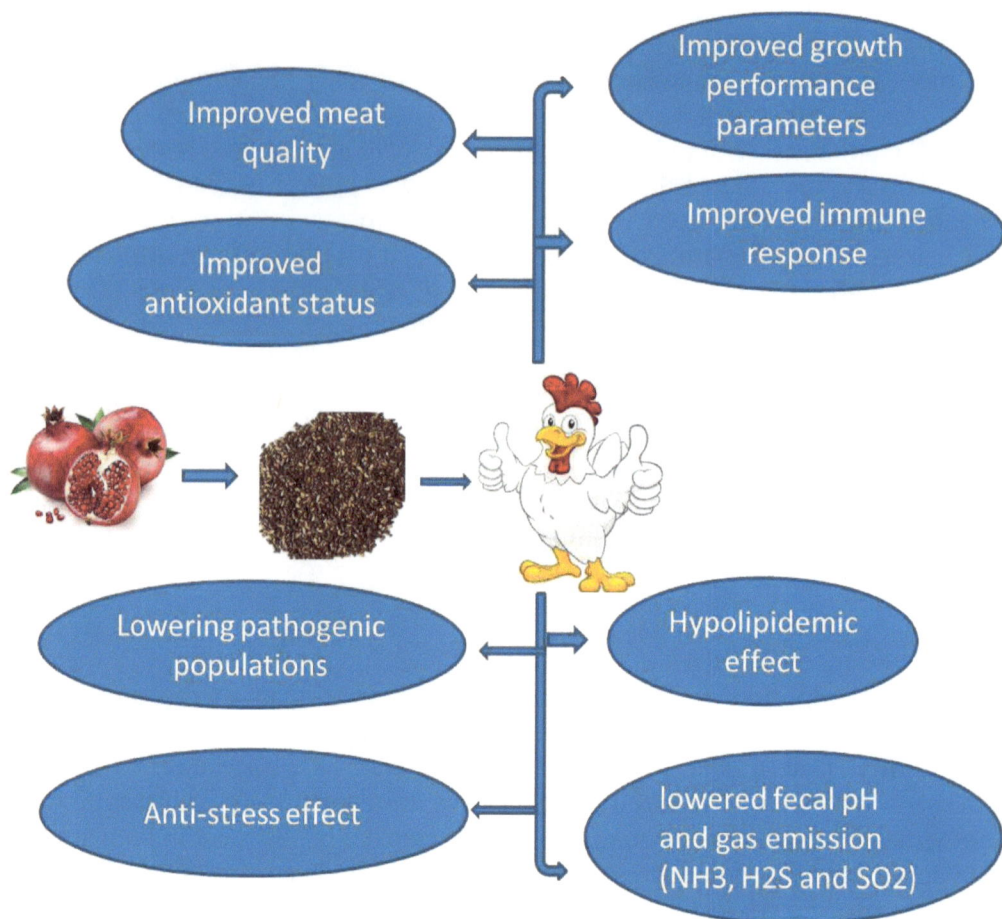

Fig. (4). Some general properties of pomegranate on poultry livestock and their performance.

CONCLUSION

Pomegranate has gained substantial attention as it is a rich source of an essential variety of phytochemical compounds. It has many promising biological properties like antioxidant, anti-inflammatory, immunomodulatory, antimicrobial, anti-tumor and wound healing enhancer. Moreover, pomegranate by-products can improve the immunity and activity of the intestinal microbes of broilers chicken. Despite several beneficial activities of pomegranate, there is currently a gap in knowledge regarding its use in the chicken diet. So, this chapter aims to broaden the knowledge of investigators, veterinary analysts, poultry nutritionists, and advisors to do more research on pomegranate to explore more beneficial impacts and recommend it as a secure natural feed added substance in poultry feed.

CONSENT FOR PUBLICATION

Not applicable.

CONFLICT OF INTEREST

The author declares no conflict of interest, financial or otherwise.

ACKNOWLEDGEMENTS

Declared none.

REFERENCES

[1]   Grashorn M. Use of phytobiotics in broiler nutrition – an alternative to infeed antibiotics? J Anim Feed Sci 2010; 19(3): 338-47.
[http://dx.doi.org/10.22358/jafs/66297/2010]

[2]   Mohammadi Gheisar M, Kim IH. Phytobiotics in poultry and swine nutrition – a review. Ital J Anim Sci 2018; 17(1): 92-9.
[http://dx.doi.org/10.1080/1828051X.2017.1350120]

[3]   Viveros A, Chamorro S, Pizarro M, Arija I, Centeno C, Brenes A. Effects of dietary polyphenol-rich grape products on intestinal microflora and gut morphology in broiler chicks. Poult Sci 2011; 90(3): 566-78.
[http://dx.doi.org/10.3382/ps.2010-00889] [PMID: 21325227]

[4]   Saeed M, Baloch A, Wang M, *et al.* Use of *Cichorium Intybus* Leaf extract as growth promoter, hepatoprotectant and immune modulent in broilers. J Anim Prod Adv 2015; 5(1): 585-91.
[http://dx.doi.org/10.5455/japa.20150118041009]

[5]   Annongu AA, Belewu MA, Joseph JK. Potentials of jatropha seeds as substitute protein in nutrition of poultry. Res J Anim Sci 2010; 4(1): 1-4.
[http://dx.doi.org/10.3923/rjnasci.2010.1.4]

[6]   Ezzat Abd El-Hack M, Alagawany M, Ragab Farag M, *et al.* Beneficial impacts of thymol essential oil on health and production of animals, fish and poultry: a review. J Essent Oil Res 2016; 28(5): 365-82.
[http://dx.doi.org/10.1080/10412905.2016.1153002]

[7]   Saeed M, El-Hack MEA, Arif M, *et al.* Impacts of distiller's dried grains with solubles as replacement of soybean meal plus vitamin E supplementation on production, egg quality and blood chemistry of laying hens. Ann Anim Sci 2017; 17(3): 849-62.
[http://dx.doi.org/10.1515/aoas-2016-0091]

[8]   Saeed M, Naveed M, Arain MA, *et al.* Quercetin: Nutritional and beneficial effects in poultry. Worlds Poult Sci J 2017; 73(2): 355-64.
[http://dx.doi.org/10.1017/S004393391700023X]

[9]   Saeed M, Abd El-Hack ME, Alagawany M, *et al.* Chicory (*Cichorium Intybus*) herb: Chemical composition, pharmacology, nutritional and healthical applications. Int J Pharmacol 2017; 13(4): 351-60.
[http://dx.doi.org/10.3923/ijp.2017.351.360]

[10]   Saeed M, Yatao X, Ur Rehman Z, *et al.* Nutritional and healthical aspects of yacon (*Smallanthus sonchifolius*) for human, animals and poultry. Int J Pharmacol 2017; 13(4): 361-9.
[http://dx.doi.org/10.3923/ijp.2017.361.369]

[11]   Ajaikumar KB, Asheef M, Babu BH, Padikkala J. The inhibition of gastric mucosal injury by *Punica granatum* L. (pomegranate) methanolic extract. J Ethnopharmacol 2005; 96(1-2): 171-6.

[http://dx.doi.org/10.1016/j.jep.2004.09.007] [PMID: 15588667]

[12]   Opara LU, Al-Ani MR, Al-Shuaibi YS. Physico-chemical Properties, Vitamin C Content, and Antimicrobial Properties of Pomegranate Fruit (Punica granatum L.). Food Bioprocess Technol 2009; 2(3): 315-21.
[http://dx.doi.org/10.1007/s11947-008-0095-5]

[13]   Shabtay A, Eitam H, Tadmor Y, *et al.* Nutritive and antioxidative potential of fresh and stored pomegranate industrial byproduct as a novel beef cattle feed. J Agric Food Chem 2008; 56(21): 10063-70.
[http://dx.doi.org/10.1021/jf8016095] [PMID: 18925742]

[14]   Galego LR, Jockusch S, Da Silva JP. Polyphenol and volatile profiles of pomegranate (*Punica granatum* L.) fruit extracts and liquors. Int J Food Sci Technol 2013; 48(4): 693-700.
[http://dx.doi.org/10.1111/ijfs.12014]

[15]   Seeram N, Adams L, Henning S, *et al. In vitro* antiproliferative, apoptotic and antioxidant activities of punicalagin, ellagic acid and a total pomegranate tannin extract are enhanced in combination with other polyphenols as found in pomegranate juice. J Nutr Biochem 2005; 16(6): 360-7.
[http://dx.doi.org/10.1016/j.jnutbio.2005.01.006] [PMID: 15936648]

[16]   Li Y, Qi Y, Huang THW, Yamahara J, Roufogalis BD. Pomegranate flower: a unique traditional antidiabetic medicine with dual PPAR-α/-γ activator properties. Diabetes Obes Metab 2008; 10(1): 10-7.
[PMID: 18095947]

[17]   Poyrazoğlu E, Gökmen V, Artık N. Organic acids and phenolic compounds in pomegranates (*Punica granatum* L.) grown in Turkey. J Food Compos Anal 2002; 15(5): 567-75.
[http://dx.doi.org/10.1016/S0889-1575(02)91071-9]

[18]   Nuamsetti T, Dechayuenyong P, Tantipaibulvut S. Antibacterial activity of pomegranate fruit peels and arils. Sci Asia 2012; 38(3): 319-22.
[http://dx.doi.org/10.2306/scienceasia1513-1874.2012.38.319]

[19]   Shiga G. Antibacterial activity of *Punica granatum* peel extracts against Shiga toxin producing E. coli. IJLBPR 2012; 1: 164-72.

[20]   Gracious Ross R, Selvasubramanian S, Jayasundar S. Immunomodulatory activity of *Punica granatum* in rabbits—a preliminary study. J Ethnopharmacol 2001; 78(1): 85-7.
[http://dx.doi.org/10.1016/S0378-8741(01)00287-2] [PMID: 11585693]

[21]   Jurenka JS. Therapeutic applications of pomegranate (*Punica granatum* L.): a review. Altern Med Rev 2008; 13(2): 128-44.
[PMID: 18590349]

[22]   Chidambara Murthy KN, Reddy VK, Veigas JM, Murthy UD. Study on wound healing activity of *Punica granatum* peel. J Med Food 2004; 7(2): 256-9.
[http://dx.doi.org/10.1089/1096620041224111] [PMID: 15298776]

[23]   Hayouni EA, Miled K, Boubaker S, *et al.* Hydroalcoholic extract based-ointment from *Punica granatum* L. peels with enhanced *in vivo* healing potential on dermal wounds. Phytomedicine 2011; 18(11): 976-84.
[http://dx.doi.org/10.1016/j.phymed.2011.02.011] [PMID: 21466954]

[24]   Al-Zoreky NS. Antimicrobial activity of pomegranate (*Punica granatum* L.) fruit peels. Int J Food Microbiol 2009; 134(3): 244-8.
[http://dx.doi.org/10.1016/j.ijfoodmicro.2009.07.002] [PMID: 19632734]

[25]   Abdollahzadeh Sh, Mashouf R, Mortazavi H, Moghaddam M, Roozbahani N, Vahedi M. Antibacterial and antifungal activities of *punica granatum* peel extracts against oral pathogens. J Dent (Tehran) 2011; 8(1): 1-6.
[PMID: 21998800]

[26]   Kawaii S, Lansky EP. Differentiation-promoting activity of pomegranate (*Punica granatum*) fruit extracts in HL-60 human promyelocytic leukemia cells. J Med Food 2004; 7(1): 13-8.
[http://dx.doi.org/10.1089/109662004322984644] [PMID: 15117547]

[27]   Carvalho Filho JM. Pomegranate seed oil (*Punica granatum* L.): a source of punicic acid (conjugated α-linolenic acid). J Human Nutri Food Sci 2014; 2(1): 1-11.

[28]   Rajani J, Karimi Torshizi MA, Rahimi S. Control of ascites mortality and improved performance and meat shelf-life in broilers using feed adjuncts with presumed antioxidant activity. Anim Feed Sci Technol 2011; 170(3-4): 239-45.
[http://dx.doi.org/10.1016/j.anifeedsci.2011.09.001]

[29]   Waheed S, Siddique N, Rahman A, Zaidi JH, Ahmad S. INAA for dietary assessment of essential and other trace elements in fourteen fruits harvested and consumed in Pakistan. J Radioanal Nucl Chem 2004; 260(3): 523-31.
[http://dx.doi.org/10.1023/B:JRNC.0000028211.23625.99]

[30]   Banihani S, Swedan S, Alguraan Z. Pomegranate and type 2 diabetes. Nutr Res 2013; 33(5): 341-8.
[http://dx.doi.org/10.1016/j.nutres.2013.03.003] [PMID: 23684435]

[31]   Mirsaeedghazi H, Emam-Djomeh Z, Ahmadkhaniha R. Effect of frozen storage on the anthocyanins and phenolic components of pomegranate juice. J Food Sci Technol 2014; 51(2): 382-6.
[http://dx.doi.org/10.1007/s13197-011-0504-z] [PMID: 24493900]

[32]   Ghasemnezhad M, Zareh S, Shiri MA, Javdani Z. The arils characterization of five different pomegranate (*Punica granatum*) genotypes stored after minimal processing technology. J Food Sci Technol 2015; 52(4): 2023-32.
[http://dx.doi.org/10.1007/s13197-013-1213-6] [PMID: 25829582]

[33]   Hmid I, Elothmani D, Hanine H, Oukabli A, Mehinagic E. Comparative study of phenolic compounds and their antioxidant attributes of eighteen pomegranate (*Punica granatum* L.) cultivars grown in Morocco. Arab J Chem 2017; 10: S2675-84.
[http://dx.doi.org/10.1016/j.arabjc.2013.10.011]

[34]   Amri Z, Lazreg-Aref H, Mekni M, *et al.* Oil characterization and lipids class composition of pomegranate seeds. BioMed research international 2017.
[http://dx.doi.org/10.1155/2017/2037341]

[35]   Yuan Z, Fang Y. Flavonol s and Flavone s Changes in Pomegranate (*Punica granatum* L.) Fruit Peel during Fruit Development. 2018.

[36]   Ercisli S, Agar G, Orhan E, Yildirim N, Hizarci Y. Interspecific variability of RAPD and fatty acid composition of some pomegranate cultivars (*Punica granatum* L.) growing in Southern Anatolia Region in Turkey. Biochem Syst Ecol 2007; 35(11): 764-9.
[http://dx.doi.org/10.1016/j.bse.2007.05.014]

[37]   Lan J, Lei F, Hua L, Wang Y, Xing D, Du L. Transport behavior of ellagic acid of pomegranate leaf tannins and its correlation with total cholesterol alteration in HepG2 cells. Biomed Chromatogr 2009; 23(5): 531-6.
[http://dx.doi.org/10.1002/bmc.1150] [PMID: 19101929]

[38]   Huang T, Peng G, Kota B, *et al.* Anti-diabetic action of flower extract: Activation of PPAR-γ and identification of an active component. Toxicol Appl Pharmacol 2005; 207(2): 160-9.
[http://dx.doi.org/10.1016/j.taap.2004.12.009] [PMID: 16102567]

[39]   Aviram M, Dornfeld L. Pomegranate juice consumption inhibits serum angiotensin converting enzyme activity and reduces systolic blood pressure. Atherosclerosis 2001; 158(1): 195-8.
[http://dx.doi.org/10.1016/S0021-9150(01)00412-9] [PMID: 11500191]

[40]   Shukla M, Gupta K, Rasheed Z, Khan KA, Haqqi TM. Bioavailable constituents/metabolites of pomegranate (*Punica granatum* L) preferentially inhibit COX2 activity *ex vivo* and IL-1beta-induced PGE2 production in human chondrocytes *in vitro*. J Inflamm (Lond) 2008; 5(1): 9.

[http://dx.doi.org/10.1186/1476-9255-5-9]

[41]   Sudheesh S, Vijayalakshmi NR. Flavonoids from *Punica granatum*—potential antiperoxidative agents. Fitoterapia 2005; 76(2): 181-6.
[http://dx.doi.org/10.1016/j.fitote.2004.11.002] [PMID: 15752628]

[42]   Youdim KA, McDonald J, Kalt W, Joseph JA. Potential role of dietary flavonoids in reducing microvascular endothelium vulnerability to oxidative and inflammatory insults Mention of trade name, proprietary product, or specific equipment does not constitute a guarantee by the US Department of Agriculture and does not imply its approval to the exclusion of other products that may be suitable. J Nutr Biochem 2002; 13(5): 282-8.
[http://dx.doi.org/10.1016/S0955-2863(01)00221-2] [PMID: 12015158]

[43]   Noda Y, Kaneyuki T, Mori A, Packer L. Antioxidant activities of pomegranate fruit extract and its anthocyanidins: delphinidin, cyanidin, and pelargonidin. J Agric Food Chem 2002; 50(1): 166-71.
[http://dx.doi.org/10.1021/jf0108765] [PMID: 11754562]

[44]   Kim H, Banerjee N, Ivanov I, *et al.* Comparison of anti-inflammatory mechanisms of mango (*Mangifera Indica* L.) and pomegranate (*Punica Granatum* L.) in a preclinical model of colitis. Mol Nutr Food Res 2016; 60(9): 1912-23.
[http://dx.doi.org/10.1002/mnfr.201501008] [PMID: 27028006]

[45]   Shah TA, Parikh M, Patel KV, Patel KG, Joshi CG, Gandhi TR. Evaluation of the effect of *Punica granatum* juice and punicalagin on NFκB modulation in inflammatory bowel disease. Mol Cell Biochem 2016; 419(1-2): 65-74.
[http://dx.doi.org/10.1007/s11010-016-2750-x] [PMID: 27352379]

[46]   Lee CJ, Chen LG, Liang WL, Wang CC. Anti-inflammatory effects of Punica granatum Linne *in vitro* and *in vivo*. Food Chem 2010; 118(2): 315-22.
[http://dx.doi.org/10.1016/j.foodchem.2009.04.123]

[47]   Esmaillzadeh A, Tahbaz F, Gaieni I, Alavi-Majd H, Azadbakht L. Cholesterol-lowering effect of concentrated pomegranate juice consumption in type II diabetic patients with hyperlipidemia. Int J Vitam Nutr Res 2006; 76(3): 147-51.
[http://dx.doi.org/10.1024/0300-9831.76.3.147] [PMID: 17048194]

[48]   Bagri P, Ali M, Aeri V, Bhowmik M, Sultana S. Antidiabetic effect of *Punica granatum* flowers: Effect on hyperlipidemia, pancreatic cells lipid peroxidation and antioxidant enzymes in experimental diabetes. Food Chem Toxicol 2009; 47(1): 50-4.
[http://dx.doi.org/10.1016/j.fct.2008.09.058] [PMID: 18950673]

[49]   Gould SWJ, Fielder MD, Kelly AF, Naughton DP. Anti-microbial activities of pomegranate rind extracts: enhancement by cupric sulphate against clinical isolates of *S. aureus*, MRSA and PVL positive CA-MSSA. BMC Complement Altern Med 2009; 9(1): 23.
[http://dx.doi.org/10.1186/1472-6882-9-23] [PMID: 19635137]

[50]   Viuda-Martos M, Fernández-López J, Pérez-Álvarez JA. Pomegranate and its Many Functional Components as Related to Human Health: A Review. Compr Rev Food Sci Food Saf 2010; 9(6): 635-54.
[http://dx.doi.org/10.1111/j.1541-4337.2010.00131.x] [PMID: 33467822]

[51]   Seeram NP, Aviram M, Zhang Y, *et al.* Comparison of antioxidant potency of commonly consumed polyphenol-rich beverages in the United States. J Agric Food Chem 2008; 56(4): 1415-22.
[http://dx.doi.org/10.1021/jf073035s] [PMID: 18220345]

[52]   Bialonska D, Kasimsetty SG, Schrader KK, Ferreira D. (*Punica granatum* L.) By-products and Ellagitannins on the Growth of Human Gut Bacteria. J Agric Food Chem 2009; 57(18): 8344-9.
[http://dx.doi.org/10.1021/jf901931b] [PMID: 19705832]

[53]   Ahmed ST, Yang CJ. Effects of dietary *Punica granatum* L. by-products on performance, immunity, intestinal and fecal microbiology, and odorous gas emissions from excreta in broilers. J Poult Sci 2017; 54(2): 157-66.

[http://dx.doi.org/10.2141/jpsa.0160116] [PMID: 32908421]

[54]   Ahmed ST, Islam MM, Bostami ABMR, Mun HS, Kim YJ, Yang CJ. Meat composition, fatty acid profile and oxidative stability of meat from broilers supplemented with pomegranate (*Punica granatum* L.) by-products. Food Chem 2015; 188: 481-8.
[http://dx.doi.org/10.1016/j.foodchem.2015.04.140] [PMID: 26041221]

[55]   Ahmed ST, Ko SY, Yang CJ. Improving the nutritional quality and shelf life of broiler meat by feeding diets supplemented with fermented pomegranate (*Punica granatum* L.) by-products. Br Poult Sci 2017; 58(6): 694-703.
[http://dx.doi.org/10.1080/00071668.2017.1363870] [PMID: 28792239]

[56]   Szymczyk B, Szczurek W. Effect of dietary pomegranate seed oil and linseed oil on broiler chickens performance and meat fatty acid profile. J Anim Feed Sci 2016; 25(1): 37-44.
[http://dx.doi.org/10.22358/jafs/65585/2016]

[57]   Saleh H, Golian A, Kermanshahi H, Mirakzehi MT. Effects of dietary α-tocopherol acetate, pomegranate peel, and pomegranate peel extract on phenolic content, fatty acid composition, and meat quality of broiler chickens. J Appl Anim Res 2017; 45(1): 629-36.
[http://dx.doi.org/10.1080/09712119.2016.1248841]

[58]   Saleh H, Golian A, Kermanshahi H, Mirakzehi MT. Antioxidant status and thigh meat quality of broiler chickens fed diet supplemented with α-tocopherolacetate, pomegranate pomace and pomegranate pomace extract. Ital J Anim Sci 2018; 17(2): 386-95.
[http://dx.doi.org/10.1080/1828051X.2017.1362966]

[59]   Akuru EA, Oyeagu CE, Mpendulo TC, Rautenbach F, Oguntibeju OO. Effect of pomegranate (*Punica granatum L*) peel powder meal dietary supplementation on antioxidant status and quality of breast meat in broilers. Heliyon 2020; 6(12): e05709.
[http://dx.doi.org/10.1016/j.heliyon.2020.e05709] [PMID: 33364487]

[60]   Ghosh S, Chatterjee PN, Maity A, Mukherjee J, Batabyal S, Chatterjee JK. Effect of supplementing pomegranate peel infusion on body growth, feed efficiency, biochemical metabolites and antioxidant status of broiler chicken. Trop Anim Health Prod 2020; 52(6): 3899-905.
[http://dx.doi.org/10.1007/s11250-020-02352-0] [PMID: 32737663]

[61]   Bostami A, Ahmed S, Islam M, *et al.* Growth performance, fecal noxious gas emission and economic efficacy in broilers fed fermented pomegranate by-products as residue of fruit industry. Int J Adv Res (Indore) 2015; 3(3): 102-14.

[62]   Yaseen AT, El-Kholy MESH, Abd El-Razik WM, Soliman MH, Soliman MH. Effect of using pomegranate peel extract as feed additive on performance, serum lipids and immunity of broiler chicks. Zagazig Vet J 2014; 42(1): 87-92.
[http://dx.doi.org/10.21608/zvjz.2014.59473]

[63]   Abbas RJ, Al-Salhie KCK, Al-Hummod SK. The effect of using different levels of pomegranate (*Punica granatum*) peel powder on productive and physiological performance of Japanese quail (*Coturnix coturnix japonica*). Livest Res Rural Dev 2017; 29(12): 2017.

[64]   Rezvani MR, Sayadpour N, Saemi F. The effect of adding pomegranate seed extract to fat-containing diets on nutrients digestibility, intestinal microflora and growth performance of broilers. Majallah-i Tahqiqat-i Dampizishki-i Iran 2018; 19(4): 310-7.
[PMID: 30774673]

[65]   Sarica S, Urkmez D. The use of grape seed-, olive leaf-and pomegranate peel-extracts as alternative natural antimicrobial feed additives in broiler diets. 2016.

[66]   Hamady G, Abdel-Moneim MA, El-Chaghaby GA, Abd-El-Ghany ZM, Hassanin MS. Effect of Pomegranate peel extract as natural growth promoter on the productive performance and intestinal microbiota of broiler chickens. African Journal of Agricultural Science and Technology 2015; 3(12): 514-9.

[67] Rama Rao SV, Raju MVLN, Prakash B, Rajkumar U, Reddy EPK. Effect of supplementing moringa (*Moringa oleifera*) leaf meal and pomegranate (*Punica granatum*) peel meal on performance, carcass attributes, immune and antioxidant responses in broiler chickens. Anim Prod Sci 2019; 59(2): 288-94.
[http://dx.doi.org/10.1071/AN17390]

[68] Kishawy AT, Omar AE, Gomaa AM. Growth performance and immunity of broilers fed rancid oil diets that supplemented with pomegranate peel extract and sage oil. Jpn J Vet Res 2016; 64 (Suppl. 2): S31-8.

[69] Kishawy ATY, Amer SA, Abd El-Hack ME, Saadeldin IM, Swelum AA. The impact of dietary linseed oil and pomegranate peel extract on broiler growth, carcass traits, serum lipid profile, and meat fatty acid, phenol, and flavonoid contents. Asian-Australas J Anim Sci 2019; 32(8): 1161-71.
[http://dx.doi.org/10.5713/ajas.18.0522] [PMID: 30744351]

[70] Manterys A, Franczyk-Zarow M, Czyzynska-Cichon I, *et al.* Haematological parameters, serum lipid profile, liver function and fatty acid profile of broiler chickens fed on diets supplemented with pomegranate seed oil and linseed oil. Br Poult Sci 2016; 57(6): 771-9.
[http://dx.doi.org/10.1080/00071668.2016.1219977] [PMID: 27636015]

[71] Hosseini-Vashan SJ, Raei-Moghadam MS. Antioxidant and immune system status, plasma lipid, abdominal fat, and growth performance of broilers exposed to heat stress and fed diets supplemented with pomegranate pulp (*Punica granatum L.*). J Appl Anim Res 2019; 47(1): 521-31.
[http://dx.doi.org/10.1080/09712119.2019.1676756]

[72] Al-Shammar KIA, Batkowska J, Zamil SJ. Role of pomegranate peels and black pepper powder and their mixture in alleviating the oxidative stress in broiler chickens. Int J Poult Sci 2019; 18(3): 122-8.
[http://dx.doi.org/10.3923/ijps.2019.122.128]

[73] Kumar KP, Reddy VR, Prakash MG. Effect of supplementing pomegranate (*Punica granatum*). peel extract on serum biochemical parameters and immune response in broilers during summer. Pharma Innov 2018; 7(1): 591-601.

[74] Abdel Baset S, Ashour EA, Abd El-Hack ME, El-Mekkawy MM. Effect of different levels of pomegranate peel powder and probiotic supplementation on growth, carcass traits, blood serum metabolites, antioxidant status and meat quality of broilers. Anim Biotechnol 2020; 1-11.
[http://dx.doi.org/10.1080/10495398.2020.1825965] [PMID: 33000991]

[75] Nagla F, Eman M, Nermeen F. Evaluating the Efficacy of Some Antibiotics and Medicinal Plant Extracts Against the Infection of *Escherichia Coli* and *Salmonella Enteritidis* and Their Effect on Poultry Meat Quality. 2016.

[76] Zhong X, Shi Y, Chen J, *et al.* Polyphenol extracts from *Punica granatum* and *Terminalia chebula* are anti-inflammatory and increase the survival rate of chickens challenged with Escherichia coli. Biol Pharm Bull 2014; 37(10): 1575-82.
[http://dx.doi.org/10.1248/bpb.b14-00163] [PMID: 25273385]

[77] Saki A, Rabet M, Zamani P, Yousefi A. The effects of different levels of pomegranate seed pulp with multi-enzyme on performance, egg quality and serum antioxidant in laying hens. Iran J Appl Anim Sci 2014; 4(4): 803-8.

[78] Lytou A, Panagou EZ, Nychas GJE. Development of a predictive model for the growth kinetics of aerobic microbial population on pomegranate marinated chicken breast fillets under isothermal and dynamic temperature conditions. Food Microbiol 2016; 55: 25-31.
[http://dx.doi.org/10.1016/j.fm.2015.11.009] [PMID: 26742613]

[79] Vaithiyanathan S, Naveena BM, Muthukumar M, Girish PS, Kondaiah N. Effect of dipping in pomegranate (*Punica granatum*) fruit juice phenolic solution on the shelf life of chicken meat under refrigerated storage (4°C). Meat Sci 2011; 88(3): 409-14.
[http://dx.doi.org/10.1016/j.meatsci.2011.01.019] [PMID: 21345604]

[80] Devatkal SK, Thorat P, Manjunatha M. Effect of vacuum packaging and pomegranate peel extract on

quality aspects of ground goat meat and nuggets. J Food Sci Technol 2014; 51(10): 2685-91.
[http://dx.doi.org/10.1007/s13197-012-0753-5] [PMID: 25328212]

[81]   Bazargani-Gilani B, Aliakbarlu J, Tajik H. Effect of pomegranate juice dipping and chitosan coating enriched with *Zataria multiflora* Boiss essential oil on the shelf-life of chicken meat during refrigerated storage. Innov Food Sci Emerg Technol 2015; 29: 280-7.
[http://dx.doi.org/10.1016/j.ifset.2015.04.007]

[82]   Kanatt SR, Chander R, Sharma A. Antioxidant and antimicrobial activity of pomegranate peel extract improves the shelf life of chicken products. Int J Food Sci Technol 2010; 45(2): 216-22.
[http://dx.doi.org/10.1111/j.1365-2621.2009.02124.x]

CHAPTER 7

Use of Chicory (*Cichorium intybus*) and its Derivatives in Poultry Nutrition

Muhammad Saeed[1], Faisal Siddique[1], Rizwana Sultan[1], Sabry A.A. El-Sayed[2], Sarah Y.A. Ahmed[3], Mayada R. Farag[4], Mohamed E. Abd El-Hack[5,*], Abdelrazeq M. Shehata[6,7] and Mahmoud Alagawany[5,*]

[1] *Cholistan University of Veterinary and Animal Sciences Bahawalpur, 63100, Pakistan*

[2] *Department of Nutrition and Clinical Nutrition, Faculty of Veterinary Medicine, Zagazig University, Zagazig, Egypt*

[3] *Department of Microbiology, Faculty of Veterinary Medicine, Zagazig University, Zagazig, Egypt*

[4] *Forensic Medicine and Toxicology Department, Veterinary Medicine Faculty, Zagazig University, Zagazig 44519, Egypt*

[5] *Poultry Department, Faculty of Agriculture, Zagazig University, Zagazig 44511, Egypt*

[6] *Department of Dairy Science & Food Technology, Institute of Agricultural Sciences, Banaras Hindu University, Varanasi 221005, India*

[7] *Department of Animal Production, Faculty of Agriculture, Al-Azhar University, Cairo 11651, Egypt*

Abstract: Chicory (*Cichorium intybus*) is a perennial herb that belongs to the Asteraceae family. Certain species are grown and used as fried, dry salad leaves, roots, or chicons as a substitute for coffee additives. It is also cultivated as forage that can be used in animal feeding. In addition, chicory has significant effects on animal and human health and has various biological activities, such as immunostimulant, antimicrobial, antioxidant, hyperlipidemic, anti-inflammatory, and antidiabetic activity. Chicory extracts protect the liver by lowering the levels of liver enzymes, *e.g.*, aspartate aminotransferase (AST), alanine aminotransferase (ALT), and alkaline phosphatase (ALP). The chicory plant plays a key role in protecting hepatocytes and other liver cells. It is used as an antimicrobial agent *in vitro* and *in vivo* against certain pathogenic bacteria species. Chicory improves immunity and feed efficacy by reducing pathogenic microorganisms in the gastrointestinal tract. *Cichorium intybus* roots were also used to alleviate slight intestinal disturbances, including the sense of flatulence, full abdomen, transient appetite loss, and indigestion. This chapter describes the role of chicory plants in promoting growth when used as feed additives in poultry feed. It also explains the mechanisms of action of chicory extracts and their role as a liver protector for poultry.

* **Corresponding author Mahmoud Alagawany and Mohamed E. Abd El-Hack:** Poultry Department, Faculty of Agriculture, Zagazig University, Zagazig 44511, Egypt; E-mails: dr.mahmoud.alagwany@gmail.com and dr.mohamed.e.abdalhaq@gmail.com

Keywords: Antioxidant, Aspartate aminotransferase, Hepatocytes, Indigestion, Protection.

INTRODUCTION

Synthetic antibiotics are used worldwide as feed additives regardless of their adverse effects on animals and humans. Since the European Union banned antibiotics as growth promoters in 1999, the quest for alternatives has been a major focus of the scientific investigation. Thus, the use of herbal feed additives is becoming more popular in animal production due to the restriction on certain antibiotics [1, 2]. Chicory (*Cichorium intybus*) is among many herbal plants that could be used in poultry feed as a natural feed additive. Because the poultry industry is one of the most extensive industries that satisfy the protein demand of consumers, the world's impressive development of the poultry industry has been attributed to the control of diseases, feed processing and genetic and managerial modifications. Diverse approaches have been employed to boost broiler efficiency in production, feed utilization and economy. Among these methods, the most used method is antibiotics usage as a growth promoter [3]. For certain countries, antibiotics used as a feed supplement were prohibited because of severe food insecurity since these antibiotics provide resistance to several bacteria in animals and humans. The demand for antibiotic-free poultry meat increases daily as we recognize the negative effects of antibiotics on human health. In addition, other factors such as technology and science advancement, hygiene, and improvements in nutrition regulation play a significant part in growing the market for healthy foods. Nowadays, poultry experts look forward to discovering other antibiotic alternatives to mitigate the health problems in poultry and consumers and boost immunity and the consistency of feed by reducing gastrointestinal tract pathogenic bacteria [4].

Various sections of herbal medicinal plants and herbal derivatives, including black cumin [5], derived products from quercetin [6] and chicory, have been used as growth promoters for poultry feed but with differing results [7, 8]. Such plants contain several properties, including growth promoters, antibacterial, antifungal, anticancer, anti-malarial, anticoccidial, gastroprotective, diuretic and immuno-stimulant. Nonetheless, a few experiments indicated a lack of knowledge on chicory use for hepatoprotection and the core mechanism for the positive effects of chicory in poultry production. Therefore, this work aims to shed light on this therapeutic herb to alleviate liver damage and substitute chicory antibiotics eventually and have future hope in the poultry industry [9].

THE DESCRIPTION OF PLANTS AND THEIR CHEMICAL COMPOSITION

Chicory is a perennial herb that belongs to the Asteraceae family (Table **1** and Fig. **1**).

Table 1. Scientific classification of chicory herb.

Kingdom:	Plantae
Clade:	Tracheophytes
Clade:	Angiosperms
Clade:	Eudicots
Clade:	Asterids
Order:	Asterales
Family:	Asteraceae
Tribe:	Cichorieae
Genus:	Cichorium
Species:	C. intybus
Binomial name	*Cichorium intybus*

Fig. (1). Chicory herb, root and dried root.

It is habitually grown with other plants such as lucerne and berseem (Fig. **2**) [10]. It has multiple applications in manufacturing and can be used as an antimicrobial and antioxidant agent in poultry [11, 12]. Compared to the therapeutic elements, kasni's nutritional profile is also good as it includes some quantities of vitamins and minerals (especially vitamin C). Different active component forms include carbohydrates (fructose, mannitol, and lactose), micro-minerals, and vitamins (Fig. **3**) [13 - 15].

Fig. (2). The edible parts in Chicory.

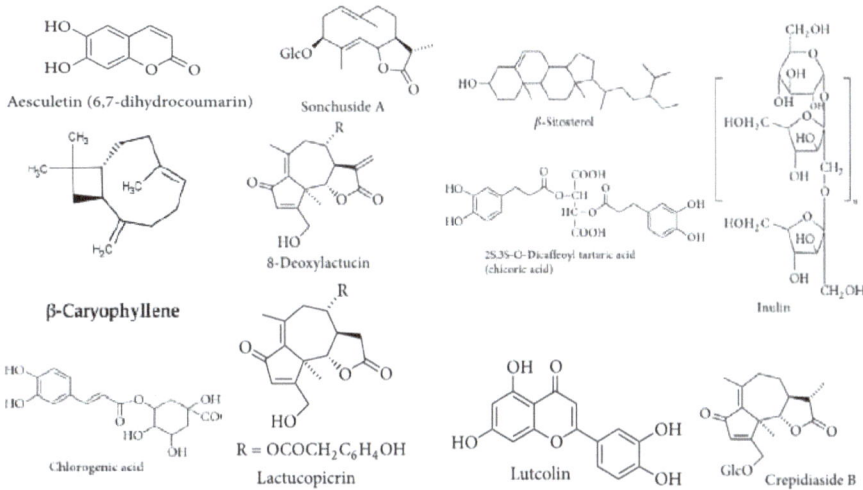

Fig. (3). Identified phytoconstituents of Chicory.

The nutritional value (per 100 g) of the chicory herb is illustrated in Table **2**.

Table 2. Nutritional value (per 100 g) of chicory herb*.

Nutrient	-
Energy	96 kJ (23 kcal)
Carbohydrates	4.7 g

(Table 2) cont.....

Nutrient	-
Dietary fiber	4 g
Sugars	0.7 g
Protein	1.7 g
Fat	0.3 g
Minerals	-
Calcium	100mg
Iron	0.9 mg
Magnesium	30 mg
Phosphorus	47 mg
Manganese	0.43 mg
Potassium	420 mg
Zinc	0.42 mg
Sodium	45 mg
Vitamins	-
Vitamin A	286 µg
beta-Caroten	3430 µg
lutein zeaxanthin	10300 µg
Thiamine (B1)	0.06 mg
Riboflavin (B2)	0.1 mg
Niacin (B3)	0.5 mg
Vitamin B6	0.105 mg
Pantothenic acid (B5)	1.159 mg
Vitamin C	24 mg
Vitamin E	2.26 mg
Folate (B9)	110 µg
Vitamin K	297.6 µg

*USDA: https://ndb.nal.usda.gov/ndb/search/list?qlookup=11152&format=Full.

Advantageous Results of Chicory with Specific Respect to its Function as a Hepatoprotective Agent

Chicory is an herbaceous plant with multiple characteristics; it acts as an antimicrobial agent (anthelmintic, anti-protozoal, antibacterial, antidiabetic) and antioxidant. We will further use chicory to help the development of effective hepatoprotective feed additives. The mechanism of action of this plant with herbs is known as a liver tonic [16].

To the best of our knowledge, the hepatoprotective effects of chicory are still ununderstood. The liver is an important organ and performs more than one physiological function. The nutritional quality of the birds is evaluated not only by what they consume but also by the biological activity of their feed. Unfortunately, the frequent use of antibiotics to promote growth in broiler production makes it very difficult to detect early signs of liver dysfunction and even long-term damage [17]. Trease and Evans [18] stated that the liver has complicated chemistry and therefore plays a significant role in the metabolism of birds. Chicory is an outstanding source of protection for hepatocytes and is very important for liver health. Clinical assessment has also shown that it can treat liver dysfunction [19]. Chicory is used as a prebiotic for *in vitro* and *in vivo* against certain pathogenic bacterial organisms. Chicory improved feed utilization and enhanced broiler immunity [20, 21]. It has been reported that the hepatoprotective function of freshly extracted chicory from ethanolic leaves was observed at different levels compared to birds treated with silymarin [22]. Chicory extracts may inhibit atrophy in skeletal muscles by reducing ceramide production and Hsp70 levels [23]. Rasmussen [24] stated that the supplementing chicory with feeds may regulate the metabolism of hepatic androsterone, which may be a means of enticing a partner.

In terms of the hepatoprotective effects of chicory, its extracts significantly decreased hepatic enzymes such as ALT, ALP, and AST [25]. However, Marzouk *et al.* [26] did not find any major effect on ALT activity by chicory leaf extract. The hepatoprotective function was observed in celery leaves and chicory supplementation on hypercholesterolemic rats, as stated by Abd El-Mageed [27]. Chicory intake reduced overall lipids, hepatic enzymes, bilirubin, and lower cholesterol [28]. By comparison, they studied the effect of kasni on dietary aflatoxicosis. They observed that when seven ppm aflatoxin B1 was present in the dietary supplement of kasni, it reduced the activity of liver enzymes (AST and ALP) [30].

Gilani and Janbaz [30] found that when aqueous methanol extract of chicory at 500 mg/kg with acetaminophen and CCL4 was added to animal feed, serum liver enzyme rates were as follows: AST (228 IU L–1), ALP (68 IU L–1) and ALT (41 IU L–1). When mice were given a 1 g/kg dose, the mortality was reduced by 30% compared to the mortality rate that was 100% in the acetaminophen group. Zafar and Ali [31] studied that while carbon tetrachloride used in combination with chicory root extract helped protect the liver by lowering liver enzymes in mice, single-use of carbon tetrachloride could damage the liver.

Gilani *et al.* [32] evaluated the protective effect of chicory extract compared to paracetamol against CCl4-induced hepatic damage. They observed that the mortality rate in mice was reduced up to 40% when esculetin was used at 1g/kg compared to paracetamol (100% mortality rate). A single dose of 1.5 ml/kg of CCl_4 damages the liver by increasing the liver enzymes ALP, ALT, and AST. However, the dose of esculetin 6 mg/kg could prevent liver damage induced by CCL4. In particular, the extract of chicory root is rich in natural antioxidants and can relieve liver damage induced by CCl4. This antioxidant activity also reduces the production of different oxidative radicals and prevents lipid peroxidation [33]. Chicory feeding plays an important role in stimulating gene expression and improving the antioxidant status by enhancing antioxidant enzymes [34, 35].

A mixture of chicory, ginger, and both has protecting effects on liver enzymes. Administration of chicory methanol extract (250 and 500 mg kg 1) or combined with ginger (250 and 500 mg/kg) showed no toxic signs, even when CCl_4 doses up to 5 g kg^{-1} were used. Similar results were reported by other researchers who confirmed the protective effect of chicory against oxytetracycline-damaged blubbery liver. They found that dietary supplements significantly reduced HDL, albumin, total protein globulin, and cholesterol compared to controls [36]. Some researchers found that toxic rats were administered with salicylic acid (25 mg/kg) in a diet consisting of dried chicory leaves (100, 200, and 300 mg/kg BWT), which significantly increased AST, ALP, and glutamate pyruvate in serum [37]. By reducing necrosis and glycogen content, the beneficial effect of Cichorium root extract was demonstrated in the animals. In rats suffering from CCl4-induced hepatitis, there was a significant increase in protein synthesis activity and the number of cells [37]. Similar results were seen in another research, which used chicory root extract at dosages of 150-450 or 200-500 mg kg–1 day–1 to treat hepatic damage in rats already suffering from CCl4-induced liver damage. Chicory (Cichorium intybus) was shown to lower microvesicular steatosis and serum indicators [38]. Dietary chicory fed to various animals demonstrated greater live weight gain (LWG) than the control groups. In rodents, aqueous extracts of chicory herbs showed glucocorticoid-mediated bone protection. This is probably because it contains flavonoids and inulin [39, 40].

BENEFICIAL APPLICATIONS OF CHICORY

Several medicinal plants have been used in ancient history for their main health condition. There has been a strong focus on plant-based therapies in traditional societies. The most effective drugs were purposefully delivered from one generation to the next. In India, chicory seeds are used in many commercial products to treat liver disease (Jigrine) [41, 42]. Cichorium intybus roots were traditionally used in Europe to treat minor GIT disorders such as indigestion, abdominal pain, flatulence, and appetite improvement [43]. In addition, chicory stems, leaves, and roots are used to treat jaundice in Africa. The chicory syrup is sometimes used as an energizer for newborns [44]. This plant juice treats tumors, uterine cancer, and malaria intended for pounds [35, 46].

PRACTICAL USAGE IN THE POULTRY SECTOR

Chicory forage is a common dietary fiber supply with advantageous features as a fiber component for poultry production [47]. Saeed *et al.* [8] indicated that Chicory leaf extract is a healthy Growth booster, immune stimulator, and hepatoprotective for broiler development. Cichorium intybus is a theoretically useful part of a fiber-rich diet that improves diet palatability in chickens [48]. Chicory fodder also includes a high amount of uronic acid derived from galactosyluronic acid in dicotyledonous plants; this acid forms the pectin building block [49]. Chicory root contains high amounts of inulin and fructooligosaccharides, which help maintain the health status of the beneficial microbes in the digestive system [50]. Inulin is one of the main prebiotic origins used in poultry [51]. Therefore, in poultry feeding, both the root and the chicory forage are essential as sources of fiber. Broiler-fed *Cichorium intybus* root powder enriched diets dramatically improved growth output by increasing the processing and absorption of food by modifying the jejunum histomorphometry, as shown by Izadi *et al.* [52]. Beta-fructans extracted from chicory greatly declined the serum cholesterol in the broiler [53, 54]. In broilers, Chicory root powder feeding (4.5 percent) reduced the concentrations of triglyceride and VLDL-cholesterol with no damaging effects on chicken birds. At the same time, overall LDL and HDL cholesterol were not affected by dietary supplements [55]. The concentration of different liver enzymes, *e.g.*, ALP, ALT, AST, LDH, and conjugated bilirubin, has been increased while uric acid concentrations with *Cichorium intybus* supplementation were beyond the normal range as shown by Najafzadeh *et al.* [56]. According to another report, chicory herb intake for 12 weeks did not affect dairy cows' blood non-esterified fatty acids and urea nitrogen [57]. Some researchers noted that chicory plants lowered serum uric acid levels and inhibited liver xanthine oxidase and xanthine dehydrogenase [58, 59]. This plant may have been produced in India, Italy, and France, but much production is done in New Zealand for poultry [60].

CONCLUSION AND FUTURE RECOMMENDATIONS

Chicory may be a practical choice for poultry's dietary and therapeutic aspects. The benefits of using chicory as a natural feed additive are promising. This is probably because the Chicory plant can lower certain hepatic enzyme levels such as AST and ALT. Additionally, as a natural hepatoprotective agent, this plant plays a crucial role in reducing multiple diseases linked to the liver. Chicory plant or its derivatives display plenty of health benefits by reducing serum and organ lipid peroxidation. Furthermore, chicory has many nutritional and pharmacological effects, including hepatoprotective, anti-inflammatory, sedative, digestive, antidiabetic, hormonal, neurological, hypolipidemic, anticancer, anti-protozoan, pain-relieving, anthelmintic and antibacterial properties. More research is needed to understand the action of molecular mechanisms based on the beneficial effects of chicory and to determine the adequate levels in chickens that act exclusively as liver tonics. Nevertheless, further studies will assess the toxicity and safety of chicory herbs or their extracts and oils and provide a new approach to evaluating their application in health and related areas.

CONSENT FOR PUBLICATION

Not applicable.

CONFLICT OF INTEREST

The author declares no conflict of interest, financial or otherwise.

ACKNOWLEDGEMENTS

Declared none.

REFERENCES

[1]   Abd El-Hack ME, Alagawany M, Farag MR, Tiwari R, Karthik K, Dhama K. Nutritional, healthical and therapeutic efficacy of black cumin (*Nigella sativa*) in animals, poultry and humans. Int J Pharmacol 2016; 12: 232-248. Abd El-Mageed NM. Hepatoprotective effect of feeding celery leaves mixed with chicory leaves and barley grains to hypercholesterolemic rats. Pharmacogn Mag 2011; 7: 151-6.

[2]   Mirza Aghazadeh A, Nabiyar E. The effect of chicory root powder on growth performance and some blood parameters of broilers fed wheat-based diets. J Appl Anim Res 2015; 43(4): 384-9. [http://dx.doi.org/10.1080/09712119.2014.978778]

[3]   Ahmad M, Qureshi R, Arshad M, Khan MA, Zafar M. Traditional herbal remedies used for the treatment of diabetes from district Attock (Pakistan). Pak J Bot 2009; 41: 2777-82.

[4]   Ahmed B, Al-Howiriny TA, Siddiqui AB. Antihepatotoxic activity of seeds of *Cichorium intybus*. J Ethnopharmacol 2003; 87(2-3): 237-40. [http://dx.doi.org/10.1016/S0378-8741(03)00145-4] [PMID: 12860315]

[5]   Ahmed N N. Alloxan diabetes-induced oxidative stress and impairment of oxidative defense system in rat brain: neuroprotective effects of *cichorium intybus*. Int J Diabetes Metab 2009; 17(3): 105-9. [http://dx.doi.org/10.1159/000497681]

[6]   Al-Snafi AE. Medical importance of *Cichorium intybus*-a review. J Pharm (Cairo) 2016; 6: 41-56.

[7]   Ali N. Effects of different levels of chicory (*Cichorium intybus* L.), zizaphora (*Zizaphora tenuior* L.), nettle (*Urtica dioica* L.) and savoury (*Satureja hortensis* L.) medicinal plants on carcass characteristics of male broilers. J Med Plants Res 2011; 5: 4354-9.

[8]   Apata DF. Growth performance, nutrient digestibility and immune response of broiler chicks fed diets supplemented with a culture of *Lactobacillus bulgaricus*. J Sci Food Agric 2008; 88(7): 1253-8. [http://dx.doi.org/10.1002/jsfa.3214]

[9]   Atta AH, Elkoly TA, Mouneir SM, Kamel G, Alwabel NA, Zaher S. Hepatoprotective effect of methanol extracts of *Zingiber officinale* and *Cichorium intybus*. Indian J Pharm Sci 2010; 72(5): 564-70.
[http://dx.doi.org/10.4103/0250-474X.78521] [PMID: 21694986]

[10]   Awad WA, Ghareeb K, Böhm J. Evaluation of the chicory inulin efficacy on ameliorating the intestinal morphology and modulating the intestinal electrophysiological properties in broiler chickens. J Anim Physiol Anim Nutr (Berl) 2011; 95(1): 65-72.
[http://dx.doi.org/10.1111/j.1439-0396.2010.00999.x] [PMID: 20579180]

[11]   Bais HP, Ravishankar GA. *Cichorium intybus* L - cultivation, processing, utility, value addition and biotechnology, with an emphasis on current status and future prospects. J Sci Food Agric 2001; 81(5): 467-84.
[http://dx.doi.org/10.1002/jsfa.817]

[12]   Belesky DP, Turner KE, Fedders JM, Ruckle JM. Mineral composition of swards containing forage chicory. Agron J 2001; 93(2): 468-75.
[http://dx.doi.org/10.2134/agronj2001.932468x]

[13]   Bhatti MY. Emerging prospects of poultry production in Pakistan at the dawn of 21st century. Vet News Views 2011; 6: 24-30.

[14]   Bischoff TA, Kelley CJ, Karchesy Y, Laurantos M, Nguyen-Dinh P, Arefi AG. Antimalarial activity of Lactucin and Lactucopicrin: sesquiterpene lactones isolated from *Cichorium intybus* L. J Ethnopharmacol 2004; 95(2-3): 455-7.
[http://dx.doi.org/10.1016/j.jep.2004.06.031] [PMID: 15507374]

[15]   Blazka ME, Wilmer JL, Holladay SD, Wilson RE, Luster MI. Role of proinflammatory cytokines in acetaminophen hepatotoxicity. Toxicol Appl Pharmacol 1995; 133(1): 43-52.
[http://dx.doi.org/10.1006/taap.1995.1125] [PMID: 7597709]

[16]   Castellini C, Cardinali R, Rebollar PG, Dal Bosco A, Jimeno V, Cossu ME. Feeding fresh chicory (*Chicoria intybus*) to young rabbits: Performance, development of gastro-intestinal tract and immune functions of appendix and Peyer's patch. Anim Feed Sci Technol 2007; 134(1-2): 56-65.
[http://dx.doi.org/10.1016/j.anifeedsci.2006.05.007]

[17]   Vasudeva N, Das S, Sharma S. *Cichorium intybus* : A concise report on its ethnomedicinal, botanical, and phytopharmacological aspects. Drug Development and Therapeutics 2016; 7(1): 1-12.
[http://dx.doi.org/10.4103/2394-6555.180157]

[18]   Dhama K, Latheef SK, Mani S, *et al.* Multiple beneficial applications and modes of action of herbs in poultry health and production: A review. Int J Pharmacol 2015; 11(3): 152-76.
[http://dx.doi.org/10.3923/ijp.2015.152.176]

[19]   Ditsch DC, Sears B. http://www.ansc.purdue.edu/SP/MG/Documents/agr190.pdf

[20]   EMA. Community herbal monograph on Cichorium intybus L., radix. London, UK: European Medicines Agency (EMA) and Committee on Herbal Medicinal Products (HMPC) 2012.

[21]   El-Hilaly J, Hmammouchi M, Lyoussi B. Ethnobotanical studies and economic evaluation of medicinal plants in Taounate province (Northern Morocco). J Ethnopharmacol 2003; 86(2-3): 149-58.
[http://dx.doi.org/10.1016/S0378-8741(03)00012-6] [PMID: 12738079]

[22]   El-Sayed YS, Lebda MA, Hassinin M, Neoman SA. Chicory (*Cichorium intybus* L.) root extract regulates the oxidative status and antioxidant gene transcripts in CCl_4-induced hepatotoxicity. PLoS ONE 2015.

[23]   Elrayeh AS, Yildiz G. Effects of inulin and {β-glucan supplementation in broiler diets on growth performance, serum cholesterol, intestinal length, and immune system. Turk J Vet Anim Sci 2012; 36: 388-94.
[http://dx.doi.org/10.3906/vet-1010-504]

[24]   Flickinger EA, Loo JV, Fahey GC Jr. Nutritional responses to the presence of inulin and oligofructose in the diets of domesticated animals: a review. Crit Rev Food Sci Nutr 2003; 43(1): 19-60.
[http://dx.doi.org/10.1080/10408690390826446] [PMID: 12587985]

[25]   Gilani AH, Janbaz KH. Evaluation of the liver protective potential of Cichorium intybus seed extract on Acetaminophen and CCl4-induced damage. Phytomedicine 1994; 1(3): 193-7.
[http://dx.doi.org/10.1016/S0944-7113(11)80064-4] [PMID: 23195938]

[26]   Gilani AH, Janbaz KH, Shah BH. Esculetin prevents liver damage induced by paracetamol and CCL$_4$. Pharmacol Res 1998; 37(1): 31-5.
[http://dx.doi.org/10.1006/phrs.1997.0262] [PMID: 9503477]

[27]   Gong J, Yin F, Hou Y, Yin Y. Review: Chinese herbs as alternatives to antibiotics in feed for swine and poultry production: Potential and challenges in application. Can J Anim Sci 2014; 94(2): 223-41.
[http://dx.doi.org/10.4141/cjas2013-144]

[28]   Guarrera PM, Forti G, Marignoli S. Ethnobotanical and ethnomedicinal uses of plants in the district of Acquapendente (Latium, Central Italy). J Ethnopharmacol 2005; 96(3): 429-44.
[http://dx.doi.org/10.1016/j.jep.2004.09.014] [PMID: 15619562]

[29]   Handa SS. Plants as drugs. East Pharmacist 1991; 34: 79-85.

[30]   Hanlidou E, Karousou R, Kleftoyanni V, Kokkini S. The herbal market of Thessaloniki (N Greece) and its relation to the ethnobotanical tradition. J Ethnopharmacol 2004; 91(2-3): 281-99.
[http://dx.doi.org/10.1016/j.jep.2004.01.007] [PMID: 15120452]

[31]   Hassan HA, Yousef MI. Ameliorating effect of chicory (*Cichorium intybus* L.)-supplemented diet against nitrosamine precursors-induced liver injury and oxidative stress in male rats. Food Chem Toxicol 2010; 48(8-9): 2163-9.
[http://dx.doi.org/10.1016/j.fct.2010.05.023] [PMID: 20478349]

[32]   Heggenstaller AH, Anex RP, Liebman M, Sundberg DN, Gibson LR. Productivity and nutrient dynamics in bioenergy double-cropping systems. Agron J 2008; 100(6): 1740-8.
[http://dx.doi.org/10.2134/agronj2008.0087]

[33]   Helal EGE, Abd El-Wahab SM, Zedan GA, Sharaf AMM. Effect of *Cichorium intybus* L. on fatty liver induced by oxytetracycline in albino rats. Egypt J Hosp Med 2011; 45(1): 522-35.
[http://dx.doi.org/10.21608/ejhm.2011.16381]

[34]   Hernández F, García V, Madrid J, Orengo J, Catalá P, Megías MD. Effect of formic acid on performance, digestibility, intestinal histomorphology and plasma metabolite levels of broiler chickens. Br Poult Sci 2006; 47(1): 50-6.
[http://dx.doi.org/10.1080/00071660500475574] [PMID: 16546797]

[35]   Holst PJ, Kemp DR, Goodacre M, Hall DG. Summer lamb production from puna chicory (*Cichorium intybus*) and lucerne (*Medicago sativa*). Anim Prod Aust 1998; 22: 145-8.

[36]   Hozayen WG, El-Desouky MA, Soliman HA, Ahmed RR, Khaliefa AK. Antiosteoporotic effect of *Petroselinum crispum, Ocimum basilicum* and *Cichorium intybus* L. in glucocorticoid-induced osteoporosis in rats. BMC Complement Altern Med 2016; 16(1): 165.
[http://dx.doi.org/10.1186/s12906-016-1140-y] [PMID: 27255519]

[37]   Huseini HF, Mahmoudabady AZ, Ziai SA, *et al.* The effects of *Cynara scolymus* L. leaf and *Cichorium intybus* L. root extracts on carbon tetrachloride induced liver toxicity in rats. Faslnamah-i Giyahan-i Daruyi 2011; 1: 33-40.

[38]   Izadi H, Arshami J, Golian A, Raji MR. Effects of chicory root powder on growth performance and histomorphometry of jejunum in broiler chicks. Vet Res Forum 2013; 4(3): 169-74.
[PMID: 25653792]

[39]   Jafari B, Rezaie A, Habibi E. Comparative effect of *Chicory* (*Cichorium intybus* L.) and *Nigella sativa* extract with an antibiotic on different parameters of broiler chickens. J Appl Environ Biol Sci 2011; 1:

525-8.

[40]   Jouad H, Haloui M, Rhiouani H, El Hilaly J, Eddouks M. Ethnobotanical survey of medicinal plants used for the treatment of diabetes, cardiac and renal diseases in the North centre region of Morocco (Fez–Boulemane). J Ethnopharmacol 2001; 77(2-3): 175-82.
[http://dx.doi.org/10.1016/S0378-8741(01)00289-6] [PMID: 11535361]

[41]   Judzentiene A, Budiene J. Volatile constituents from aerial parts and roots of *Cichorium intybus* L. (chicory) grown in Lithuania. Chemija 2008; 19: 25-8.

[42]   Khodadadi M, Mousavinasab SS, Khamesipour F, Katsande S. The effect of *Cichorium intybus* L. ethanol extraction on the pathological and biomedical indexes of the liver and kidney of broilers reared under heat stress. Rev Bras Cienc Avic 2016; 18(3): 407-12.
[http://dx.doi.org/10.1590/1806-9061-2015-0153]

[43]   Kidane A, Houdijk JGM, Athanasiadou S, Tolkamp BJ, Kyriazakis I. Effects of maternal protein nutrition and subsequent grazing on chicory (Cichorium intybus) on parasitism and performance of lambs1. J Anim Sci 2010; 88(4): 1513-21.
[http://dx.doi.org/10.2527/jas.2009-2530] [PMID: 20023143]

[44]   Kokoska L, Polesny Z, Rada V, Nepovim A, Vanek T. Screening of some Siberian medicinal plants for antimicrobial activity. J Ethnopharmacol 2002; 82(1): 51-3.
[http://dx.doi.org/10.1016/S0378-8741(02)00143-5] [PMID: 12169406]

[45]   Krylova SG, Efimova LA, Vymiatina ZK, Zueva EP. The effect of cichorium root extract on the morphofunctional state of liver in rats with carbon tetrachloride induced hepatitis model. Eksp Klin Farmakol 2006; 69(6): 34-6.
[PMID: 17209462]

[46]   Lee YH, Kim DH, Kim YS, Kim TJ. Prevention of oxidative stress-induced apoptosis of C2C12 myoblasts by a *Cichorium intybus* root extract. Biosci Biotechnol Biochem 2013; 77(2): 375-7.
[http://dx.doi.org/10.1271/bbb.120465] [PMID: 23391909]

[47]   Leporatti ML, Ivancheva S. Preliminary comparative analysis of medicinal plants used in the traditional medicine of Bulgaria and Italy. J Ethnopharmacol 2003; 87(2-3): 123-42.
[http://dx.doi.org/10.1016/S0378-8741(03)00047-3] [PMID: 12860298]

[48]   Li GY, Gao HY, Huang J, Lu J, Gu JK, Wang JH. Hepatoprotective effect of *Cichorium intybus* L., a traditional Uighur medicine, against carbon tetrachloride-induced hepatic fibrosis in rats. World J Gastroenterol 2014; 20(16): 4753-60.
[http://dx.doi.org/10.3748/wjg.v20.i16.4753] [PMID: 24782629]

[49]   Lin X, Lin CH, Zhao T, *et al.* Quercetin protects against heat stroke-induced myocardial injury in male rats: Antioxidative and antiinflammatory mechanisms. Chem Biol Interact 2017; 265: 47-54.
[http://dx.doi.org/10.1016/j.cbi.2017.01.006] [PMID: 28104348]

[50]   Lin Z, Zhang B, Liu X, Jin R, Zhu W. Effects of chicory inulin on serum metabolites of uric acid, lipids, glucose, and abdominal fat deposition in quails induced by purine-rich diets. J Med Food 2014; 17(11): 1214-21.
[http://dx.doi.org/10.1089/jmf.2013.2991] [PMID: 25314375]

[51]   Liu H, Ivarsson E, Lundh T, Lindberg JE. Chicory (*Cichorium intybus* L.) and cereals differently affect gut development in broiler chickens and young pigs. J Anim Sci Biotechnol 2013; 4(1): 50.
[http://dx.doi.org/10.1186/2049-1891-4-50] [PMID: 24341997]

[52]   Liu HY, Ivarsson E, Jönsson L, Holm L, Lundh T, Lindberg JE. Growth performance, digestibility, and gut development of broiler chickens on diets with inclusion of chicory (*Cichorium intybus* L.). Poult Sci 2011; 90(4): 815-23.
[http://dx.doi.org/10.3382/ps.2010-01181] [PMID: 21406367]

[53]   Loi MC, Maxia L, Maxia A. Ethnobotanical comparison between the villages of Escolca and Lotzorai (Sardinia, Italy). J Herbs Spices Med Plants 2005; 11(3): 67-84.

[http://dx.doi.org/10.1300/J044v11n03_07]

[54] Marzouk M, Sayed AA, Soliman AM. Hepatoprotective and antioxidant effects of *Cichorium endivia* L. leaves extract against acetaminophen toxicity on rats. J Med Sci 2011; 2: 1273-9.

[55] Miao X, Hu T, Zhang C, Wang Q, Shan C, Sun W. Effect of water-soluble extract of chicory on slaughter performance and lipid metabolism on broilers. Xibei Nongye Xuebao 2009; 18: 73-6.

[56] Mishra KS, Kishore K. Protective effects of *Cichoricum intybus* Linn (Kasni) against dietary aflatoxicosis in white albino rats. Am Eurasian J Toxicol Sci 2009; 1: 26-31.

[57] Mushtaq A, Ahmad M, Jabeen Q. Pharmacological role of *Cichorium intybus* as a hepatoprotective agent on the elevated serum marker enzymes level in albino rats intoxicated with nimesulide. Int J Curr Pharm Res 2013; 5: 25-30.

[58] Nabizadeh A. The effect of inulin on broiler chicken intestinal microflora, gut morphology, and performance. J Anim Feed Sci 2012; 21(4): 725-34.
[http://dx.doi.org/10.22358/jafs/66144/2012]

[59] Najafzadeh H, Ghadrdan AR, Jalali M, Alizadeh F. Evaluation of changes of factors related to liver function in serum of horse by administration of *Cichorium intybus*. Int J Anim Vet Adv 2011; 3: 1-5.

[60] Pieroni A. Medicinal plants and food medicines in the folk traditions of the upper Lucca Province, Italy. J Ethnopharmacol 2000; 70(3): 235-73.
[http://dx.doi.org/10.1016/S0378-8741(99)00207-X] [PMID: 10837988]

CHAPTER 8

Use of Psyllium Husk (*Plantago ovata*) in Poultry Feeding and Possible Application in Organic Production

Mahmoud Alagawany[1,*], **Rana Muhammad Bilal**[2], **Fiza Batool**[3], **Youssef A. Attia**[4,5], **Mohamed E. Abd El-Hack**[1,*], **Sameh A. Abdelnour**[6], **Mayada R. Farag**[7], **Ayman A. Swelum**[8,9] and **Mahmoud Madkour**[10]

[1] *Poultry Department, Faculty of Agriculture, Zagazig University, Zagazig 44511, Egypt*

[2] *College of Veterinary and Animal Sciences, The Islamia University of Bahawalpur Pakistan*

[3] *Department of Forestry, Faculty of Agriculture, The Islamia University of Bahawalpur, Pakistan*

[4] *Animal and Poultry Production Department, Faculty of Agriculture Damanhour University, Damanhour, Egypt*

[5] *Department of Agriculture, Faculty of Environmental Sciences, King Abdulaziz University, 21589, Jeddah, Kingdom of Saudi Arabia*

[6] *Animal Production Department, Faculty of Agriculture, Zagazig University, Zagazig 44511, Egypt*

[7] *Forensic Medicine and Toxicology Department, Veterinary Medicine Faculty, Zagazig University, Zagazig 44511, Egypt*

[8] *Department of Animal Production, College of Food and Agriculture Sciences, King Saud University, P.O. Box 2460, Riyadh 11451, Saudi Arabia*

[9] *Department of Theriogenology, Faculty of Veterinary Medicine, Zagazig University, Zagazig 44511, Egypt*

[10] *Animal Production Department, National Research Centre, Dokki, 12622 Giza, Egypt*

Abstract: Herbs or medicinal plants have gained significant attention due to their bioactive compounds that could act as antioxidants, anti-inflammatory, antimicrobial, anticancer agents, *etc.* Psyllium husk (*Plantago ovata*) is an Indian native herb. The water-loving (hydrophilic) mucilloid and water-soluble fiber derived from *Plantago ovata* have been used in traditional Indian Ayurvedic medicine as a crucial remedial mediator of constipation. Psyllium is a rich source of fiber and has many other remedial properties, including lowering the level of cholesterol, raising energy, relaxing inflammation, serving as an antidiarrhoeal, antidiabetic, laxative, and also used in hemorrhoid therapy, and as weight loss agent. The blood serum cholesterol-lowering property of the psyllium husk had drawn the researchers' main focus; thus, Psyllium is

---

[*] **Corresponding author Mahmoud Alagawany:** Poultry Department, Faculty of Agriculture, Zagazig University, Zagazig 44511, Egypt; E-mail: dr.mahmoud.alagwany@gmail.com

thought to be a plausible herbal agent helpful in treating hyperlipidemia. In various animal models, cholesterol levels are reduced by binding Psyllium husk with bile acids in the intestinal, thereby lowering its absorption rate. Screening literature has demonstrated that Psyllium husk could be utilized as an antidiarrheal mediator to cope with the diarrheal symptoms associated with poultry farming disorders. Additionally, Psyllium may also benefit various poultry species' production and growth traits. The present chapter explored Psyllium's potential responsibility for coping the hypercholesterolemia and the uses of psyllium husk as a safe feed additive in poultry farming for organic production and lowering cholesterol in meat and for production of functional foods.

Keywords: Dietary fiber, Growth, Hypocholesterolemic, Poultry, Psyllium.

INTRODUCTION

Herbal plants are recognized as an indispensable share of traditional medicine due to their phytochemical constituents. Improvements were noticed in the growth performance of birds by the dietary inclusion of several herbal plants. Antibiotics as growth agents in poultry and livestock production have been significantly limited in many countries over the past several decades. Consumers are now aware of the prevalence of antibiotics in meat and eggs and need natural foods such as plant products as substitutes. The application of herbal materials and their derivatives as growth promoters is most widely used worldwide to boost poultry and livestock productivity [1 - 5]. Several investigations have been recommended using various medicinal plant sources or their derivatives as natural feed supplements in different poultry and livestock systems [6 - 11]. Psyllium (*Plantago ovata*), an Indian native plant known as isapgol/ispaghula [13], belongs to *Plantaginaceae*. Since prehistoric times, it has been used as a traditional herbal medicine in ayurvedic and allopathic treatments [12, 13]. Its seed and husk contain high water-soluble hydrophilic mucilloid fiber levels in various primary and secondary metabolites and bioactive components [14]. In India, one of the most crucial principle producers of *Plantago ovata* is present and widely grown in various subtropical and tropical areas like China, India, Egypt, Iran, Japan, Korea, *etc* [15].

Dietary fiber and its fractions are among the different nutraceutical essential foods that have earned the most importance over the last few decades. Consumption of dietary fiber foods has been documented to boost the long-term maintenance of down atherogenic Low Density Lipoprotein (LDL)-cholesterol [13, 16]. According to the American Heart Association's advice, dietary fiber and other lifestyle practices could decrease the peril of cardiovascular ailment. They suggested augmenting the soluble fiber level in the diet to more than 25 g / d to effectively reduce total cholesterol and LDL levels [17].

Psyllium husk is generally utilized as a sedative agent [18], which reduces the concentration of serum cholesterol and LDL in hypercholesterolemic children and adults [18 - 24], raises the levels of High Density Lipoprotein (HDL) cholesterol, as well as reduces the level of glucose in diabetic patients [18]. Also, it has been suggested as a useful dietary assistant for psychotherapy in reducing total and LDL-cholesterol levels in adults [19], specifically those who do not respond effectively to low-fat cholesterol and low-fat diets [20]. Many reported studies on different animal models, including humans, have confirmed Psyllium hypocholesterolemic therapy claims [24, 25].

The water-soluble dietary fiber is an effective therapeutic agent as a functional nutritional product against various ailments such as physiological diabetes mellitus, obesity [26], diarrhea [27], constipation [28], intestinal inflammation [29], cardiovascular diseases [30], hypercholesterolemia [31], and also act as prebiotic [32] and antioxidant [33]. The pharmacological and therapeutic activities of psyllium polysaccharides have been thoroughly investigated by Madgulkar *et al.* [34].

Chicken eggs are a vital part of the human diet and are considered healthy food with high-quality protein [4]. Since chicken eggs have a high cholesterol level (about 213-280 mg), health consultants are frequently advised to reduce the consumption of eggs from regular diets, especially for individuals with hyper-cholesterol-related syndromes [35]. Besides, customers are hesitant to embrace its benefits because they have concerns about its high cholesterol content, which affects their health. Scientists are now searching for new strategies to supply functional eggs with lower cholesterol concentrations [36]. The addition of psyllium husk in poultry diets is a successful technique to minimize cholesterol in the blood serum and egg yolk [37]. Several dietary fiber items are currently introduced into poultry diets to reduce the cholesterol levels in eggs. The present work aimed to address psyllium fiber as a beneficial agent for lowering serum total cholesterol and serum-LDL-cholesterol and egg or meat in poultry, its essential component in organic and low-cholesterol egg and meat for functional foods.

Geographical Source of Psyllium

The morphology of the psyllium plant is presented in Figs (**1** and **2**).

Psyllium is cultivated in several parts of India, including Maharashtra, Rajasthan, Punjab, and Gujarat, while in Pakistan, it is found in Sindh and West Pakistan [38]. It is also grown in European countries such as France and Spain to cover the European market requirements. The seeds are produced all over Southern Europe and North Africa [38, 39]. Psyllium husk is used as industrial powders distributed to numerous countries such as the UK, Spain, USA, Italy, Norway, France, Korea, Indonesia, Germany, Mexico, Canada, Japan, Sweden, Australia, and Denmark (Fig. **3**).

Fig. (1). Morphology of psyllium plant.

Fig. (2). Morphology of psyllium seeds.

Fig. (3). Whole plant (**a**) and powder (**b**) of psyllium seeds.

Biological Benefits of Psyllium Husk

Psyllium is a rich fiber source and has several biological properties such as relieving constipation, functioning as a laxative, an antidiabetic and antidiarrheal, being used in hemorrhoid remedy, serving as a weight loss agent, soothing inflammation, and increasing energy and lowering cholesterol effects [33, 34].

The Anti-cholesterol Activity of Psyllium

Cholesterol is a significant lipid molecule that shows a central function in producing several crucial hormones like steroids and maintaining cell membranes (Fig. **3**). However, it has been shown that high cholesterol levels (hypercholesterolemia) are a significant cause of coronary heart disease that could lead to a heart attack [24]. Bile secretions are produced from cholesterol in the liver and released into the most critical cholesterol absorption route known as the intestinal cavity. The bile secretion's main component is reabsorbed in the ileum and recycled back to the liver for further excretion by enterohepatic circulation into the small intestine and consequently decreased bile acid production [36].

Psyllium consists of high soluble fiber levels, which are associated with bile secretions in the small intestine, produces multifaceted bile absorption from the small intestines, and enhances bile acid synthesis and secretion to replace lost acid [23]. Cholesterol is then derived from the bloodstream to generate bile acids, thus decreasing blood cholesterol levels. This adsorption of bile acids by several low binding sites on the polysaccharides' structure would result in increased excretory products of bile acids, which would increase the cholesterol metabolism into bile acids in the liver and then decrease the amount of serum cholesterol [40]. Men who suffered from blood cholesterol were observed to lower their total serum cholesterol by 4 percent [21]. This decrease was mainly due to the stimulation of bile acid secretion by Psyllium hydrophilic mucilloids. Previous studies have shown that bile secretions fusion is induced in several species by *Psyllium ovate* [25, 41] and humans. The fusion of *Psyllium ovate* was also induced.

A traditional method was used to reduce serum cholesterol to adjust the hepatic cholesterol synthesizer. At the same time, psyllium husk affects fat and cholesterol uptake, which helps lower the cholesterol level.

Psyllium as a Potent Hypocholesterolemic Agent in Humans

In the routine diet of hypercholesterolemia prevention, healthcare consultants indicated that Psyllium is the most hazardous factor for heart ailment. Several previous reports have found that Psyllium has a possible role in lowering cholesterol levels by lowering total cholesterol and LDL [24]. During treatment with mild to moderate hyper cholesterol [19, 20], Psyllium can be used as an additional complement to the low-fat diet due to its protection.

Sprecher *et al.* [42] have found a 3.5 and 5.1% decrease in total cholesterol and LDL levels in eight weeks of study. Healthy men and women between 21 and 70 years old with key hyper-cholesterol are helped by dietary supplementation psyllium. Moreyra *et al.* [43] also reported cholesterol reduction in a mixture of 10 mg simvastatin and 15 g psyllium in patients with hyperlipidemia aged 18 and 80 years compared to 20 mg simvastatin community in the cholesterol-reduction treatments. In contrast to older patients in a population less than 60 years, psyllium husk was more successful at reducing serum cholesterol levels [44]. The effect was more significant for men than for women [24].

Psyllium Husk as a Potent Hypocholesterolemic Agent in Animal and Poultry

Several studies described the total cholesterol and LDL levels in different animal models that are reduced by *Psyllium ovata* [23]. Treatments of diabetic rats and hyper cholesterol-induced albino rats with Psyllium decreased serum glucose and cholesterol levels, confirming antidiabetes and hypocholesterols [38]. Rats have been shown to provide ten% defatted psyllium husk with a half-purified diet with almost stabilized liver and serum triglyceride concentrations, including 0.5 percent of cholesterol and lower serum cholesterol levels, in a research conducted by Kritchevsky *et al.* [45].

Psyllium husk has been added to milk substitutes as a healthy natural supplement to encourage growth in neonatal milk calves, showing a constructive effect on physiological properties, with improved growth, and health status due to increased gastrointestinal canal size and lower intestinal fermentation. Also, Anderson *et al.* [19] have demonstrated a 4 percent reduction in overall serum cholesterol and a 7 percent decrease in LDL compared with placebo in the diet that contains low fat and 10.2 g of Psyllium per day. In another study, 176 aging individuals with continuous use of Psyllium for 12 months caused a significant decrease in total serum cholesterol [46].

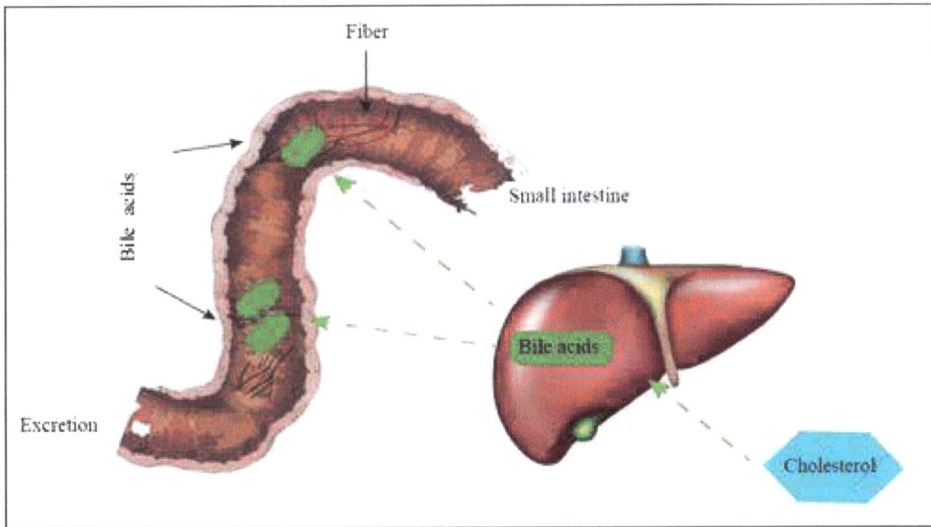

Fig. (4). The mechanism for the cholesterol-lowering effect of psyllium husk fiber [47].

Role of Psyllium in the Therapy of other Human Ailments

Digestive Function and Metabolism

Psyllium is a useful stool-bulging device. It has the peculiar property of relieving constipation and diarrhea. Hydrophilic molecules that retain the water by forming hydrogen connexions are polysaccharides derived from Psyllium. Psyllium soluble fibers have a particular character, capable of connecting water by gelatinizing when the water has been trapped within the three-dimensional network of molecules, where insoluble fibers carry sponge-like water [29]. As was well established, the water- storing ability of solvent fibers modifies the weight and length of the intestinal transport of stools, which induces suspended gastric emptying by hydration of polysaccharides that form a viscous gel [40].

A placebo-controlled analysis of those with fluid stool fecal incontinence with the addition of Psyllium showed a 50% reduction in the frequency of incontinent stools where the water-insoluble solids and overall water keeping capability of stools for psyllium community were the highest [48]. The relaxation of stools could be affected intensively by small increases in water [37]. The findings of human studies suggested the effect of Psyllium on diarrheal-related disorders as a secondary medication for the treatment of fecal incontinence and constipation. However, if constipation occurs, it decreases the passage time through increased stool volatility [35].

Hemorrhoids

In systematic study research, scientists deliberated on treating symptomatic hemorrhoids to reduce the prolonged stress caused by chronic constipation and used fiber laxatives to improve hemorrhoid symptoms, particularly bleeding [49]. Several accumulating pieces of evidence revealed that Psyllium had been used for the medication of hemorrhoids and associated ailments. A significant effect on the decrease in clogged hemorrhoidal cushions and improvement of internal bleeding hemorrhoids have been observed to complement dietary fibers in patients' diets with bleeding internal hemorrhoids [50]. Adequate added fiber in the form of 5-6 teaspoons of psyllium husk with a daily amount of 600 ml of water combined with accurate medication goals could prevent bleeding and hemorrhoids and the need for surgery in most patients with progression hemorrhoids [51].

Salween and Basson [52] have indicated that a bowel preparation for non-constipated patients with suspected hemorrhoidal bleeding is unsuitable for the psyllium base fiber supplement and should not be added only a few days before the use of polyethylene glycol preparations is discontinued. In the case of pregnant women, it was proposed that the ingestion of psyllium powder during the third trimester of pregnancy could substantially prevent anal hemorrhoid cracking and constipation [53]. Still, the safety of psyllium usage during pregnancy might be constrained [54].

Clinical Effect

Several investigators identified the use of psyllium husk for constipation, hypercholesterolemia, colon cancer, diabetes, diarrhea, inflammatory bowel disease, and irritable bowel syndrome, as stated by Monica [55] and Majmudar *et al.* [56].

CONCLUSION

Psyllium can be used to treat the medical effects of ulcerative colitis, colon cancer, irritable bowel syndrome, diarrhea, inflammatory bowel diseases, constipation, hypercholesterolemia, and diabetes in humans. The potential of

Psyllium to sustain human health needs more exploitation and successful practical use. Moreover, psyllium husk might be utilized as a natural healthy additive feeding source in poultry and animal feed. Its commercial potential for producing organic eggs and meat, which is needed in the future, should be significantly utilized. Future research may find far more useful uses of this valuable plant for humans, animals, and poultry.

CONSENT FOR PUBLICATION

Not applicable.

CONFLICT OF INTEREST

The author declares no conflict of interest, financial or otherwise.

ACKNOWLEDGEMENTS

Declared none.

REFERENCES

[1]   Gong J, Yin F, Hou Y, Yin Y. Review: Chinese herbs as alternatives to antibiotics in feed for swine and poultry production: Potential and challenges in application. Can J Anim Sci 2014; 94(2): 223-41.
        [http://dx.doi.org/10.4141/cjas2013-144]

[2]   Dhama K, Latheef SK, Mani S, *et al.* Multiple beneficial applications and modes of action of herbs in poultry health and production.AReview. Int J Pharmacol 2015; 11(3): 152-76.
        [http://dx.doi.org/10.3923/ijp.2015.152.176]

[3]   El-Hack MEA, Alagawany M, Farag MR, Tiwari R, Karthik K, Dhama K. Nutritional, healthical and therapeutic efficacy of black cumin (*Nigella sativa*) in animals, poultry and humans. Int J Pharmacol 2016; 12(3): 232-48.
        [http://dx.doi.org/10.3923/ijp.2016.232.248]

[4]   Abd El-Hac ME, Alagawany M, Ragab Fara M, Dhama K. Use of maize distiller's dried grains with solubles (DDGS) in laying hen diets: trends and advances. Asian J Anim Vet Adv 2015; 10(11): 690-707.
        [http://dx.doi.org/10.3923/ajava.2015.690.707]

[5]   Abdelnour S, Alagawany M, El-Hack MEA, Sheiha AM, Saadeldin IM, Swelum AA. Growth, carcass traits, blood hematology, serum metabolites, immunity, and oxidative indices of growing rabbits fed diets supplemented with red or black pepper oils. Animals (Basel) 2018; 8(10): 168.
        [http://dx.doi.org/10.3390/ani8100168] [PMID: 30279392]

[6]   Durrani FR, Chand N, Jan M, Sultan A, Durrani Z, Akhtar S. Immunomodulatory and growth promoting effect of neem leaves infusion in broiler chicks. Sarhad J Agric 2008; 24: 655-9.

[7]   Rodríguez-Cabezas ME, Gálvez J, Camuesco D, *et al.* Intestinal anti-inflammatory activity of dietary fiber (Plantago ovata seeds) in HLA-B27 transgenic rats. Clin Nutr 2003; 22(5): 463-71.
        [http://dx.doi.org/10.1016/S0261-5614(03)00045-1] [PMID: 14512034]

[8]   Saeed M, Naveed M, Arain MA, *et al.* Quercetin: Nutritional and beneficial effects in poultry. Worlds Poult Sci J 2017; 73(2): 355-64. a
        [http://dx.doi.org/10.1017/S004393391700023X]

[9]   Saeed M, Ur Rehman Z, Arian MA, Soomro RN. Nutritional and Healthical Aspects of Yacon (*Smallanthus sonchifolius*) for Human, Animals and Poultry. Int J Pharma 2017.

[10]   Saeed M, Abd EL-Hack ME, Alagawany M, Arain MA. Chicory (*Cichorium intybus*) Herb: Chemical Composition, Pharmacology, Nutritional and Healthical Applications. 2017.

[11]   Saeed M, Abd EL-Hack, Arif M, *et al.* Impacts of distiller's dried grains with solubles as replacement of soybean meal plus vitamin e supplementation on production, egg quality and blood chemistry of laying hens. Ann Anim Sci 2017.
       [http://dx.doi.org/10.1515/aoas-2016-0091]

[12]   Kendall CWC. The health benefits of Psyllium. Insert to the Canadian Journal of Dietetic Practice and Research 2004; 65: 3.

[13]   Deokar G, Kshirsagar S, Deore PA, Kakulte H. Pharmaceutical benefits of Plantago ovate (Isabgol seed): a review. Pharmaceutical and Biological Evaluations 2016; 3(1): 32-41.

[14]   Talukder P, Talapatra S, Ghoshal N, Raychaudhuri SS. Antioxidantactivity and HPLC analysis of phenolic compounds during *in vitro* callus cultureofPlantago ovata Forsk and effect of exogenous additives on accumulation of phenolic compounds. J Sci Food Agric 2015; 96: 232-44.
       [http://dx.doi.org/10.1002/jsfa.7086] [PMID: 25640456]

[15]   Chevallier A. The Encyclopedia of Medicinal Plants. London: Dorling Kindersley 1996.

[16]   Davidson MH, Maki KC, Kong JC, *et al.* Long-term effects of consuming foods containing psyllium seed husk on serum lipids in subjects with hypercholesterolemia. Am J Clin Nutr 1998; 67(3): 367-76.
       [http://dx.doi.org/10.1093/ajcn/67.3.367] [PMID: 9497178]

[17]   Krauss RM, Eckel RH, Howard B, *et al.* AHA dietary guidelines revision 2000 a statement for healthcare professionals from the Nutrition Committee of the American Heart Association. Circulation 2000; 102(18): 2284-99.
       [http://dx.doi.org/10.1161/01.CIR.102.18.2284] [PMID: 11056107]

[18]   Amanullah MAK, Badani LA, Ali Z, Rahojo MIJ, Sadiq N, Bugti GA. Comparative study on yield and yield components of Psyllium Plantago ovata varieties in Balochistan. J Pharmacogn Phytochem 2016; 5(6): 270-2.

[19]   Anderson JW, Allgood LD, Lawrence A, *et al.* Cholesterol-lowering effects of psyllium intake adjunctive to diet therapy in men and women with hypercholesterolemia: meta-analysis of 8 controlled trials. Am J Clin Nutr 2000; 71(2): 472-9. a
       [http://dx.doi.org/10.1093/ajcn/71.2.472] [PMID: 10648260]

[20]   Anderson JW, Davidson MH, Blonde L, *et al.* Long-term cholesterol-lowering effects of psyllium as an adjunct to diet therapy in the treatment of hypercholesterolemia. Am J Clin Nutr 2000; 71(6): 1433-8. b
       [http://dx.doi.org/10.1093/ajcn/71.6.1433] [PMID: 10837282]

[21]   Anderson JW, Zettwoch N, Feldman T, Tietyen-Clark J, Oeltgen P, Bishop CW. Cholesterol-lowering effects of psyllium hydrophilic mucilloid for hypercholesterolemic men. Arch Intern Med 1988; 148(2): 292-6.
       [http://dx.doi.org/10.1001/archinte.1988.00380020036007] [PMID: 3277558]

[22]   Anderson JW, Allgood LD, Turner J, Oeltgen PR, Daggy BP. Effects of psyllium on glucose and serum lipid responses in men with type 2 diabetes and hypercholesterolemia. Am J Clin Nutr 1999; 70(4): 466-73.
       [http://dx.doi.org/10.1093/ajcn/70.4.466] [PMID: 10500014]

[23]   Terpstra AHM, Lapré JA, de Vries HT, Beynen AC. Hypocholesterolemic effect of dietary psyllium in female rats. Ann Nutr Metab 2000; 44(5-6): 223-8.
       [http://dx.doi.org/10.1159/000046688] [PMID: 11146328]

[24]   Aleixandre A, Miguel M. Dietary fiber and blood pressure control. Food Funct 2016; 7(4): 1864-71.

[http://dx.doi.org/10.1039/C5FO00950B] [PMID: 26923351]

[25]   Everson GT, Daggy BP, McKinley C, Story JA. Effects of psyllium hydrophilic mucilloid on LDL-cholesterol and bile acid synthesis in hypercholesterolemic men. J Lipid Res 1992; 33(8): 1183-92.
        [http://dx.doi.org/10.1016/S0022-2275(20)40770-9] [PMID: 1431597]

[26]   Hartvigsen ML, Gregersen S, Lærke HN, Holst JJ, Bach Knudsen KE, Hermansen K. Effects of concentrated arabinoxylan and β-glucan compared with refined wheat and whole grain rye on glucose and appetite in subjects with the metabolic syndrome: a randomized study. Eur J Clin Nutr 2014; 68(1): 84-90.
        [http://dx.doi.org/10.1038/ejcn.2013.236] [PMID: 24253758]

[27]   Mehmood MH, Aziz N, Ghayur MN, Gilani AH. Pharmacological basis for the medicinal use of psyllium husk (Ispaghula) in constipation and diarrhea. Dig Dis Sci 2011; 56(5): 1460-71.
        [http://dx.doi.org/10.1007/s10620-010-1466-0] [PMID: 21082352]

[28]   Gilani AH, Aziz N, Khan MA, Khan S, Zaman V. Laxative effect of ispaghula: physical or chemical effect? Phytother Res 1998; 12(S1): S63-5.
        [http://dx.doi.org/10.1002/(SICI)1099-1573(1998)12:1+<S63::AID-PTR252>3.0.CO;2-7]

[29]   Rodriguez , Cabezas ME, Galvez J, Camuesco D, *et al.* Intestinalanti-inflammatory activity of dietary fiber (Plantagoovataseeds)in HLA-B27 transgenic rats. ClinNutr 2003; 22: 463-71.

[30]   de Bock M, Jose G B Derraik, Christine M Brennan, *et al.* Psyllium Supplementation in Adolescents Improves Fat Distribution & Lipid Profile: A Randomized, Participant Blinded, Placebo-Controlled, Crossover Trial. PLoS One 2012; 7(7): e41735.

[31]   Wei ZH, Wang H, Chen XY, *et al.* Time- and dose-dependent effect of Psyllium on serum lipids in mild-to-moderate hypercholesterolemia: a metaanalysis of controlled clinical trials Time- and dosedependent effect of Psyllium on serum lipids. Eur J Clin Nutr 2009; 63: 821-7.
        [http://dx.doi.org/10.1038/ejcn.2008.49] [PMID: 18985059]

[32]   Elli M, Cattivelli D, Soldi S, Bonatti M, Morelli L. Evaluation of prebiotic potential of refined psyllium (Plantago ovata) fiber in healthy women. J Clin Gastroenterol 2008; 42 (Suppl. 3): S174-6.
        [http://dx.doi.org/10.1097/MCG.0b013e31817f183a] [PMID: 18685505]

[33]   Patel MK, Mishra A, Jha B. Non-targeted Metabolite Profiling and Scavenging Activity Unveil the Nutraceutical Potential of Psyllium (Plantago ovata Forsk). Front Plant Sci 2016; 7: 431.
        [http://dx.doi.org/10.3389/fpls.2016.00431] [PMID: 27092153]

[34]   Madgulkar AR, Monica r. P. Rao AR, Deepa Warrier. Characterization of Psyllium (plantago ovata) polysaccharide and its uses. Polysaccharides 2014.
        [http://dx.doi.org/10.1007/978-3-319-03751-6_49-1]

[35]   Oparil S, Schmieder RE. New approaches in the treatment of hypertension. Circ Res 2015; 116(6): 1074-95.
        [http://dx.doi.org/10.1161/CIRCRESAHA.116.303603] [PMID: 25767291]

[36]   Rakita S, Spasevski N, Čolović D, *et al.* The influence of laying hens' diet enriched with omega-3 fatty acids, paprika and marigold on physical properties of eggs. Journal on Processing and Energy in Agriculture 2016; 20(2): 58-62.

[37]   Mukhtar N, Mehmood R, Khan SH, Ashrif NM, Mirza MW. Effect of Psyllium Husk Fiber on Growth Performance, Egg Quality Traits and Lipid Profile in Layers under High Ambient Temperature. J World Poult Res 2017; 7(1): 15-22.

[38]   Ahmed I, Muhammad Naeem, Abdul Shakoor, Zaheer Ahmed, Hafiz Muhammad Nasir Iqbal. Investigation of Antidiabetic and Hypocholesterolemic Potential of PsylliumHusk Fiber (Plantago psyllium). Diabetic and Hypercholesterolemic Albino RatsInternational Journal of Medical, Health, Biomedical, Bioengineering and Pharmaceutical Engineering 2010; 4: 1. scholar.waset.org/ 1999.9/13773

[39]   Gupta AK. Handbook on herbs: cultivation & processing. Delhi: Asia Pacific Business Press Inc 2005;

pp. 329-30.

[40]   Blackwood AD, Salter J, Dettmar PW, Chaplin MF. Dietary fibre, physicochemical properties and their relationship to health. J R Soc Promot Health 2000; 120(4): 242-7.
[http://dx.doi.org/10.1177/146642400012000412] [PMID: 11197452]

[41]   Matheson HB, Colón IS, Story JA. Cholesterol 7 alpha-hydroxylase activity is increased by dietary modification with psyllium hydrocolloid, pectin, cholesterol and cholestyramine in rats. J Nutr 1995; 125(3): 454-8.
[PMID: 7876920]

[42]   Sprecher DL, Harris BV, Goldberg AC, *et al.* Efficacy of psyllium in reducing serum cholesterol levels in hypercholesterolemic patients on high- or low-fat diets. Ann Intern Med 1993; 119(7_Part_1): 545-54.
[http://dx.doi.org/10.7326/0003-4819-119-7_Part_1-199310010-00001] [PMID: 8363164]

[43]   Moreyra AE, Wilson AC, Koraym A. Effect of combining psyllium fiber with simvastatin in lowering cholesterol. Arch Intern Med 2005; 165(10): 1161-6.
[http://dx.doi.org/10.1001/archinte.165.10.1161] [PMID: 15911730]

[44]   Schectman G, Hiatt J, Hartz A. Evaluation of the effectiveness of lipid-lowering therapy (bile acid sequestrants, niacin, psyllium and lovastatin) for treating hypercholesterolemia in veterans. Am J Cardiol 1993; 71(10): 759-65.
[http://dx.doi.org/10.1016/0002-9149(93)90820-3] [PMID: 8456750]

[45]   Kritchevsky D, Tepper SA, Klurfeld DM. Influence of psyllium preparations on plasma and liver lipids of cholesterol-fed rats. Artery 1995; 21(6): 303-11.
[PMID: 8833230]

[46]   Vega-López S, Vidal-Quintanar RL, Fernandez ML. Sex and hormonal status influence plasma lipid responses to psyllium. Am J Clin Nutr 2001; 74(4): 435-41.
[http://dx.doi.org/10.1093/ajcn/74.4.435] [PMID: 11566640]

[47]   Xing LC, Santhi D, Shar AG, *et al.* Psyllium Husk (*Plantago ovata*) as a Potent Hypocholesterolemic Agent in Animal, Human and Poultry. Int J Pharmacol 2017; 13(7): 690-7.
[http://dx.doi.org/10.3923/ijp.2017.690.697]

[48]   Bliss DZ, Jung HJ, Savik K, *et al.* Supplementation with dietary fiber improves fecal incontinence. Nurs Res 2001; 50(4): 203-13.
[http://dx.doi.org/10.1097/00006199-200107000-00004] [PMID: 11480529]

[49]   Alonso-Coello P, Guyatt G, Heels-Ansdell D, *et al.* Laxatives for the treatment of hemorrhoids. Cochrane Database Syst Rev 2005; (4): CD004649.
[PMID: 16235372]

[50]   Ashwini r Madgulkar. Characterization of Psyllium (plantago ovata) polysaccharide and its uses. Polysaccharides 1-17.2014;

[51]   Garg P, Singh P. Adequate dietary fiber supplement and TONE can help avoid surgery in most patients with advanced hemorrhoids. Minerva Gastroenterol 2017; 63(2).
[http://dx.doi.org/10.23736/S1121-421X.17.02364-9]

[52]   Salwen WA, Basson MD. Effect of four-day psyllium supplementation on bowel preparation for colonoscopy:A prospective double blind randomized trial [ISRCTN76623768]. BMC Gastroenterol 2004; 4(1): 2.
[http://dx.doi.org/10.1186/1471-230X-4-2] [PMID: 15005812]

[53]   Ghahramani L, Hosseini SV, Rahimikazerooni S, *et al.* The effect of oral Psyllium herbal laxative powder in prevention of hemorrhoids and anal fissure during pregnancy, a randomized double blind clinical trial. Annals of Colorectal Research 2013; 1(1): 23-7.
[http://dx.doi.org/10.17795/acr-11488]

[54]   Bor R, Farkas K, Balint A, Molnar T. Psyllium Powder Laxatives Are Effective Treatment of

Constipation in Pregnancy, but What Is About Safety? Annals of Colorectal Research 2014; 2(1): e15204.
[http://dx.doi.org/10.17795/acr-15204]

[55]   Rao MRP, Warrier DU, Gaikwad SR, Shevate PM. Phosphorylation of psyllium seed polysaccharide and its characterization. Int J Biol Macromol 2016; 85: 317-26.
[http://dx.doi.org/10.1016/j.ijbiomac.2015.12.043]

[56]   Majmudar H, Mourya V, Devdhe S, Chandak R. Pharmaceutical applications of ispaghula husk: mucilage. Int J Pharm Sci Rev Res 2008; 18(1): 49-55.

CHAPTER 9

Dandelion Herb: Chemical Composition and Use in Poultry Nutrition

Mahmoud Alagawany[1,*], Mohamed E. Abd El-Hack[1,*], Mayada R. Farag[2,*], Sameh A. Abdelnour[3], Kuldeep Dhama[4], Ayman A. Swelum[5,6] and Alessandro Di Cerbo[7]

[1] *Poultry Department, Faculty of Agriculture, Zagazig University, Zagazig 44511, Egypt*

[2] *Forensic Medicine and Toxicology Department, Veterinary Medicine Faculty, Zagazig University, Zagazig 44511, Egypt*

[3] *Animal Production Department, Faculty of Agriculture, Zagazig University, Zagazig 44511, Egypt*

[4] *Division of Pathology, ICAR-Indian Veterinary Research Institute, Izatnagar, Bareilly- 243 122, Uttar Pradesh, India*

[5] *Department of Animal Production, College of Food and Agriculture Sciences, King Saud University, P.O. Box 2460, Riyadh 11451, Saudi Arabia*

[6] *Department of Theriogenology, Faculty of Veterinary Medicine, Zagazig University, Zagazig 44511, Egypt*

[7] *School of Biosciences and Veterinary Medicine, University of Camerino, Matelica, Italy*

Abstract: *Taraxacum officinale,* also known as dandelion herb, is a popular medicinal and therapeutic herb used for many years and is mostly raised in Europe, Asia, North and South America. It contains several nutrients and bioactive substances, especially the leaves and roots of this herb, which are a rich source of fiber, lecithin, choline, and micronutrients such as minerals (potassium, magnesium, calcium, zinc, *etc.*, iron) and vitamins (A, C, K, and B-complex). The root has been commonly used for digestive and liver problems due to its stimulatory effects on the production of bile and detoxification functions. The leaves of dandelion have stimulatory functions on the digestive system and possess diuretic effects. Furthermore, several studies have shown that dandelion leaves can enhance the growth and productivity of poultry. Various functions on the intestinal mucosa have been reported, including the effects on the architecture of villi, villus height/crypt depth ratio, and cellular infiltration. This herb also has various beneficial functions, such as immunomodulatory effects, stimulation of the digestive system and insulin activation, enhancing the metabolism of androgens, and acting as a probiotic, antiangiogenic, antineoplastic and demulcent. Moreover, the dandelion herb can treat indigestions and hepatitis B infection. Due to the lack of

* **Corresponding author Mohamed E. Abd El-Hack and Mahmoud Alagawany:** Poultry Department, Faculty of Agriculture, Zagazig University, Zagazig 44511, Egypt and Poultry Department, Faculty of Agriculture, Zagazig University, Zagazig 44511, Egypt; E-mails: dr.mohamed.e.abdalhaq@gmail.com and mmalagwany@zu.edu.eg

studies on the effects of dandelion, further research has to be conducted to exploit the medicinal properties of this herb for its beneficial health impact on humans, pet and livestock animals (*e.g.*, poultry) nutrition.

Keywords: Beneficial effects, Chemical composition, Dandelion, Pharmaceutical, Poultry.

INTRODUCTION

Recently, therapeutic herbs are becoming increasingly popular worldwide for their medicinal and health-promoting properties in humans and animals, particularly poultry [1 - 5]. This review focuses on the medicinal plant known as dandelion (*Taraxacum officinale*). Dandelion, a perennial plant, belongs to the Asteraceae (Compositae) family. Dandelion is commonly utilized in several conventional and current herbal therapeutic procedures, especially in Asia, North America, Africa and Europe [6 - 10]. It is described as a weedy species and can be found in various climatic conditions, on roadsides, shores and areas where the soil is moist [11]. Dandelion is known in English terms as Blowball, Swine's snout, Cankerwort, Lion'stooth [12] and Arabic terms like Khas Berri and Hindiba [13]. The Latin name for dandelion is *Taraxacum*, which could be a derivate of the Arabic word "Tharakhchakon" [11] or the Greek word "Tarraxos" [13]. However, the most common name for this herb is "Dandelion", which originates from the French expression "Dent de lion", meaning "Lion's tooth", due to the serrated leaves of this plant [13]. The root of the dandelion has been most commonly used in digestive disorders, promoting the digestion process and functions of the liver, while the leaves have diuretic and gastrointestinal effects [14]. Although not all researchers agree, several studies have pointed out numerous beneficial effects of dandelion, including immunomodulatory, digestion, and insulin stimulation. It has been proven to have probiotic, demulcent, antiangiogenic, and antineoplastic functions [8, 15]. As previously mentioned, Dandelion has been used for decades as a traditional therapeutic herb due to its various medicinal properties, especially in Chinese traditional medicine. From the pharmacological point of view, the effects of dandelion are due to the presence of some bioactive substances, like flavonoids, tannins, saponins, lactones, and alkaloids [8, 15, 16]. Williams *et al.* [17] have isolated several flavonoids from dandelion, including caffeic acid, chlorogenic acid, luteolin, and luteolin 7-glucoside. The leaf of dandelion contains fibers, proteins, minerals such as calcium, phosphorus, potassium, magnesium, iron, and vitamins including A, C, and B-complex [18, 19].

A ban on antibiotics as feed additives in poultry nutrition is realized because of the increased occurrence of pathogens resistance against therapeutic antibiotics used in poultry nutrition [20, 21]. Due to the limited usage of antibiotics in diets, efforts are being made to find alternative strategies to enhance the health status, growth performance, boost immunity, and increase the productivity and economic benefits of poultry farms [22 - 24]. A plausible solution to these problems could be the more frequent use of phytogenics, a group of natural growth promoters and antimicrobial agents, in herbs, plant extracts, cold-pressed or essential oils and phytochemicals [23, 25 - 27]. Such supplements are added to the diets of poultry to raise the productivity and economic feasibility by enhancing digestibility, bioavailability and absorption of nutrients while at the same time eliminating pathogenic microorganisms from the gastrointestinal system [1, 28, 29]. Showing very promising results, dandelion is one such phytogenic agent, which could have beneficial effects when supplemented with poultry diets [10, 30 - 32].

Structure and Chemical Composition

Generally, the dandelion herb is cultured mainly in India, and its rhizome, root and leaf have a wide application in medicine. The major components in dandelion are sesquiterpene lactones (having anti-cancer anti- and inflammatory effects) [33, 34], phenylpropanoids (which are not toxic and have analgesic, anti-inflammatory, and hypotensive functions), polysaccharides, and triterpenoid saponins.

The main sesquiterpene lactones that are commonly existing, include glycosides, containing taraxacolides, dihydrolactucin, ixerin, ainslioside, taraxacosides, and taraxinic acids. As a part of the sporopollenin structure, phenylpropanoids are plentifully present in dandelion herb and consist of several bioactive acids such as inulin, caffeic acid, cichoric acid, chlorogenic acid, monocaffeoyltartaric acid, and 4-caffoeylquinic acid, (a class of fibres known as fructans) [7, 8, 15, 35]. The root of the dandelion herb is a rich source of triterpenes, including taraxol, taraxerol, taraxasterol, β-amyrin; ψ-taraxasterol, sterols (stigmasterols, β-sitosterol) and inulin [7, 36, 37]. The stem, root and flower of dandelion contain high concentrations of flavonoids, saponins, tannins, alkaloids and phenols, whereas flower extracts have the highest concentration of flavonoids [13, 16, 38]. Furthermore, phenols and steroids are also represented [7]. This herb is a very rich source of minerals, such as calcium (Ca), magnesium (Mg), sodium (Na), potassium (K), iron (Fe), zinc (Zn), and copper (Cu) [13, 39, 40]. Besides, macro- and micro-elements, the leaf of dandelion contains higher concentrations of β-carotene compared to carrots and greater amounts of Fe and Ca in comparison to spinach [39]. Qureshi *et al.* [41] reported that dandelion contains 90.66% dry matter, 13.81% crude fibre, 11.40% crude protein, 9.34% moisture, and 3.3% ether extract. The mineral composition of dandelion contains Ca, Mg, Na, Cu and Zn as 7.3, 2, 0.013, 0.05 and 0.14 mg per 100 g, respectively [40].

Beneficial Roles of Dandelion for Health

Effect of Dandelion on Performance, Carcass and Meat Quality

Some beneficial functions of the dandelion herb and its derivatives are shown in Table **1**. The dietary supplementation of dandelion showed remarkable improvements in body weight gain (BWG), growth performance and feed efficiency in poultry [32, 42]. Furthermore, in a study conducted by Yan *et al.* [31], dietary supplementation of dandelion resulted in the enhancement in BWG, feed conversion rate (FCR), and nutrient utilization. Qureshi *et al.* [41] investigated the impact of dietary supplementation with dandelion leaves on broilers and pointed out that the growth performance parameters improved throughout the growing period.

Table 1. Some beneficial applications of dandelion herb and its fractions.

Activities	Modes of Actions	Literatures
Growth enhancer	Dandelion herb alone or with other herbal plants improved FI, FCR, BW, meat quality, and mortality rate.	[30, 41, 44]
Immune stimulant	The dandelion herb improves macrophage function and increases the lymphoid organs' weights. Dandelion increases nitric oxide production through inducible nitric oxide synthase (iNOS) induction.	[71, 72]
Anti-parasitic and antibacterial effects	The extract of dandelion decreased the number of pathogens like *S. aureus* and *E. coli* and improved the number of beneficial bacteria. Also, this herb has also been found effective against endoparasites like coliform.	[31, 32, 57]
Antioxidant effect	↓free radicals, singlet oxygen, and hydroxyl radicals. ↑the antioxidant enzymes activities such as SOD, GSH-Px and CAT.	[50, 64, 65]
Hepatoprotective effect	Dandelion and its derivatives improve bile secretion and enhance liver functions and enzymes.	[50, 52]
Anti-inflammatory effect	The dandelion herb also has an anti-inflammatory impact against inflammation. The ethanol extracts of dried aerial parts of this herb inhibit inflammation and angiogenesis in rodents.	[67, 69]
Anti-cancer effect	This herb plays a crucial role against many types of cancer. Also, it could inhibit tumor growth by inhibiting the tumor's development.	[54]
Digestive stimulant and prebiotic activity	The contents of fiber in the dandelion herb have a high ability to eliminate toxins from the body and balance gut microbiota. Dandelion can be used as a prebiotic feed additive due to its high inulin content.	[33]
Hypoglycaemic and hypolipidemic effects	The bioactive components isolated from the dandelion herb stimulate the pancreatic beta-cell release of insulin. These also reduce the levels of blood triglyceride and cholesterol.	[12, 32, 42]

Except for the dressing percentage and small intestine length, dietary supplementation with dandelion had no effects on carcass parameters compared to the control group [43]. Compared to the 0.25% dandelion diet, birds that were supplemented with a 0.5% dandelion diet had increased live body weight (LBW) and body weight gain (BWG), and inferior feed conversion ratio (FCR) and mortality rate. On the other side, no remarkable effects were noticed on carcass features (viscera and dressing %) and stress coefficient while dietary supplementing broilers with dandelion compared to the untreated group [30].

In a study conducted on broilers at 52 days of age, Schleicher *et al.* [44] reported that herbs such as chamomile and dandelion had adverse effects on LBW and FCR. Moreover, they noticed that the mortality rate was decreased, but the reason was unclear. Oh *et al.* [45] investigated the influence of dandelion and probiotics based on the effects of dandelion on growth and meat quality in broiler chickens without finding any significant change in LBW, BWG, FCR and feed consumption in comparison with the control group. On the other side, the dietary supplementation with dandelion extracts decreased the pH value and the concentrations of volatile basic nitrogen and thiobarbituric acid reactive substance in chicken meat [46].

Hepatoprotective and Anti-cancer Activities

Recently, the incidence of liver disease has an increasing trend. Due to their efficiency and economic benefits, therapeutic herbs have been more frequently used to treat liver disease and cancers. Several herbs were proven to have critical hepatoprotective effects, but not all of them have been pharmacologically evaluated. The hepatoprotective effects of dandelion have been extensively studied and proven [47 - 50]. In previous trials, the therapeutic effects of dandelion were investigated after the administration of some hepatotoxic substances, resulting in a regeneration of hepatocytes and lower serum levels of

liver enzymes, such as alanine aminotransferase (ALT), alkaline phosphatase (ALP) and aspirate aminotransferase (AST). Due to the cholagogic activity of the dandelion herb, root extracts enhance the production and release of bile after oral administration [51]. After treatment with dandelion, the concentrations of antioxidant enzymes, such as superoxide dismutase (SOD), glutathione peroxidase (GSH-Px) and catalase (CAT), remarkably increased in the liver [50].

Qureshi *et al.* [52] reported that after dietary supplementation with dandelion leaves, the regenerative capacity of the liver significantly increased in broilers compared to the control group. Besides, Tabassum *et al.* [53] suggested that the beneficial effect of dandelion on liver function might be due to the presence of some compounds of this herb such as triterpenes, flavonoids (luteolin and apigenin), sesquiterpene lactones, vitamins (A, C, riboflavin and thiamine), and fatty acids (myristic) able to increase the secretion of bile and enhance the functionality of the liver [13]. Moreover, the extracts of dandelion have been reported to have antiteratogenic and anti-cancer effects [54].

Antibacterial and Antiparasitic Activities

Bacterial resistance to drugs has become an enormous problem for public health. As previously mentioned, medicinal herbs have become a valuable alternative to antibacterial drugs. Various researchers indicated that dandelion has antibacterial effects against *Saphylococcus aureus* and *Escherichia coli* in a dose-dependent manner [13, 32, 55, 56]. In particular, compounds such as phenols, triterpenes, tannins, steroids saponins, reducing sugar, anthracenosides, and phlobatannins are known to have antimicrobial effects. Furthermore, dandelion has therapeutic effects on several pathogens and can be used as a traditional therapeutic agent in several pathological states [13, 55]. Yan *et al.* [31] reported a decrease in the bacterial coliform count in feces. Also, dandelion has shown antiparasitic activities in poultry [57].

Antioxidant and Anti-inflammatory Activities

A possible reason for the therapeutic effects of various herbs could be the presence of antioxidant substances. The harmful effects of free radicals can be alleviated with antioxidants such as diterpenes, flavonoids, and phenolic acids [58, 59]. These phenolic components have antioxidant properties because of their scavenging active oxygen species and free radicals such as singlet oxygen and superoxide and hydroxyl radicals [60, 61]. Several studies have proven that dandelion has antioxidant activities [35, 62 - 65]. Also, it has been claimed that the flower of dandelion has the highest antioxidant capacity among other parts [64, 65]. Hu and Kitts [66] reported that the antioxidant activity of dandelion flowers could be due to the presence of luteolin and luteolin-7-o-glycoside, which improve the free radical scavenging activity.

Jeon *et al.* [67] reported that the extracts of dandelion have anti-inflammatory activities. The sesquiterpene lactones from the dandelion root have been proven to have anti-inflammatory effects [68]. It has been suggested that phenylpropanoids could be the main substances responsible for such activities of the root and flower of dandelion [69]. Jeon *et al.* [67] demonstrated that ethanol extracts of dried aerial parts of dandelion impeded the inflammatory process and inhibited angiogenesis in rodents.

Hypoglycaemic and Hypolipidemic Effects

In several *in vivo* experiments on animals, the root and leaf of dandelion have been reported to have hypoglycaemic effects [32, 42], whereas the mechanism of action is not completely clear. It has been suggested that the bioactive components of dandelion stimulate the pancreas to secrete insulin, which could explain the hypoglycaemic effects [70]. On the other side, several studies reported that dandelion lowers cholesterol and triglyceride serum levels, exhibiting a hypolipidemic effect [13, 32], which can be attributed to the occurrence of several biological components such as tannins alkaloids, glycosides, phenolic compounds, flavonoids and in dandelion by stimulating the secretion of bile. After dietary supplementation of dandelion, no remarkable differences in total cholesterol, LDL, HDL and total cholesterol serum levels were noticed [45].

Immune System Enhancer

The dandelion plant has essential functions in stimulating the non-specific immune response. The dandelion plant induces immunomodulatory effects by impeding tumor necrosis factor (TNF) [71, 72]. Another interesting study by Lee *et al.* [73] observed that a combinatory use of dandelion and recombinant interferon-gamma (rIFN-γ) remarkably inducted the production of TNF-α, interleukin (IL)-12p70, and IL-10. Such combined application increased the production of nitric oxide (NO) through inducible nitric oxide synthase (iNOS) in peritoneal macrophages. Such findings indicate that dandelion greatly influences several lymphocyte populations or body tissues, suggesting that dandelion can modify immune reactions. Several polysaccharides in the dandelion herb might be major intermediaries in immune communications [67]. Despite the mentioned facts, the exact role of the dandelion is still unclear [74].

Digestion Stimulant

The dandelion plant and its extracts exhibit several natural stimulatory effects on the gastrointestinal system. Such effects could be due to resins (α- and β- amyrin, triterpene and Epi-lupeol), which stimulate reflex excitation of taste receptors and increase the production of saliva, bile, and gastric juices, in this way stimulating appetite and ingestion [7]. Furthermore, the high fiber content in this plant enhances the excretion of toxins from the body and balances gut microbiota. The leaves of dandelion encompass substances that stimulate appetite, eudesmanolides and lactones, which might improve the secretion of gastric juice.

Effect of Dandelion on Hematological and Biochemical Blood Parameters

The supplementation of dandelion in broiler chickens resulted in improved values of packed cell volume (PCV) and hemoglobin (Hb) concentration, while the heterophil/lymphocyte ratio (H/L) decreased [42, 75]. Such effects could be the repercussion of the anti-stress attributes of dandelion [42], which are related to its antioxidant capability [64, 65]. By stimulating the production of hormones in the adrenal gland, stress can increase the H/L ratio [76]. Based on the previous findings, the H/L ratio can be a reliable indicator for assessing poultry's health status and stress [77]. Park *et al.* [78] described that dietary inclusion with dandelion in broilers led to a decrease in the concentrations of total cholesterol, triglyceride, glutamic pyruvic transaminase (GPT), and glutamic oxaloacetate transaminase (GOT) in comparison to the control group.

CONCLUSION

The present review article highlights dandelion's most important and bioactive substances, a multipurpose medicinal and therapeutic herb with different biological functions and health-promoting effects. Several researchers have investigated the applications of dandelion in the poultry industry and indicated that the supplementary use of this plant has beneficial outcomes related to growth and production. Despite that many functions of dandelion in enhancing production and improving health are known, further studies are needed especially to exploit its optimum usage in improving poultry production and safeguarding health.

CONSENT FOR PUBLICATION

Not applicable.

CONFLICT OF INTEREST

The author declares no conflict of interest, financial or otherwise.

ACKNOWLEDGEMENTS

Declared none.

REFERENCES

[1]   Mahima RA, Rahal A, Deb R, *et al.* Immunomodulatory and therapeutic potentials of herbal, traditional/indigenous and ethnoveterinary medicines. Pak J Biol Sci 2012; 15(16): 754-74.
[http://dx.doi.org/10.3923/pjbs.2012.754.774] [PMID: 24175417]

[2]   Dhama K, Latheef SK, Mani S, *et al.* Multiple beneficial applications and modes of action of herbs in poultry health and production: A review. Int J Pharmacol 2015; 11(3): 152-76.
[http://dx.doi.org/10.3923/ijp.2015.152.176]

[3]   Tiwari R, Latheef SK, Ahmed I, *et al.* Herbal Immunomodulators - A Remedial Panacea for Designing and Developing Effective Drugs and Medicines: Current Scenario and Future Prospects. Curr Drug Metab 2018; 19(3): 264-301.
[http://dx.doi.org/10.2174/1389200219666180129125436] [PMID: 29380694]

[4]   Alagawany M, Elnesr SS, Farag MR, *et al.* Use of licorice (Glycyrrhiza glabra) herb as a feed additive in poultry: current knowledge and prospects. Animals (Basel) 2019; 9(8): 536.
[http://dx.doi.org/10.3390/ani9080536] [PMID: 31394812]

[5]   Shu Z, Wu T, Shahen M, *et al.* System-pharmacology dissection of traditional chinese herbs sini decoction for treatment of cardiovascular diseases. An Acad Bras Cienc 2019; 91(3): e20180424.
[http://dx.doi.org/10.1590/0001-3765201920180424] [PMID: 31553364]

[6]   Sweeney B, Vora M, Ulbricht C, Basch E. Evidence-based systematic review of dandelion (Taraxacum officinale) by natural standard research collaboration. J Herb Pharmacother 2005; 5(1): 79-93.
[http://dx.doi.org/10.1080/J157v05n01_09] [PMID: 16093238]

[7]   Yarnell E, Abascal JD. Dandelion (Taraxacum officinale and T mongolicum). Integr Med (Encinitas) 2009; 8: 310-6.

[8]   Martinez M, Poirrier P, Chamy R, *et al.* Taraxacum officinale and related species—An ethnopharmacological review and its potential as a commercial medicinal plant. J Ethnopharmacol 2015; 169: 244-62.
[http://dx.doi.org/10.1016/j.jep.2015.03.067] [PMID: 25858507]

[9]   Qureshi S, Adil S, Abd El-Hack ME, Alagawany M, Farag MR. Beneficial uses of dandelion herb (*Taraxacum officinale*) in poultry nutrition. Worlds Poult Sci J 2017; 73(3): 591-602.
[http://dx.doi.org/10.1017/S0043933917000459]

[10]   Anonymous A. Drugs and Lactation Database (LactMed). Bethesda (MD): National Library of Medicine (US). 2018.

[11]   Morley TI. Spring Flora of Minnesota. Minneapolis, MN: University of Minnesota press 1974; p. 255.

[12]   Peter L. Solving Weed Problems. Guilford, Conn.: Lyons Press 2001; pp. 210-1.

[13]   Jassim AKN, Farhan SA, Noori OM. () Identification of Dandelion *Taraxacum officinale* leaves components and study its extracts effect on different microorganisms. J Al-Nahrain University Science 2012; 15(3): 7-14.
[http://dx.doi.org/10.22401/JNUS.15.3.02]

[14]   Grieve M. A Modern Herbal. New York: Dover Publications 1931.

[15]   González-Castejón M, Visioli F, Rodriguez-Casado A. Diverse biological activities of dandelion. Nutr Rev 2012; 70(9): 534-47.
[http://dx.doi.org/10.1111/j.1753-4887.2012.00509.x] [PMID: 22946853]

[16]   Mir MA, Sawhney SS, Jassal MS. Qualitative and quantitative analysis of phytochemicals of *Taraxacum officinale.* Wudpecker J Pharm Pharmacol 2013; 2: 1-5.

[17]   Williams CA, Goldstone F, Greenham J. Flavonoids, cinnamic acids and coumarins from the different tissues and medicinal preparations of *Taraxacum officinale.* Phytochemistry 1996; 42(1): 121-7.
[http://dx.doi.org/10.1016/0031-9422(95)00865-9] [PMID: 8728061]

[18]   Schmidt M. The delightful dandelion. J Organic Garden 1979; 26: 112-7.

[19]   Jackson BS. The lowly dandelion deserves more respect. Can Geogr Mag 1982; 102: 54-9.

[20]   Khan RU, Naz S, Nikousefat Z, Tufarelli V, Laudadio V. *Thymus vulgaris* : alternative to antibiotics in poultry feed. Worlds Poult Sci J 2012; 68(3): 401-8.
[http://dx.doi.org/10.1017/S0043933912000517]

[21]   Diaz-Sanchez S, D'Souza D, Biswas D, Hanning I. Botanical alternatives to antibiotics for use in

organic poultry production. Poult Sci 2015; 94(6): 1419-30.
[http://dx.doi.org/10.3382/ps/pev014] [PMID: 25743421]

[22]   Dhama K, Tiwari R, Khan RU, *et al.* Growth promoters and novel feed additives improving poultry production and health, bioactive principles and beneficial applications: The trends and advances A Review. Int J Pharmacol 2014; 10(3): 129-59.
[http://dx.doi.org/10.3923/ijp.2014.129.159]

[23]   El-Hack MEA, Mahgoub SA, Alagawany M, Dhama K. Influences of dietary supplementation of antimicrobial cold pressed oils mixture on growth performance and intestinal microflora of growing Japanese quails. Int J Pharmacol 2015; 11(7): 689-96.
[http://dx.doi.org/10.3923/ijp.2015.689.696]

[24]   Yadav AS, Kolluri G, Gopi M, Karthik K, Malik YS, Dhama K. Exploring alternatives to antibiotics as health promoting agents in poultry- a review. J Exp Biol Agric Sci 2016; 4(3S): 368-83.
[http://dx.doi.org/10.18006/2016.4(3S).368.383]

[25]   Windisch W, Schedle K, Plitzner C, Kroismayr A. Use of phytogenic products as feed additives for swine and poultry1. J Anim Sci 2008; 86(14) (Suppl. 14): E140-8.
[http://dx.doi.org/10.2527/jas.2007-0459] [PMID: 18073277]

[26]   Dhama K, Tiwari R, Chakrabort S, *et al.* Evidence based antibacterial potentials of medicinal plants and herbs countering bacterial pathogens especially in the era of emerging drug resistance: An integrated update. Int J Pharmacol 2013; 10(1): 1-43.
[http://dx.doi.org/10.3923/ijp.2014.1.43]

[27]   Farag MR, Alagawany M, El-Hack MEA, Tufarelli V. Alleviative effect of some phytochemicals on cyadox-induced oxidative damage in rabbit erythrocytes. Jpn J Vet Res 2016; 64(3): 171-82.
[PMID: 29786988]

[28]   Alagawany M, Ashour EA, Reda FM. () Effect of dietary supplementation of garlic (*Allium sativum*) and turmeric (*Curcuma longa*) on growth performance, carcass traits, blood profile and oxidative status in growing rabbits. Ann Anim Sci 2016; 16(2): 489-505.
[http://dx.doi.org/10.1515/aoas-2015-0079]

[29]   Alagawany M, Abd El-Hack ME, Farag MR, Sachan S, Karthik K, Dhama K. The use of probiotics as eco-friendly alternatives for antibiotics in poultry nutrition. Environ Sci Pollut Res Int 2018; 25(11): 10611-8.
[http://dx.doi.org/10.1007/s11356-018-1687-x] [PMID: 29532377]

[30]   Al-Kassi GAM, Witwit NM. A comparative study on diet supplementation with a mixture of herbal plants and dandelion as a source of prebiotics on the performance of broilers. Pak J Nutr 2009; 9(1): 67-71.
[http://dx.doi.org/10.3923/pjn.2010.67.71]

[31]   Yan L, Zhang ZF, Park JC, Kim IH. Evaluation of *Houttuynia cordata* and *Taraxacum officinale* on growth performance, nutrient digestibility, blood characteristics, and fecal microbial shedding in diet for weaning pigs. Asian-Australas J Anim Sci 2012; 25(10): 1439-44.
[http://dx.doi.org/10.5713/ajas.2012.12215] [PMID: 25049500]

[32]   Qureshi S, Banday MT, Adil S, Shakeel I, Munshi ZH. Effect of dandelion leaves and fenugreek seeds with or without enzyme addition on performance and blood biochemistry of broiler chicken and evaluation of their *in vitro* antibacterial activity. Indian J Anim Sci 2015; 85: 1248-54.

[33]   Faber K. [The dandelion Taraxacum officinale Weber]. Pharmazie 1958; 13(7): 423-36.
[PMID: 13578653]

[34]   Lis B, Olas B. Pro-health activity of dandelion (Taraxacum officinale L.) and its food products – history and present. J Funct Foods 2019; 59: 40-8.
[http://dx.doi.org/10.1016/j.jff.2019.05.012]

[35]   Hudec J, Burdová M, Kobida L, *et al.* Antioxidant capacity changes and phenolic profile of Echinacea

purpurea, nettle (Urtica dioica L.), and dandelion (Taraxacum officinale) after application of polyamine and phenolic biosynthesis regulators. J Agric Food Chem 2007; 55(14): 5689-96.
[http://dx.doi.org/10.1021/jf070777c] [PMID: 17579437]

[36]   Rutherford PP, Deacon AC. β-Fructofuranosidases from roots of dandelion (*Taraxacum officinale* Weber). Biochem J 1972; 126(3): 569-73.
[http://dx.doi.org/10.1042/bj1260569] [PMID: 5075268]

[37]   Kristo ST, Terdz P, Simændi B, Køry B. Preparation of bioactive constituents from *Taraxacum officinale* L. by means of various extraction methods. Journal of Oil, Soap. Cosmetics 2000; 49: 93-7.

[38]   Hu C, Kitts DD. Dandelion (Taraxacum officinale) flower extract suppresses both reactive oxygen species and nitric oxide and prevents lipid oxidation *in vitro*. Phytomedicine 2005; 12(8): 588-97.
[http://dx.doi.org/10.1016/j.phymed.2003.12.012] [PMID: 16121519]

[39]   Bruneton J. Pharmacognosy, Phytochemistry, Medicinal Plants. Paris, France: Lavoisier Publishers 1995; p. 915.

[40]   Khan R, Khan MA, Sultan S. () Nutritional quality of sixteen terrestial weeds for the formulation of cost-effective animal feed. J Anim Plant Sci 2013; 23: 75-9.

[41]   Qureshi S, Banday MT, Shakeel I, Adil S, Khan AA. Effect of raw and enzyme-treated dandelion leaves and fenugreek seed supplemented diet on gut microflora of broiler chicken. Appl Biol Res 2016; 18(1): 76-9.
[http://dx.doi.org/10.5958/0974-4517.2016.00012.4]

[42]   Al-Kassie GAM, Al-Jumaa YMF, Jameel YJ. Effect of probiotic (*Aspergillus niger*) and prebiotic (*Taraxacum officinale*) on blood picture and biochemical properties of broiler chicks. Int J Poult Sci 2008; 7(12): 1182-4.
[http://dx.doi.org/10.3923/ijps.2008.1182.1184]

[43]   Qureshi S, Banday MT, Shakeel I, Adil S. Feeding value of raw or enzyme treated dandelion leaves and fenugreek seeds alone or in combination in meat type chicken. Pak J Nutr 2015; 15(1): 9-14. b
[http://dx.doi.org/10.3923/pjn.2016.9.14]

[44]   Schleicher A, Fritz Z, Kinal S. Zastosowanie wybranych ziół w mieszankach treściwych dla kurcząt rzeźnych. Rocz Nauk Zootech 1998; 25(3): 213-24.

[45]   Oh JI, Kim GM, Ko SY, Bae IH, Lee SS, Yang CJ. Effect of dietary dandelion (*Taraxzcum coreanum*) and dandelion fermented probiotics on productivity and meat quality of broilers. Hangug Gageum Haghoeji 2007; 34(4): 319-27.
[http://dx.doi.org/10.5536/KJPS.2007.34.4.319]

[46]   Park CI, Kim YJ. Effects of dietary supplementation of mulberry leaves and dandelion extracts on storage of chicken meat. Hangug Gageum Haghoeji 2010; 37(4): 313-21.
[http://dx.doi.org/10.5536/KJPS.2010.37.4.313]

[47]   Park C, Zhou Y, Song Y. Hepatoprotective effect of dandelion(*Taraxacum officinale*) against acute liver injury induced by Carbon tetrachloride in Sprague☐Dawley rats. FASEB J 2007; 21(6): 862.
[http://dx.doi.org/10.1096/fasebj.21.6.A1122-d]

[48]   Fallah H, Zareei M, Ziai SA. The effects of *Taraxacum officinale* l. and *Berberis vulgaris* l. root extracts on carbon tetrachloride induced liver toxicity in rat. J Med Plant 2010; 9: 6.

[49]   Al-Malki A, Kamel AA, Gamal A, Hassan A. Hepatoprotective effect of *Dandelion (Taraxacum officinale)* against induced chronic liver cirrhosis. J Med Plants Res 2013; 7: 26-35.

[50]   Gulfraz M, Ahamd D, Ahmad MS, *et al.* Effect of leaf extracts of *Taraxacum officinale* on CCl_4 induced hepatotoxicity in rats, *in vivo* study. Pak J Pharm Sci 2014; 27(4): 825-9.
[PMID: 25015447]

[51]   Vogel G. Natural substances with effects on the liver. In: Wagner H, Wolff P, Eds. new natural products and plant drugs with pharmacological, biological or therapeutic activity. Heidelberg:

Springer-Verlag 1977.
[http://dx.doi.org/10.1007/978-3-642-66682-7_11]

[52] Qureshi S, Banday MT, Shakeel I, *et al.* Histomorphological studies of broiler chicken fed diets supplemented with either raw or enzyme treated dandelion leaves and fenugreek seeds. Vet World 2016; 9(3): 269-75.
[http://dx.doi.org/10.14202/vetworld.2016.269-275] [PMID: 27057110]

[53] Tabassum N, Shah MY, Qazi MA. Prophlactic activity of extract of *Taraxacum officinale* against hepatocellular injury induced in mice. Pharmacologyonline 2010; 2: 344-52.

[54] Takasaki M, Konoshima T, Tokuda H, *et al.* Anti-carcinogenic activity of Taraxacum plant. I. Biol Pharm Bull 1999; 22(6): 602-5.
[http://dx.doi.org/10.1248/bpb.22.602] [PMID: 10408234]

[55] Oseni LA, Issah Y. Screening ethanolic and aqueous leaf extracts of *Taraxacum officinale* for *in vitro* bacteria growth inhibition. J Pharm Biomed 2012; 20: 29-32.

[56] Ionescu D, Predan G, Rizea A, Dune A. (Antimicrobial activity of some hydroalcoholic extracts of artichoke *(Cynara scolymus)*, burdock *(Arctium lappa)* and dandelion *(Taraxacum officinale)*. Bull Transilv Univ Bras 2013; 6: 122-7.

[57] Lans C, Turner N. Organic parasite control for poultry and rabbits in British Columbia, Canada. J Ethnobiol Ethnomed 2011; 7(1): 21.
[http://dx.doi.org/10.1186/1746-4269-7-21] [PMID: 21756341]

[58] Yanishlieva NV, Marinova E, Pokorný J. Natural antioxidants from herbs and spices. Eur J Lipid Sci Technol 2006; 108(9): 776-93.
[http://dx.doi.org/10.1002/ejlt.200600127]

[59] Rahal A, Kumar A, Singh V, *et al.* Oxidative stress, prooxidants and antioxidants: the interplay. BioMed Res Int 2014.
[http://dx.doi.org/10.1155/2014/761264]

[60] Hall CA, Cuppett SL. Structure-activities of natural antioxidants. Antioxidant Methodology *in-vivo* and *in-vitro* Concepts. In: Auroma OI, Cuppett SL, Eds. Champaign: AOCS Press 1997; pp. 141-70.

[61] Farag MR, Alagawany M, Tufarelli V. *In vitro* antioxidant activities of resveratrol, cinnamaldehyde and their synergistic effect against cyadox-induced cytotoxicity in rabbit erythrocytes. Drug Chem Toxicol 2017; 40(2): 196-205.
[http://dx.doi.org/10.1080/01480545.2016.1193866] [PMID: 27314888]

[62] Kim HM, Lee EH, Shin TY, Lee KN, Lee JS. *Taraxacum officinale* restores inhibition of nitric oxide production by cadmium in mouse peritoneal macrophages. Immunopharmacol Immunotoxicol 1998; 20(2): 283-97.
[http://dx.doi.org/10.3109/08923979809038545] [PMID: 9653673]

[63] Kim HM, Oh CH, Chung CK. Activation of inducible nitric oxide synthase by *Taraxacum officinale* in mouse peritoneal macrophages. Gen Pharmacol 1999; 32(6): 683-8.
[http://dx.doi.org/10.1016/S0306-3623(98)00227-4] [PMID: 10401993]

[64] Hu C, Kitts DD. Antioxidant, prooxidant, and cytotoxic activities of solvent-fractionated dandelion *(Taraxacum officinale)* flower extracts *in vitro*. J Agric Food Chem 2003; 51(1): 301-10.
[http://dx.doi.org/10.1021/jf0258858] [PMID: 12502425]

[65] Ghaima KK, Noor MH, Safaa AA. Antibacterial and antioxidant activities of ethyl acetate extract of *Nettle (Urtica dioica)* and dandelion *(Taraxacum officinale)*. J Appl Pharm Sci 2013; 3: 96-9.

[66] Hu C, Kitts DD. Luteolin and luteolin-7-O-glucoside from dandelion flower suppress iNOS and COX-2 in RAW264.7 cells. Mol Cell Biochem 2004; 265(1/2): 107-13.
[http://dx.doi.org/10.1023/B:MCBI.0000044364.73144.fe] [PMID: 15543940]

[67] Jeon HJ, Kang HJ, Jung HJ, *et al.* Anti-inflammatory activity of *Taraxacum officinale*. J

Ethnopharmacol 2008; 115(1): 82-8.
[http://dx.doi.org/10.1016/j.jep.2007.09.006] [PMID: 17949929]

[68]   Kashiwada Y, Takanaka K, Tsukada H. Sesquiterpene glucosides from anti-leukotriene B4 release fraction of *Taraxacum officinale*. J Asian Nat Prod Res 2001; 191-7.

[69]   Hu C, Kitts DD. Dandelion (*Taraxacum officinale*) flower extract suppresses both reactive oxygen species and nitric oxide and prevents lipid oxidation *in vitro*. Phytomedicine 2005; 12(8): 588-97.
[http://dx.doi.org/10.1016/j.phymed.2003.12.012] [PMID: 16121519]

[70]   Hussain Z, Waheed A, Qureshi RA, *et al.* The effect of medicinal plants of Islamabad and Murree region of Pakistan on insulin secretion from INS-1 cells. Phytother Res 2004; 18(1): 73-7.
[http://dx.doi.org/10.1002/ptr.1372] [PMID: 14750205]

[71]   Kim HM, Shin HY, Lim KH, *et al. Taraxacum officinale* inhibits tumor necrosis factor-alpha production from rat astrocytes. Immunopharmacol Immunotoxicol 2000; 22(3): 519-30.
[http://dx.doi.org/10.3109/08923970009026009] [PMID: 10946829]

[72]   Koo HN, Hong SH, Song BK, Kim CH, Yoo YH, Kim HM. Taraxacum officinale induces cytotoxicity through TNF-α and IL-1α secretion in Hep G2 cells. Life Sci 2004; 74(9): 1149-57.
[http://dx.doi.org/10.1016/j.lfs.2003.07.030] [PMID: 14687655]

[73]   Lee BR, Lee JH, An HJ. Effects of *Taraxacum officinale* on fatigue and immunological parameters in mice. Molecules 2012; 17(11): 13253-65.
[http://dx.doi.org/10.3390/molecules171113253] [PMID: 23135630]

[74]   Lee SH, Park JB, Park HJ, Park YJ, Sin JI. Biological properties of different types and parts of the dandelions: comparisons of antioxidative, immune cell proliferative and tumor cell growth inhibitory activities. Food Sci Nutr 2005; 10: 172-8.

[75]   Berezi EP, Monago C, Adelagun ROA. Haematological profile of rats treated with aqueous extracts of common dandelion leaf (*Taraxacum officinale* Weber) against carbon tetrachloride (CCl4) toxicity. Int J Biotechnol Biochem 2013; 2: 263-7.

[76]   Gross WB, Siegel HS. Evaluation of the heterophil/lymphocyte ratio as a measure of stress in chickens. Avian Dis 1983; 27(4): 972-9.
[http://dx.doi.org/10.2307/1590198] [PMID: 6360120]

[77]   McFARLANE JM, Curtis S. Multiple concurrent stressors in chicks. 3. Effects on plasma corticosterone and the heterophil:lymphocyte ratio. Poult Sci 1989; 68(4): 522-7.
[http://dx.doi.org/10.3382/ps.0680522] [PMID: 2748500]

[78]   Park CI, Shon JC, Kim YJ. Effect of dietary supplementation of mulberry leaves and dandelion extracts on performance and blood characteristics of chickens. Hangug Gageum Haghoeji 2010; 37(2): 173-80.
[http://dx.doi.org/10.5536/KJPS.2010.37.2.173]

CHAPTER 10

Probiotics in Poultry Nutrition as a Natural Alternative for Antibiotics

Mohamed E. Abd El-Hack[1,*], **Mahmoud Alagawany**[1], **Nahed A. El-Shall**[2], **Abdelrazeq M. Shehata**[3,4], **Abdel-Moneim E. Abdel-Moneim**[5] and **Mohammed A. E. Naiel**[6]

[1] *Department of Poultry, Faculty of Agriculture, Zagazig University, Zagazig 44519, Egypt*

[2] *Department of poultry and fish diseases, Faculty of Veterinary Medicine, Alexandria University, Edfina, Elbehira 22758, Egypt*

[3] *Department of Animal Production, Faculty of Agriculture, Al-Azhar University, Cairo 11651, Egypt*

[4] *Department of Dairy Science & Food Technology, Institute of Agricultural Sciences, Banaras Hindu University, Varanasi 221005, India*

[5] *Biological Applications Department, Nuclear Research Center, Egyptian Atomic Energy Authority, 13759, Egypt*

[6] *Department of Animal Production, Faculty of Agriculture, Zagazig University, Zagazig 44519, Egypt*

Abstract: Since the early 1950s, antibiotics have been used in poultry for improving feed efficiency and growth performance. Nevertheless, various side effects have appeared, such as antibiotic resistance, antibiotic residues in eggs and meat, and imbalance of beneficial intestinal bacteria. Consequently, it is essential to find other alternatives that include probiotics that improve poultry production. Probiotics are live microorganisms administered in adequate doses and improve host health. Probiotics are available to be used as feed additives, increasing the availability of the nutrients for enhanced growth by digesting the feed properly. Immunity and meat and egg quality can be improved by supplementation of probiotics in poultry feed. Furthermore, the major reason for using probiotics as feed additives is that they can compete with various infectious diseases causing pathogens in poultry's gastrointestinal tract. Hence, this chapter focuses on the types and mechanisms of action of probiotics and their benefits, by feed supplementation, for poultry health and production.

Keywords: Growth promoters, Health benefits, Nutrition, Poultry, Probiotics.

* **Corresponding author Mohamed E. Abd El-Hack:** Department of Poultry, Faculty of Agriculture, Zagazig University, Zagazig 44519, Egypt; E-mail:mezzat@zu.edu.eg

Mohamed E. Abd El-Hack & Mahmoud Alagawany (Eds.)
All rights reserved-© 2022 Bentham Science Publishers

INTRODUCTION

Eight decades ago, antibiotics were used as growth promoters in the animal's diet, when it was observed that supplementation of *Streptomyces aureofaciens* harbor chlortetracycline residues in the bird or animal's feed increased their growth rate [1]. Nevertheless, microbial resistance to antibiotics that used for the treatment of diseases has emerged, destruction of beneficial microbiota in the gut was observed, as well as reduced growth rate has resulted from the use of antibiotics as feed additives which may be due to the increasing incidence of subclinical necrotic enteritis [NE] and dysbacteriosis [2]. Hence, European Union [EU] has prohibited using antibiotics as growth promoters or food additives since 2006.

According to the Organization of Food and Agriculture and the Economic Co-operation and Development Organization, the global meat consumption by 2023 was estimated to be an average 36.3 kg in retail weight, a 2.4 kg increase from 2013. Furthermore, around 72% of the increased meat consumption is expected to be obtained from poultry. Chicken is the most nutritious, low-priced, low in fat, adjustable, and high-quality protein source [3], and therefore, the demand for antibiotic-free poultry is increased. Research works are interested in providing alternatives like avian egg antibodies, cytokines, toll-like receptors, probiotics, and others for growth-promoting, preventing diseases, and stimulating the host immunity [4, 5].

Probiotics are defined as "Live strains of strictly selected microorganisms which, when administered in adequate amounts, confer a health benefit on the host" [6]. Also, Abd El-Hack *et al.* [7], defined probiotics as "live microbial feed additives which beneficially affect the host animal *via* enhancing the balance in the gut and consequently improving feed efficiency, nutrient absorption, growth rate and economic aspects of poultry". Interestingly, probiotics have many beneficial aspects and can enhance growth in human beings, animals, poultry, and fish [8 - 11]. Probiotics include several species such as bacteria, yeast, or fungi, and the most prevalent used probiotics are *Bifidobacterium, Bacillus subtilis, Streptococcus,* and *lactobacillus,* which are also capable of reducing many pathogenic bacteria like *Escherichia coli, Clostridium perfringens, Salmonella typhimurium, Staphylococcus aureus, etc* [12]. This article highlights the types and sources of probiotics, mechanisms of action, and their beneficial effects on health, immunity, and production in poultry.

PROBIOTICS TYPES AND SOURCES

The types of probiotics include bacteria, most often [*Lactobacillus, Bacillus, Bifidobacterium, and Enterococcus*] [13 - 16], yeast [*Saccharomyces cerevisiae* and S*accharomyces boulardii*] [17, 18] and fungi such as *Aspergillus* and

Candida species [19]. Most bacteria belong to *Lactobacillus* or *Bifidobacterium* species and present as normal inhabitants of the GIT [Autochthonous probiotics]. In contrast, other species that can be isolated from outside the host like fermented products and soil are can be defined as Allochthonous probiotics [20]. Interestingly, *Lactobacillus plantarum* and *Leuconostoc mesenteroides* can be isolated from non-conventional sources such as fresh fruits and vegetables and can be used as probiotics [21].

Whatever the probiotic source, they must be safe and can't induce any adverse effects on the host or contribute to infectious diseases. Moreover, they should be identified on the basis of genotype and phenotype and check their bile and gastric juice tolerance. In addition, a probiotic used as a feed supplement in animals or poultry may contain a single strain or a mixture of two or more species [22]. These products are produced by fermentation with strain-specific pH and temperature and dried by spray or freeze-drying process [23].

There are various forms of commercial probiotics such as liquid, gel, powder, paste, and granules available in capsules, sachets, tablets, *etc*. Interestingly, dry probiotic form has higher survival life during storage and better resistance to gastric juices. Moreover, hydroxypropyl methylcellulose phthalate 55, when used as a matrix for tablet probiotic form, increased its viability in poultry [12, 24]. A probiotic must be provided in adequate amounts in the feed for animals or poultry that recommended 10^9 cfu/kg of feed for most probiotic products [25]. However, Mountzouris *et al.* [15], added a commercial multi-strain probiotic [PoultryStar ME] in poultry feed at 10^8 cfu/kg. They observed that the growth rate of broiler chickens was enhanced without an observable modification of caecal microflora composition while increasing the dose to 10^9 cfu/kg, the caecal coliform populations were reduced. Not all microorganisms can be probiotic as probiotic strains must be isolated from healthy individuals and appropriately selected with specific criteria to be able to give the desired effect and meet all required conditions to be commercially applied according to the recommendations of the European Food Safety Authority [26] and Food and Agriculture Organization [23] as in Fig. (**1**).

"Direct-fed Microbial [DFM] products" has been used by Food and Drug Administration [FDA] to define probiotics that are used in animal feed that means "products that are purported to contain live [viable] microorganisms [bacteria and/or yeast]" [27]. The Association of American Feed Control Officials [AAFCO] publications listed the microorganisms that can be used as DFM products and have been approved by FDA (Table **1**).

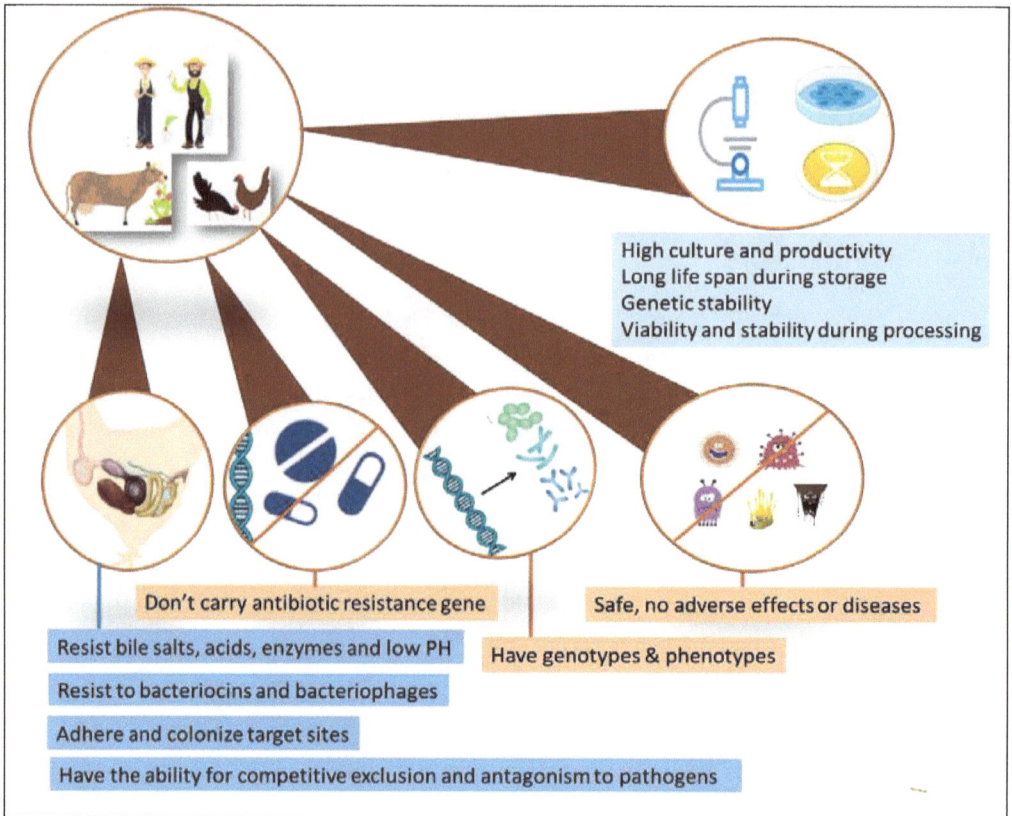

Fig. (1). Criteria and properties for selection of probiotic strains according to [FAO, 2002; EFSA, 2005].

Table 1. The official probiotic list of AAFCO that can be used in direct feed microbial products [DFM] in the animal feed [Source:Yadav *et al.* [38].

Lactobacillus spp.		Bifidobacterium spp.	Enterococcus spp. E. cremoris
L. acidophilus	L. fermentum	B. thermophilum	E. diacetylactis
L. brevis	L. helveticus	B. adolescentis	E. faecium
L. bulgaricus	L. lactis	B. animalis	E. intermedius
L. casei	L. plantarum	B. bifidum	E. lactis
L. cellobiosus	L. reuterii	B. infantis	E. thermophilus
L. curvatus	L. farciminis [swine only]	B. longum	
L. delbruekii	L. buchneri [cattle only]		

(Table 1) cont.....

Bacillus spp. B. coagulans B. lentus B. licheniformis B. pumilus B. subtilis	Bacteroides spp. B. amylophilus B. capillosus B. ruminocola B. suis	Pediococcus spp. P. acidilacticii P. cerevisiae [damnosus] P. pentosaceus	Propionibacterium spp. P. freudenreichii P. shermanii P. acidpropionici [cattle only]
Aspergillus spp. A. niger A. oryzae		*Saccharomyces cerevisiae: Yeast*	*Leuconostoc mesenteroides*

MECHANISM OF PROBIOTIC ACTION

As an alternative to antibiotic growth promoters, probiotics follow various mechanisms to exert their action; closely related probiotic strains may show different modes of action [28, 29].They may act through reduction of the numbers of pathogenic microorganisms in the GIT *via* producing antimicrobial substances, such as hydrogen peroxide and bacteriocins [30, 31], short-chain fatty acids, such as acetic and lactic acid [14], cyclic lipopeptide compounds, *e.g.*, fengycin, surfactin, iturin A and bacillomycin D [32 - 35] and/or polyketides, *e.g.* difficidin, macrolactin, chlorotetaine, and bacillaene [33, 36].

Probiotics also can exclude pathogenic bacteria by competitive adhesion to epithelial binding sites [31], and therefore the time of a probiotic administration is critical for the outcome of its use; since it is most effective when applied to the host before the pathogens enter the intestine naturally or experimentally and multiply [28]. Aguiar *et al.* [37], investigated three *B. subtilis* isolates *in vivo* to be used as a probiotic to reduce colonization of *C. jejuni* based on their motility. They concluded that the better motility of probiotic isolates and the faster reach to colonization sites facilitated, lowering *C. jejuni* colonization.

Another mode by which probiotics can act is the prevention of chronic inflammation of the gastrointestinal epithelium, which is the barrier between pathogenic microorganisms and the entire body through stimulation of the innate immunity [39] that includes immunoglobulins [IgA], antimicrobial peptides, mucus, and the epithelial junction adhesion complex [40, 41].Moreover, the adaptive immune response is stimulated upon using probiotics as the population of intestinal intraepithelial lymphocytes [IEL] was increased in the GIT of chickens that indicated by increased CD3$^+$, CD4$^+$, CD8$^+$ T-lymphocytes, IL-2, and IFN-γ production in the GIT of chickens [17, 42]. Anti- and pro-inflammatory cytokines production can be regulated by probiotics, in addition, several models of immune stimulation, including serum antibody [IgA and IgM], natural killer and macrophages [NK] cells, dendritic cells and phenotype, apoptosis, nitric oxide, and accent of AP-1 and NF-kB pathway [31, 43 - 45].

Probiotics can promote growth through regenerating intestinal mucosa [46] and enhancing the secretion of digestive enzymes [47, 48], resulting in proper digestion. Zhang and Kim [45] reported a 5% increase in body weight gain with an observation of increased ileal digestibility of essential amino acids. Furthermore, Chawla *et al.* [49], and Raghuwanshi *et al.* [50], recorded improved bioavailability of calcium, iron, copper, manganese, and magnesium in the chicken. The villus height: crypt ratio in the poultry intestine was increased, and a subsequent increase of the nutrient absorption was also observed [51, 52]. On another side, probiotic bacteria can also reduce cholesterol levels, lowering cardiovascular disease risk [53]. In a study by Rather *et al.* [54], *Lactobacillus plantarum* YML009 showed antiviral activity against H1N1 virus, but its mode of action is still unclear.

ASPECTS OF PROBIOTIC APPLICATIONS IN THE POULTRY

Growth Performance

Probiotics have shown positive effect on the growth performance of broiler chickens [51, 55, 56].This effect may be resulted from improving the digestibility of protein and starch through increasing protease and amylase activity, which in turn result in better absorption and availability of nutrients, thus enhancing broiler growth [57, 58]. Another possible manner is improving intestinal health through reducing numbers of pathogenic *E. coli* [55, 59 - 61] or *Clostridium* spp [62]. among intestinal microflora, though *Lactobacillus, Bacillus* spp., or yeast populations may be changed or not. However, there are reports of no change of either *E. coli* or lactic acid bacterial numbers by supplementation of probiotics [63].

The enhanced performance and increased productivity of poultry *et al.* using probiotics as feed additives may be due to increased feed consumption and improved feed conversion ratio [30]. Still, interestingly, Yadav *et al.* [64], recorded a significant decrease in feed intake by supplementation of three strains of *Bacillus subtilis* but with significant improvement of weight gain and FCR. The probiotic [*Lactobacillus acidophilus, Lactobacillus plantarum, Pediococcus pentosaceus, Streptococcus faecium, Sacchalomyces cervisiae, Bacillus subtilis,* and *Bacillus lichenifermis*] in combination with prebiotics in ratio 9:1 [w/w] for 42 days had decreased feed intake significantly and improved FCR while weight gain wasn't affected [63]. Furthermore, Zarei *et al.* [61]; Mookiah *et al.* [14], and He *et al.* [65], reported no significant effect on feed intake due to probiotic supplementation, but weight gain and FCR were improved. Hence, it is unnecessary for probiotics to increase feed intake to improve weight gain and FCR as they may act through better digestibility and absorption of nutrients. On the other side, dietary supplementation of *Bacillus subtilis* spore at 0.2 g/kg [66] and *Lactobacillus bulgaricus* at 20, 40, and 60 mg/kg feed [67] increased the apparent digestion coefficients of amino acids and crude protein. Consequently, it reduced their requirements and broiler's feed cost.

In ovo supplementation of 1×10^6 cfu of a mixture of *Lactobacillus casei, Lactobacillus acidophilus, Bifidobacterium bifidium,* and *Enterococcus faecium* increased weight gain of chicks only at first four days post-hatching without significant differences for feed intake, FCR, hatchability, or mortality [68]. In contrast, some probiotic products did not improve the growth performance of broiler chickens [69, 70], even when used in combination with organic acids [60] or vitamins and minerals [71].

Increased villus height and improved intestinal histomorphology led to increasing nutrient absorption and improving growth performance by feeding with probiotics. Nevertheless, Raksasiri *et al.* [63]; Rodjan *et al.* [60]; de Souza *et al.* [70], and Kazemi *et al.* [62], reported that no differences in growth performance in broilers fed different strains of probiotics compared to control while villus height and/or crypt depth was increased. So, it is a complex process, and the variable effect of different strains of probiotics or even the same strains is questionable, and their exact mechanism of action is inconsistent.

Recent studies suggested that symbiotic [a] combination of probiotics with prebiotics could be more effective [48, 72]. On the other hand, there are contrast results by Mookiah *et al.* [14], who reported that neither growth performance nor numbers of caecal microflora were affected in chickens fed probiotic or symbiotic. In addition, Raksasiri *et al.* [63], supplemented Jerusalem artichoke and BACTOSAC-P® [*Lactobacillus plantarum, Lactobacillus acidophilus, Pediococcus pentosaceus, Streptococcus faecium, Bacillus lichenifermis, Bacillus subtilis* and *Sacchalomyces cervisiae*] at ratio 1:9 [w/w] at the sources of prebiotic and probiotic, respectively, this treatment improved only FCR and reduced ammonia concentration in the intestinal tract with no effect on populations of *E. coli* and lactic acid bacteria as well as crypt depth of the intestine.

Another approach of probiotic combinations is multi-enzymes [xylanase, amylase, protease, and/or phytase] with probiotic [*Bacillus* spp.] that were evaluated by Singh *et al.* [73], Murugesan *et al.* [74], and Wealleans *et al.* [75], who and revealed that this combination could effectively optimize the nutrient digestibility in broilers that may be attributed to improved gut health and a conducive environment provided by probiotics for better enzyme activity to improve nutrient digestibility [76] that could be a complementary process [77]. In addition, Sugiharto *et al.* [71], concluded that supplementation of commercial broilers with vitamins and minerals in combination with 0.5% of multi-strain probiotic in the feed had increased the relative weight of pancreas and ileum and reduced serum concentration of uric acid in addition to heart relative weight, but revealed a non-significant effect on growth performance, carcass pH, and drip loss of breast muscles of broiler chickens.

Antibiotic Alternatives to Counter Infectious Pathogens

Recently, probiotics as feed additive alternatives to antibiotics has been documented [78]. Probiotics have shown the ability to enhance intestinal microbial balance and stimulate the natural defence against pathogenic bacteria [79, 80]. Moreover, another mode of action of probiotics, especially lactic acid bacteria, is produce antimicrobial substances, including bacteriocins and organic acids [81]. However, many factors can affect the efficacy of probiotics, such as the included strains, the interaction between various strains, and the synergistic combination of probiotics and other products [82]. Thus, using multi-strain probiotics has been proved to be the optimal way of potentiating the beneficial impacts of probiotics through improving the growth enhancer bacteria with pathogenic bacteria antagonism in birds' gastrointestinal tract [83].

Furthermore, Rathnapraba [84] investigated the efficacy of probiotics [500 g/tone of feed] alone or in combination with bacitracin methylene disalicylate [BMD] [each 500 g/ton of feed] on broiler chickens with experimentally-induced necrotic enteritis [NE] [*C. perfringens*] that proceeded by inoculation of *Eimeria necatrix* oocysts growth performance and intestinal health against in broilers. Supplementation of probiotics either alone or in combination with BMD significantly increased weight gain, FCR, and ileal villi length, and significantly reduced duodenal *E. coli* and *C. perfringens* counts. In the same context, the best FCR and reduced mortality rates post challenge of fifteen days aged broiler chickens with *Clostridium perfringens* was found in 0.5, 0.12 g Kg-1 of *Bacillus subtilis* [2×10^7 CFU/g] and Photobiotic compound [benzophenanthridine alkaloids, sanguinarine, and protopine] treated groups, respectively, alone or in combined form. Moreover, *Bacillus subtilis* supplemented diet increased villus length, total villus area, small intestine, and ileum weight and reduced histopathological lesions of the intestine and liver [85].

In another study, 0.5, 0.12 g Kg^{-1} of *Bacillus subtilis* [2×10^7 CFU/g] and photobiotic compound [benzophenanthridine alkaloids, sanguinarine, and protopine] showed a significant increase in some meat quality and carcass traits, such as dressing, spleen, and thymus percentages as well as the lowest temperature and pH values and the highest hardness and chewiness texture values were recorded in fifteen days aged broiler chickens inoculated with *Clostridium perfringens* [86]. Pradikta *et al.* [87], compared the effect of powder and liquid probiotic feed supplements [*Lactobacillus* sp 1,4 x 10^{10} cfu/ml] towards intestinal microflora of Isa Brown layer chickens and concluded that both powder and liquid *Lactobacillus* sp probiotic provide a positive effect on the Total Plate Count of Lactic Acid Bacteria, *Salmonella* sp, and *E. coli* with a more beneficial effect for the powder form than the liquid one.

Several studies revealed that probiotic supplementation has reduced *E. coli* population and total coliform counts and increased beneficial bacteria in the intestine of chickens [60, 88, 89]. Moreover, probiotic additives have succeeded in reduction of *Salmonella enteritidis, S. typhimurium, S. gallinarum,* and *Campylobacter jejuni [C. jejuni]* numbers in the gut [90 - 93]. Birds fed *Bacillus subtilis* C-3102 [94] or *Lactobacillus gasseri* SBT2055 [95] inhibited the adhesion, invasion, and colonization of *C. jejuni.* In addition, PrimaLac® supplementation protected broiler chickens from *C. jejuni* challenge that may be attributed to the release of organic acids and proteinaceous molecules by probiotic bacteria that kill the pathogenic *C. jejuni* [96]. Furthermore, gut colonization of *Salmonella* enteritidis S1400 in poultry has been reduced by administering a combination of *Lactobacillus salivarius* and *Enterococcus faecium* PXN33 [97]. Turkeys fed genetically restructured probiotic strain [*E. coli*] had reduced *Salmonella enterica* numbers in the gastrointestinal tract due to Microcin J25, an antimicrobial peptide secreted by this probiotic strain [98]. In addition, the beneficial application of probiotics in controlling infection of poultry with *Listeria monocytogenes* was reported [99].

On another side, the efficacy of simultaneous administration of probiotics with live vaccines can be used to enhance protection produced by the latter and diminish their possible adverse effects, such as what was observed by El-Shall *et al.* [22], who administered broiler chickens a commercial probiotic composed of [*Lactobacillus acidophilus, Lactobacillus plantarum, Pediococcus pentosaceus, Saccharomyces cerevisiae, Bacillus subtilis,* and *Bacillus licheniformis*] in drinking water simultaneously with live *Salmonella* vaccine [Avipro®*Salmonella* Vac E]. Chickens were challenged with *Salmonella enteritidis* at four weeks of age. Probiotics diminished the negative effect of live *Salmonella* vaccine on growth performance, reducing fecal shedding of challenged bacterium and its re-isolation from liver, spleen, heart, and cecum.

More else, coccidiosis is included in this concern as Ritzi *et al.* [100], found that growth performance and resistance of birds to Eimeria species was improved by dietary supplementation of probiotics either in water or feed. Pender *et al.* [2016b] administered 1×10^6 cfu of probiotic bacteria [PrimaLac®] *in ovo* at the 18thday of embryonic age, which protected hatched chicks from mixed Eimeria species challenge at third-day post-hatching. Smart ProLive® containing *Bacillus subtilis* and *Pediococcus acidilactici* at a 1x10^7 CFU/mL was administered in drinking water for *E. tenella* experimentally infected broiler chickens aged 2 weeks, continuously from the d 14 to the experimental end [d 35] and significantly improved caecal and ileal villus height and crypt depth were reported comparing to salinomycin as well as improvement of body weight, feed intake, and FCR that did not differ significantly from the non-infected control group were observed [101]. Moreover, Pender *et al.* [68], had attributed the anticoccidial efficacy of probiotics to their modulating effect on caecal and ileal immune response genes. Hence, in case of forbidding anticoccidial drugs use in broiler rearing, probiotics can be used as an alternative.

Even anti aflatoxin B1 [AFB1] effect was reported by probiotic *in vitro* [70] and also *in vivo* by Śliżewska *et al.* [102], who obtained an improved weight gain of broiler chickens than from that induced by 5 mg AFB1and residues in the liver and kidneys as well as increased the excretion of AFB1 by a probiotic containing 4.5 ×10^{10}*Lactobacilli* [*L. plantarum, L. reuteri, L. rhamnosus, L. paracasei,* and *L. pentosus*] and 4 ×10^6 *Saccharomyces cerevisiae* yeasts.

Egg Production

Egg production and egg quality criteria were enhanced in laying hens fed diets containing probiotics [103 - 106]. Ribeiro *et al.* [107], found that probiotic supplementation [*Bacillus subtilis* at 8×105 cfu/g feed and multi-strain probiotics at 0.4%] to layer diet increased egg production, enhanced egg quality, and reduced costs of feed. On the other side, no significant effect on egg production [108, 109] or egg size [108] were induced by probiotics supplementation. Nevertheless, *Lactobacillus* supplementation increased egg weight and size throughout the laying period [110]. Furthermore, Aalaei *et al.* [111], investigated the effect of 0.1g/kg of multi-strain probiotic [containing *Lactobacillus acidophilus* 2.5 × 10^7 cfu/g, *Lactobacillus casei* 2.5 × 10^7 cfu/g, *Bifidobacterium thermophilum* 2.5 × 10^7 cfu/g and *Enterococcus faecium* 2.5 × 10^7 cfu/g] and/or 0.1g/kg of single-strain probiotic [*Pediococcus acidilactici* 1 × 10^{10} cfu/g] on performance and gastrointestinal health of 51 weeks old broiler breeder hens [Ross 308] in a 10-week trial. None of the body weight, egg production, fertility, hatchability, number of yellow follicles, yolk weight and color index, eggshell thickness, and weight was influenced; and gastrointestinal tract function was not improved. However, reduced ileum coliforms count was observed. In laying hens, Ramasamy *et al.* [110], found that the cholesterol content of eggs produced by hens fed a diet supplemented with probiotics [*Lactobacillus* culture] was reduced by 10.4% compared to those of the control hens at 28 weeks of age. Tang *et al.* [112], confirmed that supplementation of probiotic [0.1% PrimaLac®] to layer diets significantly [P < 0.05] decreased the egg yolk cholesterol and total saturated fatty acids at 28 weeks of age and increased total unsaturated fatty acids at 28, 32 and 36 weeks of age.

Health and Immunity

Probiotics could be an antibiotic alternative in poultry nutrition with a further advantage: improving the health status and immunity of birds [103]. They have positive effects on the mucosal immune system and their barrier functions and the intestinal luminal environment [113]. In addition, they can be used as mucosal adjuvants to stimulate the vaccine-specific immune responses in the host. Moreover, probiotics have various beneficial effects through their antioxidant, anti-allergic, anti-inflammatory, anti-diabetic, anti-mutagenic, anti-cancerous, and antiviral potentials [114].

In this regard, using probiotics as a feed additive for chickens can improve the bird's immune status by reducing harmful intestinal microbes. Consequently, the use of antibiotics has to be reduced. The used *Bacillus coagulans* and *Lactobacillus* probiotics increased cecal LAB count and reduced *E. coli*. Moreover, no Salmonella sp. was detected in 5-week-old broilers [69]. They recorded an increase of cecal short volatile fatty acids [Acetic, Butyric, and Valeric acids] that are lipophilic, which penetrate the bacterial cell wall and produce H$^+$ ions, which in turn destroy the bacteria [115].

Giannenas *et al.* [116], found that probiotic products including *Enterococcus faecium, Bifidobacterium animalis, Lactobacillus reuteri animalis, Bacillus subtilis*, or a mixture of multi-species probiotics at 5×10^8 cfu/kg feed have improved growth performance and broilers' intestinal health. The use of *Clostridium butyricum* at different concentrations in chicken's feed by Yang *et al.* [117], and Liao *et al.* [118], has induced a balance of the intestinal microbiota and improved the antioxidation and immune functions of broiler chickens. An enhanced intestinal T-cell immune system was observed using a probiotic product containing *Lactobacillus fermentum* and *Saccharomyces cerevisiae* at 0.1% or 0.2% in the feed of broilers during 1st three weeks of age [17].

Antibody production for Newcastle disease [ND] and infectious bursal disease [IBD] had been augmented by PrimaLac® administration in poultry diets [119]. Furthermore, an enhanced mucosal immunity against ND was obtained by feeding *Echinacea purpurea* and protexin® probiotics to turkey poults [120]. Also, improved humoral immunity and ileal amino acid digestibility of broilers were induced by dietary administration with *Bacillus subtilis, Clostridium butyricum, and Lactobacillus acidophilus* [45].

Even under stress factors, probiotics have shown beneficial effects; Landy and Kavyani [121] examined the effect of multi-strain probiotic [0.9 g Primalac®/kg diet] supplementation in broiler chickens reared under heat stress, and the results revealed that this product induced high caecal populations of *Lactobacilli* spp and low coliforms, increased broiler performance as well as significantly improved the humoral antibody responses to vaccination against Newcastle disease [ND] virus [HI], *Infectious Bronchitis* and Gumboro disease [ELISA]. Similar results regarding growth performance were obtained by dietary supplementation with 1 g/kg of *B. subtilis* [122] through improving beneficial bacteria colonization in the gut and regeneration of the villus-crypt structure. Hu *et al.* [123], found that humoral immune response to Lasota Newcastle disease vaccine in broiler chickens was enhanced through probiotic treatment at 200g/kg of drinking water. It was due to increased expression of IL-7 mRNA in Harder's gland, caecal tonsils, duodenum, and ileal Peyer's patches.

However, Rewatkar *et al.* [124], exhibited non-significant high HI titers for ND by adding dietary oregano essential oil [*Origanum vulgare*] and probiotics [*Saccharomyces cerevisiae*] in broiler chickens in spite of high protein concentration in blood serum, low cholesterol, and BUN value were observed. Moreover, ErdoĞMuŞ *et al.* [101], reported no significant difference for infectious bronchitis or Newcastle disease antibodies obtained by adding Smart ProLive® containing *Bacillus subtilis* and *Pediococcus acidilactici* at 1×10^7 CFU/mL in drinking water. It may be explained by that the time interval between the last ND vaccine and measuring its HI titers was one week in the study of Landy and Kavyani [121] while it was 10 and 14 days in the study by ErdoĞMuŞ *et al.* [101], and Rewatkar *et al.* [124], respectively.

Regarding coccidiosis, Lee *et al.* [125], observed significantly reduced lesions in GIT post *Eimeria maxima* infection in chickens fed a diet supplemented with *Bacillus* based direct-fed microbial at 5×10^6 cfu compared with control birds. This may be attributed to the improvement of immunity as evidenced by increased serum nitric oxide levels in birds supplemented with probiotics. Application of probiotics [Poultry Star® gel] with coccidia vaccine had enhanced the protection against challenges with *Eimeria* species [10]. Rajput *et al.* [126], reported that *Saccharomyces boulardii* and *Bacillus subtilis* B10 could modulate the intestinal ultrastructure through increasing mRNA expression of occluding, cloudin2 and cloudin3. On the contrary, Aalaei *et al.* [127], concluded multi-strain [Lactofeed] and mono-strain [Pediguard] probiotics, either each alone or in combination with each other, had no positive effect on T-cell-mediated immune response [expression of toll-like receptors [TLR]] in 51 weeks aged broiler breeder hens and the immune response to PHA-P injection, serum glutathione peroxidase activity, malondialdehyde, and cholesterol concentration and blood haematology of broiler breeder and are not advisable for breeder nutrition.

CONCLUSION

Probiotics can be considered among the best alternatives for antibiotic usage in the poultry industry. They have many health and production benefits, including growth promotion and immunomodulation, which in turn resulted in safe meat and egg production and increased economic benefits. Nevertheless, variations in the findings of various studies with some claim no extra benefit of probiotic usage can be seen. Hence, focusing on finding the optimal dose, correct strain/s of probiotic microbial, duration of use, form, and delivery method for certain required action should be highlighted to obtain maximum benefit. Furthermore, if there are more field trials for their potential use, an additional area of benefit can be explored that can also be obtained if a better understanding of their mode of action is performed. Although a significant amount of work is available showing many possible benefits of using probiotics as feed additives for poultry production, they have already been commercially used. However, more research is still needed to develop some standard protocol for their application.

CONSENT FOR PUBLICATION

Not applicable.

CONFLICT OF INTEREST

The author declares no conflict of interest, financial or otherwise.

ACKNOWLEDGEMENT

All the authors of the manuscript thank and acknowledge their respective Institutes, Universities, and Organizations for necessary help in compilation of this manuscript.

REFERENCES

[1]   Eckert NH, Lee JT, Hyatt D, *et al.* Influence of probiotic administration by feed or water on growth parameters of broilers reared on medicated and nonmedicated diets. J Appl Poult Res 2010; 19(1): 59-67.
[http://dx.doi.org/10.3382/japr.2009-00084]

[2]   Palamidi I, Fegeros K, Mohnl M, *et al.* Probiotic form effects on growth performance, digestive function, and immune related biomarkers in broilers. Poult Sci 2016; 95(7): 1598-608.
[http://dx.doi.org/10.3382/ps/pew052] [PMID: 26944970]

[3]   Kutepatil O. Poultry Probiotics Market to a Mass Huge Profits as an Alternative to Antibiotics. Poultry, Fisheries & Wildlife Sciences 2018; 6(1).
[http://dx.doi.org/10.4172/2375-446X.1000190]

[4]   Tiwari R, Chakrabort S, Dhama K, Wani MY, Kumar A, Kapoor S. Wonder world of phages: potential biocontrol agents safeguarding biosphere and health of animals and humans- current scenario and perspectives. Pak J Biol Sci 2014; 17(3): 316-28.
[http://dx.doi.org/10.3923/pjbs.2014.316.328] [PMID: 24897785]

[5]   Yadav AS, Kolluri G, Gopi M, Karthik K, Malik YS, Dhama K. Exploring alternatives to antibiotics as health promoting agents in poultry- a review. J Exp Biol Agric Sci 2016; 4(3S): 368-83.
[http://dx.doi.org/10.18006/2016.4(3S).368.383]

[6]   Hill C, Guarner F, Reid G, *et al.* The International Scientific Association for Probiotics and Prebiotics consensus statement on the scope and appropriate use of the term probiotic. Nat Rev Gastroenterol Hepatol 2014; 11(8): 506-14.
[http://dx.doi.org/10.1038/nrgastro.2014.66] [PMID: 24912386]

[7]   Abd El-Hack ME, Mahgoub SA, Alagawany M, Ashour EA. Improving productive performance and mitigating harmful emissions from laying hen excreta *via* feeding on graded levels of corn DDGS with or without *Bacillus subtilis* probiotic. J Anim Physiol Anim Nutr (Berl) 2017; 101(5): 904-13.
[http://dx.doi.org/10.1111/jpn.12522] [PMID: 27184423]

[8]   Alagawany M, Abd El-Hack ME, Arif M, Ashour EA. Individual and combined effects of crude protein, methionine, and probiotic levels on laying hen productive performance and nitrogen pollution in the manure. Environ Sci Pollut Res Int 2016; 23(22): 22906-13.
[http://dx.doi.org/10.1007/s11356-016-7511-6] [PMID: 27572695]

[9]   Popova T. Effect of probiotics in poultry for improving meat quality. Curr Opin Food Sci 2017; 14: 72-7.
[http://dx.doi.org/10.1016/j.cofs.2017.01.008]

[10]   Ritzi MM, Abdelrahman W, van-Heerden K, Mohnl M, Barrett NW, Dalloul RA. Combination of probiotics and coccidiosis vaccine enhances protection against an Eimeria challenge. Vet Res 2016.
[http://dx.doi.org/10.1186/s13567-016-0397-y]

[11]   Zorriehzahra MJ, Delshad ST, Adel M, *et al.* Probiotics as beneficial microbes in aquaculture: an update on their multiple modes of action: a review. Vet Q 2016; 36(4): 228-41.
[http://dx.doi.org/10.1080/01652176.2016.1172132] [PMID: 27075688]

[12]   Iannitti T, Palmieri B. Therapeutical use of probiotic formulations in clinical practice. Clin Nutr 2010; 29(6): 701-25.
[http://dx.doi.org/10.1016/j.clnu.2010.05.004] [PMID: 20576332]

[13]   Abdelqader A, Irshaid R, Al-Fataftah AR. Effects of dietary probiotic inclusion on performance, eggshell quality, cecal microflora composition, and tibia traits of laying hens in the late phase of production. Trop Anim Health Prod 2013; 45(4): 1017-24.
[http://dx.doi.org/10.1007/s11250-012-0326-7] [PMID: 23271415]

[14]   Mookiah S, Sieo CC, Ramasamy K, Abdullah N, Ho YW. Effects of dietary prebiotics, probiotic and synbiotics on performance, caecal bacterial populations and caecal fermentation concentrations of broiler chickens. J Sci Food Agric 2014; 94(2): 341-8.
[http://dx.doi.org/10.1002/jsfa.6365] [PMID: 24037967]

[15]   Mountzouris KC, Tsitrsikos P, Palamidi I, *et al.* Effects of probiotic inclusion levels in broiler nutrition on growth performance, nutrient digestibility, plasma immunoglobulins, and cecal microflora composition. Poult Sci 2010; 89(1): 58-67.
[http://dx.doi.org/10.3382/ps.2009-00308] [PMID: 20008803]

[16]   Pedroso A, Hurley-Bacon A, Zedek A, *et al.* Can probiotics improve the environmental microbiome and resistome of commercial poultry production? Int J Environ Res Public Health 2013; 10(10): 4534-59.
[http://dx.doi.org/10.3390/ijerph10104534] [PMID: 24071920]

[17]   Bai SP, Wu AM, Ding XM, *et al.* Effects of probiotic-supplemented diets on growth performance and

intestinal immune characteristics of broiler chickens. Poult Sci 2013; 92(3): 663-70.
[http://dx.doi.org/10.3382/ps.2012-02813] [PMID: 23436517]

[18]   Rahman MS, Mustari A, Salauddin M, Rahman MM. Effects of probiotics and enzymes on growth performance and haematobiochemical parameters in broilers. J Bangladesh Agric Univ 2014; 11(1): 111-8.
[http://dx.doi.org/10.3329/jbau.v11i1.18221]

[19]   Daşkıran M, Önol AG, Cengiz Ö, *et al.* Influence of dietary probiotic inclusion on growth performance, blood parameters, and intestinal microflora of male broiler chickens exposed to posthatch holding time. J Appl Poult Res 2012; 21(3): 612-22.
[http://dx.doi.org/10.3382/japr.2011-00512]

[20]   Evaluation of certain food additives. Twenty-third Report of the Joint FAO/WHO Expert Committee on Food Additives. Food Cosmet Toxicol 1981; 19: 381.
[http://dx.doi.org/10.1016/0015-6264(81)90400-4]

[21]   Sornplang P, Piyadeatsoontorn S. Probiotic isolates from unconventional sources: a review. J Anim Sci Technol 2016.
[http://dx.doi.org/10.1186/s40781-016-0108-2]

[22]   El-Shall NA, Awad AM, El-Hack MEA, *et al.* The simultaneous administration of a probiotic or prebiotic with live Salmonella vaccine improves growth performance and reduces fecal shedding of the bacterium in Salmonella-challenged broilers. Animals (Basel) 2019; 10(1): 70.
[http://dx.doi.org/10.3390/ani10010070] [PMID: 31906020]

[23]   Sheep FAO. 1982. Animal Genetic Resources Information 1984; 2: 45-6.
[http://dx.doi.org/10.1017/S1014233900003771]

[24]   Jiang T, Li HS, Han GG, *et al.* Oral Delivery of Probiotics Using pH-Sensitive Tablets. J Microbiol Biotechnol 2017; 27(4): 739-46.
[http://dx.doi.org/10.4014/jmb.1606.06071] [PMID: 28081355]

[25]   Seidavi A, Tavakoli M, Asroosh F, Scanes CG, El-Hack A, Mohamed E, *et al.* Antioxidant and antimicrobial activities of phytonutrients as antibiotic substitutes in poultry feed. Environ Sci Pollut Res Int 2021; 1-26.
[PMID: 34811612]

[26]   Opinion of the Scientific Committee on a request from EFSA related to a generic approach to the safety assessment by EFSA of microorganisms used in food/feed and the production of food/feed additives. EFSA J 2005; 3(6): 226.
[http://dx.doi.org/10.2903/j.efsa.2005.226]

[27]   US-FDA. CPG Sec. 689.100 Direct-Fed Microbial Products. 2015. http://wwwfdagov/ICECI/ComplianceManuals/CompliancePolicyGuidanceManual/ucm074707htm
[http://dx.doi.org/10.1211/PJ.2015.20069450]

[28]   Lodemann U. Effects of Probiotics on Intestinal Transport and Epithelial Barrier Function Bioactive Foods in Promoting Health. Elsevier 2010; pp. 303-33.

[29]   Roselli M, Finamore A, Britti MS, *et al.* The novel porcine Lactobacillus sobrius strain protects intestinal cells from enterotoxigenic Escherichia coli K88 infection and prevents membrane barrier damage. J Nutr 2007; 137(12): 2709-16.
[http://dx.doi.org/10.1093/jn/137.12.2709] [PMID: 18029488]

[30]   Shim YH, Ingale SL, Kim JS, *et al.* A multi-microbe probiotic formulation processed at low and high drying temperatures: effects on growth performance, nutrient retention and caecal microbiology of broilers. Br Poult Sci 2012; 53(4): 482-90.
[http://dx.doi.org/10.1080/00071668.2012.690508] [PMID: 23130583]

[31]   Tiwari G, Tiwari R, Pandey S, Pandey P. Promising future of probiotics for human health: Current scenario. Chronicles of Young Scientists 2012; 3(1): 17.

[http://dx.doi.org/10.4103/2229-5186.94308]

[32]   Arrebola E, Jacobs R, Korsten L. Iturin A is the principal inhibitor in the biocontrol activity of *Bacillus amyloliquefaciens* PPCB004 against postharvest fungal pathogens. J Appl Microbiol 2010; 108(2): 386-95.
[http://dx.doi.org/10.1111/j.1365-2672.2009.04438.x] [PMID: 19674188]

[33]   Chen XH, Koumoutsi A, Scholz R, *et al.* Genome analysis of Bacillus amyloliquefaciens FZB42 reveals its potential for biocontrol of plant pathogens. J Biotechnol 2009; 140(1-2): 27-37.
[http://dx.doi.org/10.1016/j.jbiotec.2008.10.011] [PMID: 19041913]

[34]   Ongena M, Jacques P. Bacillus lipopeptides: versatile weapons for plant disease biocontrol. Trends Microbiol 2008; 16(3): 115-25.
[http://dx.doi.org/10.1016/j.tim.2007.12.009] [PMID: 18289856]

[35]   Sun L, Lu Z, Bie X, Lu F, Yang S. Isolation and characterization of a co-producer of fengycins and surfactins, endophytic Bacillus amyloliquefaciens ES-2, from Scutellaria baicalensis Georgi. World J Microbiol Biotechnol 2006; 22(12): 1259-66.
[http://dx.doi.org/10.1007/s11274-006-9170-0]

[36]   Rapp C, Jung G, Katzer W, Loeffler W. Chlorotetain fromBacillus subtilis, an Antifungal Dipeptide with an Unusual Chlorine-containing Amino Acid. Angew Chem Int Ed Engl 1988; 27(12): 1733-4.
[http://dx.doi.org/10.1002/anie.198817331]

[37]   Aguiar VF, Donoghue AM, Arsi K, *et al.* Targeting motility properties of bacteria in the development of probiotic cultures against Campylobacter jejuni in broiler chickens. Foodborne Pathog Dis 2013; 10(5): 435-41.
[http://dx.doi.org/10.1089/fpd.2012.1302] [PMID: 23531121]

[38]   Yadav S, Athol V, Peter J, Wayne L. MakkarEd HP Probiotics in animal nutrition: production, impact and regulation. Rome Italy: FAO 2016.

[39]   Galdeano CM, Perdigón G. The probiotic bacterium Lactobacillus casei induces activation of the gut mucosal immune system through innate immunity. Clin Vaccine Immunol 2006; 13(2): 219-26.
[http://dx.doi.org/10.1128/CVI.13.2.219-226.2006] [PMID: 16467329]

[40]   Baumgart DC, Dignass AU. Intestinal barrier function. Curr Opin Clin Nutr Metab Care 2002; 5(6): 685-94.
[http://dx.doi.org/10.1097/00075197-200211000-00012] [PMID: 12394645]

[41]   Ohland CL, MacNaughton WK. Probiotic bacteria and intestinal epithelial barrier function. Am J Physiol Gastrointest Liver Physiol 2010; 298(6): G807-19.
[http://dx.doi.org/10.1152/ajpgi.00243.2009] [PMID: 20299599]

[42]   Sato K, Takahashi K, Tohno M, *et al.* Immunomodulation in gut-associated lymphoid tissue of neonatal chicks by immunobiotic diets. Poult Sci 2009; 88(12): 2532-8.
[http://dx.doi.org/10.3382/ps.2009-00291] [PMID: 19903951]

[43]   Cao GT, Zeng XF, Chen AG, *et al.* Effects of a probiotic, Enterococcus faecium, on growth performance, intestinal morphology, immune response, and cecal microflora in broiler chickens challenged with Escherichia coli K88. Poult Sci 2013; 92(11): 2949-55.
[http://dx.doi.org/10.3382/ps.2013-03366] [PMID: 24135599]

[44]   Roselli M, Finamore A, Britti MS, Bosi P, Oswald I, Mengheri E. Alternatives to in-feed antibiotics in pigs: Evaluation of probiotics, zinc or organic acids as protective agents for the intestinal mucosa. A comparison of *in vitro* and *in vivo* results. Anim Res 2005; 54(3): 203-18.
[http://dx.doi.org/10.1051/animres:2005012]

[45]   Zhang ZF, Kim IH. Effects of multistrain probiotics on growth performance, apparent ileal nutrient digestibility, blood characteristics, cecal microbial shedding, and excreta odor contents in broilers. Poult Sci 2014; 93(2): 364-70.
[http://dx.doi.org/10.3382/ps.2013-03314] [PMID: 24570458]

[46] Perdigon G, Alvarez S, Rachid M, Agüero G, Gobbato N. Immune system stimulation by probiotics. J Dairy Sci 1995; 78(7): 1597-606.
      [http://dx.doi.org/10.3168/jds.S0022-0302(95)76784-4] [PMID: 7593855]

[47] Jin LZ, Ho YW, Abdullah N, Jalaludin S. Digestive and bacterial enzyme activities in broilers fed diets supplemented with Lactobacillus cultures. Poult Sci 2000; 79(6): 886-91.
      [http://dx.doi.org/10.1093/ps/79.6.886] [PMID: 10875772]

[48] Li X, Qiang L, Xu C. Effects of Supplementation of Fructooligosaccharide and/or <i>Bacillus Subtilis</i> to Diets on Performance and on Intestinal Microflora in Broilers. Arch Tierzucht 2008; 51(1): 64-70.
      [http://dx.doi.org/10.5194/aab-51-64-2008]

[49] Chawla S, Katoch S, Sharma K, Sharma V. Biological response of broiler supplemented with varying dose of direct fed microbial. Vet World 2013; 6(8): 521.
      [http://dx.doi.org/10.5455/vetworld.2013.521-524]

[50] Raghuwanshi S, Misra S, Sharma R, Ps B. Probiotics: Nutritional Therapeutic Tool. J Probiotics Health 2018; 6(1)
      [http://dx.doi.org/10.4172/2329-8901.1000194]

[51] Afsharmanesh M, Sadaghi B. Effects of dietary alternatives (probiotic, green tea powder, and Kombucha tea) as antimicrobial growth promoters on growth, ileal nutrient digestibility, blood parameters, and immune response of broiler chickens. Comp Clin Pathol 2014; 23(3): 717-24.
      [http://dx.doi.org/10.1007/s00580-013-1676-x]

[52] Biloni A, Quintana CF, Menconi A, *et al.* Evaluation of effects of EarlyBird associated with FloraMax-B11 on Salmonella Enteritidis, intestinal morphology, and performance of broiler chickens. Poult Sci 2013; 92(9): 2337-46.
      [http://dx.doi.org/10.3382/ps.2013-03279] [PMID: 23960116]

[53] Jones ML, Tomaro-Duchesneau C, Martoni CJ, Prakash S. Cholesterol lowering with bile salt hydrolase-active probiotic bacteria, mechanism of action, clinical evidence, and future direction for heart health applications. Expert Opin Biol Ther 2013; 13(5): 631-42.
      [http://dx.doi.org/10.1517/14712598.2013.758706] [PMID: 23350815]

[54] Rather IA, Choi KH, Bajpai VK, Park YH. Antiviral mode of action of <i>Lactobacillus plantarum</i> YML009 on Influenza virus H1N1. Bangladesh J Pharmacol 2015; 10(2): 475.
      [http://dx.doi.org/10.3329/bjp.v10i2.23068]

[55] Ahmed ST, Islam MM, Mun HS, Sim HJ, Kim YJ, Yang CJ. Effects ofBacillus amyloliquefaciens as a probiotic strain on growth performance, cecal microflora, and fecal noxious gas emissions of broiler chickens. Poult Sci 2014; 93(8): 1963-71.
      [http://dx.doi.org/10.3382/ps.2013-03718] [PMID: 24902704]

[56] Hrnčár C, Bujko J. Effect of different levels of green tea (Camellia sinensis) on productive performance, carcass characteristics and organs of broiler chickens. Potravinárstvo 2017; 11(1): 623-8.
      [http://dx.doi.org/10.5219/809]

[57] Sen S, Ingale SL, Kim YW, *et al.* Effect of supplementation of Bacillus subtilis LS 1-2 to broiler diets on growth performance, nutrient retention, caecal microbiology and small intestinal morphology. Res Vet Sci 2012; 93(1): 264-8.
      [http://dx.doi.org/10.1016/j.rvsc.2011.05.021] [PMID: 21757212]

[58] Wang Y, Gu Q. Effect of probiotic on growth performance and digestive enzyme activity of Arbor Acres broilers. Res Vet Sci 2010; 89(2): 163-7.
      [http://dx.doi.org/10.1016/j.rvsc.2010.03.009] [PMID: 20350733]

[59] Kim YJ, Bostami ABMR, Islam MM, Mun HS, Ko SY, Yang CJ. Effect of Fermented Ginkgo biloba and Camelia sinensis-Based Probiotics on Growth Performance, Immunity and Caecal Microbiology in Broilers. Int J Poult Sci 2016; 15(2): 62-71.

[http://dx.doi.org/10.3923/ijps.2016.62.71]

[60]   Rodjan P, Soisuwan K, Thongprajukaew K, *et al*. Effect of organic acids or probiotics alone or in combination on growth performance, nutrient digestibility, enzyme activities, intestinal morphology and gut microflora in broiler chickens. J Anim Physiol Anim Nutr (Berl) 2018; 102(2): e931-40. [http://dx.doi.org/10.1111/jpn.12858] [PMID: 29250860]

[61]   Zarei A, Lavvaf A, Motamedi Motlagh M. Effects of probiotic and whey powder supplementation on growth performance, microflora population, and ileum morphology in broilers. J Appl Anim Res 2018; 46(1): 840-4. [http://dx.doi.org/10.1080/09712119.2017.1410482]

[62]   Kazemi SA, Ahmadi H, Karimi Torshizi MA. Evaluating two multistrain probiotics on growth performance, intestinal morphology, lipid oxidation and ileal microflora in chickens. J Anim Physiol Anim Nutr (Berl) 2019; 103(5): 1399-407. [http://dx.doi.org/10.1111/jpn.13124] [PMID: 31141245]

[63]   Raksasiri B, Paengkoum P, Paengkoum S, Poonsuk K. The effect of supplementation of synbiotic in broiler diets on production performance, intestinal histo-morphology and carcass quality. Agric Technol Thail 2018; 14(7): 1743-54.

[64]   Yadav M, Dubey M, Yadav M, Shankar KS. Effect of Supplementation of Probiotic (*Bacillus subtilis*) on Growth Performance and Carcass Traits of Broiler Chickens. Int J Curr Microbiol Appl Sci 2018; 7(8): 4840-9. [http://dx.doi.org/10.20546/ijcmas.2018.708.510]

[65]   He T, Long S, Mahfuz S, *et al*. Effects of Probiotics as Antibiotics Substitutes on Growth Performance, Serum Biochemical Parameters, Intestinal Morphology, and Barrier Function of Broilers. Animals (Basel) 2019; 9(11): 985. [http://dx.doi.org/10.3390/ani9110985] [PMID: 31752114]

[66]   Zaghari M, Zahroojian N, Riahi M, Parhizkar S. Effect of *Bacillus Subtilis* Spore (GalliPro ®) Nutrients Equivalency Value on Broiler Chicken Performance. Ital J Anim Sci 2015; 14(1): 3555. [http://dx.doi.org/10.4081/ijas.2015.3555]

[67]   Apata DF. Growth performance, nutrient digestibility and immune response of broiler chicks fed diets supplemented with a culture ofLactobacillus bulgaricus. J Sci Food Agric 2008; 88(7): 1253-8. [http://dx.doi.org/10.1002/jsfa.3214]

[68]   Pender CM, Kim S, Potter TD, Ritzi MM, Young M, Dalloul RA. Effects of *in ovo* supplementation of probiotics on performance and immunocompetence of broiler chicks to an *Eimeria* challenge. Benef Microbes 2016; 7(5): 699-705. [http://dx.doi.org/10.3920/BM2016.0080] [PMID: 27726419]

[69]   Al-Khalaifa H, Al-Nasser A, Al-Surayee T, *et al*. Effect of dietary probiotics and prebiotics on the performance of broiler chickens. Poult Sci 2019; 98(10): 4465-79. [http://dx.doi.org/10.3382/ps/pez282] [PMID: 31180128]

[70]   de Souza LFA, Araújo DN, Stefani LM, *et al*. Probiotics on performance, intestinal morphology and carcass characteristics of broiler chickens raised with lower or higher environmental challenge. Austral J Vet Sci 2018; 50(1): 35-41. [http://dx.doi.org/10.4067/S0719-81322018000100107]

[71]   Sugiharto S, Isroli I, Yudiarti T, Widiastuti E. The effect of supplementation of multistrain probiotic preparation in combination with vitamins and minerals to the basal diet on the growth performance, carcass traits, and physiological response of broilers. Vet World 2018; 11(2): 240-7. [http://dx.doi.org/10.14202/vetworld.2018.240-247] [PMID: 29657411]

[72]   Utami MMD, Wahyono ND. Supplementation of probiotic and prebiotic on the performance of broilers. IOP Conf Ser Earth Environ Sci 2018; 207: 012024. [http://dx.doi.org/10.1088/1755-1315/207/1/012024]

[73] Singh AK, Tiwari UP, Berrocoso JD, Dersjant-Li Y, Awati A, Jha R. Effects of a combination of xylanase, amylase and protease, and probiotics on major nutrients including amino acids and non-starch polysaccharides utilization in broilers fed different level of fibers. Poult Sci 2019; 98(11): 5571-81.
[http://dx.doi.org/10.3382/ps/pez310] [PMID: 31198939]

[74] Murugesan GR, Romero LF, Persia ME. Effects of protease, phytase and a Bacillus sp. direct-fed microbial on nutrient and energy digestibility, ileal brush border digestive enzyme activity and cecal short-chain fatty acid concentration in broiler chickens. PLoS One 2014; 9(7): e101888-e.10.1371.
[http://dx.doi.org/10.1371/journal.pone.0101888]

[75] Wealleans AL, Walsh MC, Romero LF, Ravindran V. Comparative effects of two multi-enzyme combinations and a Bacillus probiotic on growth performance, digestibility of energy and nutrients, disappearance of non-starch polysaccharides, and gut microflora in broiler chickens. Poult Sci 2017; 96(12): 4287-97.
[http://dx.doi.org/10.3382/ps/pex226] [PMID: 29053809]

[76] Salim HM, Kang HK, Akter N, *et al.* Supplementation of direct-fed microbials as an alternative to antibiotic on growth performance, immune response, cecal microbial population, and ileal morphology of broiler chickens. Poult Sci 2013; 92(8): 2084-90.
[http://dx.doi.org/10.3382/ps.2012-02947] [PMID: 23873556]

[77] Momtazan R, Moravej H, Zaghari M, Taheri H. A note on the effects of a combination of an enzyme complex and probiotic in the diet on performance of broiler chickens. Ir J Agric Food Res 2011; 249-54.

[78] Sharifi SD, Dibamehr A, Lotfollahian H, Baurhoo B. Effects of flavomycin and probiotic supplementation to diets containing different sources of fat on growth performance, intestinal morphology, apparent metabolizable energy, and fat digestibility in broiler chickens. Poult Sci 2012; 91(4): 918-27.
[http://dx.doi.org/10.3382/ps.2011-01844] [PMID: 22399731]

[79] Newaj-Fyzul A, Al-Harbi AH, Austin B. Review: Developments in the use of probiotics for disease control in aquaculture. Aquaculture 2014; 431: 1-11.
[http://dx.doi.org/10.1016/j.aquaculture.2013.08.026]

[80] Ohashi Y, Ushida K. Health-beneficial effects of probiotics: Its mode of action. Anim Sci J 2009; 80(4): 361-71.
[http://dx.doi.org/10.1111/j.1740-0929.2009.00645.x] [PMID: 20163595]

[81] Aliakbarpour HR, Chamani M, Rahimi G, Sadeghi AA, Qujeq D. The *Bacillus subtilis* and Lactic Acid Bacteria Probiotics Influences Intestinal Mucin Gene Expression, Histomorphology and Growth Performance in Broilers. Asian-Australas J Anim Sci 2012; 25(9): 1285-93.
[http://dx.doi.org/10.5713/ajas.2012.12110] [PMID: 25049692]

[82] Song J, Xiao K, Ke YL, *et al.* Effect of a probiotic mixture on intestinal microflora, morphology, and barrier integrity of broilers subjected to heat stress. Poult Sci 2014; 93(3): 581-8.
[http://dx.doi.org/10.3382/ps.2013-03455] [PMID: 24604851]

[83] Patel SG, Raval AP, Bhagwat SR, Sadrasaniya DA, Patel AP, Joshi SS. Effects of Probiotics Supplementation on Growth Performance, Feed Conversion Ratio and Economics of Broilers. Journal of Animal Research 2015; 5(1): 155.
[http://dx.doi.org/10.5958/2277-940X.2015.00026.1]

[84] Rathnapraba S. Evaluation of the Probiotic Mixture and Bacitracin Methylene Disalicylate Supplementation on Intestinal Health of *Clostridium perfringens* Induced Necrotic Enteritis in Broiler Chickens. Int J Curr Microbiol Appl Sci 2021; 10(2): 265-70.
[http://dx.doi.org/10.20546/ijcmas.2021.1002.032]

[85] Hussein EOS, Ahmed SH, Abudabos AM, *et al.* Effect of Antibiotic, Phytobiotic and Probiotic Supplementation on Growth, Blood Indices and Intestine Health in Broiler Chicks Challenged with

Clostridium perfringens. Animals (Basel) 2020; 10(3): 507.
[http://dx.doi.org/10.3390/ani10030507] [PMID: 32197455]

[86]   Hussein EOS, Ahmed SH, Abudabos AM, *et al.* Ameliorative Effects of Antibiotic-, Probiotic- and Phytobiotic-Supplemented Diets on the Performance, Intestinal Health, Carcass Traits, and Meat Quality of *Clostridium perfringens*-Infected Broilers. Animals (Basel) 2020; 10(4): 669.
[http://dx.doi.org/10.3390/ani10040669] [PMID: 32290578]

[87]   Pradikta RW, Sjofjan O, Djunaidi IH. Evaluation on Addition of Powder And Liquid Probiotic In Poultry Feed Towards Intestinal Microflora of Layer. IOSR J Agric Vet Sci 2014; 7(11): 43-7.
[http://dx.doi.org/10.9790/2380-07113]

[88]   Dibaji SM, Seidavi A, Asadpour L, Moreira da Silva F. Effect of a synbiotic on the intestinal microflora of chickens. J Appl Poult Res 2014; 23(1): 1-6.
[http://dx.doi.org/10.3382/japr.2012-00709]

[89]   Faseleh Jahromi M, Wesam Altaher Y, Shokryazdan P, *et al.* Dietary supplementation of a mixture of Lactobacillus strains enhances performance of broiler chickens raised under heat stress conditions. Int J Biometeorol 2016; 60(7): 1099-110.
[http://dx.doi.org/10.1007/s00484-015-1103-x] [PMID: 26593972]

[90]   Ghareeb K, Awad WA, Mohnl M, *et al.* Evaluating the efficacy of an avian-specific probiotic to reduce the colonization ofCampylobacter jejuni in broiler chickens. Poult Sci 2012; 91(8): 1825-32.
[http://dx.doi.org/10.3382/ps.2012-02168] [PMID: 22802174]

[91]   Oh JK, Pajarillo EAB, Chae JP, Kim IH, Yang DS, Kang D-K. Effects of Bacillus subtilis CSL2 on the composition and functional diversity of the faecal microbiota of broiler chickens challenged with Salmonella Gallinarum. J Anim Sci Biotechnol 2017.
[http://dx.doi.org/10.1186/s40104-016-0130-8]

[92]   Park JH, Kim IH. The effects of the supplementation of Bacillus subtilis RX7 and B2A strains on the performance, blood profiles, intestinal Salmonella concentration, noxious gas emission, organ weight and breast meat quality of broiler challenged with Salmonella typhimuri. J Anim Physiol Anim Nutr (Berl) 2015; 99(2): 326-34.
[http://dx.doi.org/10.1111/jpn.12248] [PMID: 25244020]

[93]   Saint-Cyr MJ, Guyard-Nicodème M, Messaoudi S, Chemaly M, Cappelier J-M, Dousset X, *et al.* Recent Advances in Screening of Anti-Campylobacter Activity in Probiotics for Use in Poultry. Front Microbiol 2016.
[http://dx.doi.org/10.3389/fmicb.2016.00553]

[94]   Fritts CA, Kersey JH, Motl MA, *et al.* Bacillus subtilis C-3102 (Calsporin) Improves Live Performance and Microbiological Status of Broiler Chickens. J Appl Poult Res 2000; 9(2): 149-55.
[http://dx.doi.org/10.1093/japr/9.2.149]

[95]   Nishiyama K, Seto Y, Yoshioka K, Kakuda T, Takai S, Yamamoto Y, *et al.* Lactobacillus gasseri SBT2055 reduces infection by and colonization of Campylobacter jejuni. PLoS One 2014.
[http://dx.doi.org/10.1371/journal.pone.0108827]

[96]   Ebrahimi H, Rahimi S, Khaki P, Grimes JL, Kathariou S. The effects of probiotics, organic acid, and a medicinal plant on the immune systemand gastrointestinal microflora in broilers challenged with Campylobacter jejuni. Turk J Vet Anim Sci 2016; 40: 329-36.
[http://dx.doi.org/10.3906/vet-1502-68]

[97]   Carter A, Adams M, La Ragione RM, Woodward MJ. Colonisation of poultry by Salmonella Enteritidis S1400 is reduced by combined administration of Lactobacillus salivarius 59 and Enterococcus faecium PXN-33. Vet Microbiol 2017; 199: 100-7.
[http://dx.doi.org/10.1016/j.vetmic.2016.12.029] [PMID: 28110775]

[98]   Forkus B, Ritter S, Vlysidis M, Geldart K, Kaznessis YN. Antimicrobial Probiotics Reduce Salmonella enterica in Turkey Gastrointestinal Tracts. Sci Rep 2017.
[http://dx.doi.org/10.1038/srep40695]

[99]   Dhama K, Karthik K, Tiwari R, *et al.* Listeriosis in animals, its public health significance (food-borne zoonosis) and advances in diagnosis and control: a comprehensive review. Vet Q 2015; 35(4): 211-35.
[http://dx.doi.org/10.1080/01652176.2015.1063023] [PMID: 26073265]

[100]   Ritzi MM, Abdelrahman W, Mohnl M, Dalloul RA. Effects of probiotics and application methods on performance and response of broiler chickens to an Eimeria challenge. Poult Sci 2014; 93(11): 2772-8.
[http://dx.doi.org/10.3382/ps.2014-04207] [PMID: 25214558]

[101]   Erdoğmuş SZ, Gülmez N, Findik A, Şah H, Gülmez M. Probiyotiklerin Eimeria tenella İle Enfekte Broiler Piliçlerin Sağlık Durumu ve Verim Performansı Üzerine Etkileri. Kafkas Univ Vet Fak Derg 2018.
[http://dx.doi.org/10.9775/kvfd.2018.20889]

[102]   Śliżewska K, Cukrowska B, Smulikowska S, Cielecka-Kuszyk J. The Effect of Probiotic Supplementation on Performance and the Histopathological Changes in Liver and Kidneys in Broiler Chickens Fed Diets with Aflatoxin B1. Toxins (Basel) 2019; 11(2): 112.
[http://dx.doi.org/10.3390/toxins11020112] [PMID: 30781814]

[103]   Cox CM, Dalloul RA. Immunomodulatory role of probiotics in poultry and potential *in ovo* application. Benef Microbes 2015; 6(1): 45-52.
[http://dx.doi.org/10.3920/BM2014.0062] [PMID: 25213028]

[104]   Sobczak A, Kozłowski K. The effect of a probiotic preparation containing Bacillus subtilis ATCC PTA-6737 on egg production and physiological parameters of laying hens. Ann Anim Sci 2015; 15(3): 711-23.
[http://dx.doi.org/10.1515/aoas-2015-0040]

[105]   Yörük MA, Gül M, Hayirli A, Macit M. The effects of supplementation of humate and probiotic on egg production and quality parameters during the late laying period in hens. Poult Sci 2004; 83(1): 84-8.
[http://dx.doi.org/10.1093/ps/83.1.84] [PMID: 14761088]

[106]   Youssef AW, Hassan HMA, Ali HM, Mohamed MA. Effect of Probiotics, Prebiotics and Organic Acids on Layer Performance and Egg Quality. Asian J Polit Sci 2013; 7(2): 65-74.
[http://dx.doi.org/10.3923/ajpsaj.2013.65.74]

[107]   Ribeiro V, Albino LFT, Rostagno HS, Barreto SLT, Hannas MI, Harrington D, *et al.* Corrigendum to "Effects of the dietary supplementation of Bacillus subtilis levels on performance, egg quality and excreta moisture of layers" [Anim. Feed Sci. Technol. 195 [2014] 142–146]. Anim Feed Sci Technol 2015; 209: 286.
[http://dx.doi.org/10.1016/j.anifeedsci.2015.09.009]

[108]   Afsari M. Effects of Dietary Inclusion of Olive Pulp Supplemented with Probiotics on Productive Performance, Egg Quality and Blood Parameters of Laying Hens. Annu Res Rev Biol 2014; 4(1): 198-211.
[http://dx.doi.org/10.9734/ARRB/2014/5212]

[109]   Ramasamy K, Abdullah N, Wong MCVL, Karuthan C, Ho YW. Bile salt deconjugation and cholesterol removal from media by *Lactobacillus* strains used as probiotics in chickens. J Sci Food Agric 2010; 90(1): 65-9.
[http://dx.doi.org/10.1002/jsfa.3780] [PMID: 20355013]

[110]   Ramasamy K, Abdullah N, Jalaludin S, Wong M, Ho YW. Effects of *Lactobacillus* cultures on performance of laying hens, and total cholesterol, lipid and fatty acid composition of egg yolk. J Sci Food Agric 2009; 89(3): 482-6.
[http://dx.doi.org/10.1002/jsfa.3477]

[111]   Aalaei M, Khatibjoo A, Zaghari M, Taherpour K, Akbari Gharaei M, Soltani M. Comparison of single- and multi-strain probiotics effects on broiler breeder performance, egg production, egg quality and hatchability. Br Poult Sci 2018; 59(5): 531-8.
[http://dx.doi.org/10.1080/00071668.2018.1496400] [PMID: 29976078]

[112] Tang SGH, Sieo CC, Kalavathy R, *et al.* Chemical Compositions of Egg Yolks and Egg Quality of Laying Hens Fed Prebiotic, Probiotic, and Synbiotic Diets. J Food Sci 2015; 80(8): C1686-95.
[http://dx.doi.org/10.1111/1750-3841.12947] [PMID: 26174350]

[113] Arif M, Iram A, Bhutta MAK, *et al.* The biodegradation role of Saccharomyces cerevisiae against harmful effects of mycotoxin contaminated diets on broiler performance, immunity status, and carcass characteristics. Animals (Basel) 2020; 10(2): 238.
[http://dx.doi.org/10.3390/ani10020238] [PMID: 32028628]

[114] Alloui MN, Szczurek W, Świątkiewicz S. The Usefulness of Prebiotics and Probiotics in Modern Poultry Nutrition: a Review / Przydatność prebiotyków i probiotyków w nowoczesnym żywieniu drobiu – przegląd. Ann Anim Sci 2013; 13(1): 17-32.
[http://dx.doi.org/10.2478/v10220-012-0055-x]

[115] Kuruti K, Nakkasunchi S, Begum S, Juntupally S, Arelli V, Anupoju GR. Rapid generation of volatile fatty acids (VFA) through anaerobic acidification of livestock organic waste at low hydraulic residence time (HRT). Bioresour Technol 2017; 238: 188-93.
[http://dx.doi.org/10.1016/j.biortech.2017.04.005] [PMID: 28433907]

[116] Giannenas I, Papadopoulos E, Tsalie E, *et al.* Assessment of dietary supplementation with probiotics on performance, intestinal morphology and microflora of chickens infected with Eimeria tenella. Vet Parasitol 2012; 188(1-2): 31-40.
[http://dx.doi.org/10.1016/j.vetpar.2012.02.017] [PMID: 22459110]

[117] Yang CM, Cao GT, Ferket PR, *et al.* Effects of probiotic, Clostridium butyricum, on growth performance, immune function, and cecal microflora in broiler chickens. Poult Sci 2012; 91(9): 2121-9.
[http://dx.doi.org/10.3382/ps.2011-02131] [PMID: 22912445]

[118] Liao XD, Ma G, Cai J, *et al.* Effects ofClostridium butyricum on growth performance, antioxidation, and immune function of broilers. Poult Sci 2015; 94(4): 662-7.
[http://dx.doi.org/10.3382/ps/pev038] [PMID: 25717087]

[119] Murarolli VDA, Burbarelli MFC, Polycarpo GV, Ribeiro PAP, Moro MEG, Albuquerque R. Prebiotic, probiotic and symbiotic as alternative to Antibiotics on the Performance and Immune Response of Broiler Chickens. Rev Bras Cienc Avic 2014; 16(3): 279-84.
[http://dx.doi.org/10.1590/1516-635x1603279-284]

[120] Hasanzadeh M, Tolouei T, Nikbakht G, Alkaragoly H, Rezaei Far A, Ghahri H. Efficacy of Echinacea purpurea and protexin on systemic and mucosal immune response to Newcastle diseases virus vaccination [VG/GA strain] in commercial turkey poults. Iran J Vet Med 2017; 11(1): 85-95.

[121] Landy N, Kavyani A. Effects of using a multi-strain probiotic on performance, immune responses and cecal microflora composition in broiler chickens reared under cyclic heat stress condition. Iran J Appl Anim Sci 2013; 3(4): 703-8.

[122] Al-Fataftah AR, Abdelqader A. Effects of dietary Bacillus subtilis on heat-stressed broilers performance, intestinal morphology and microflora composition. Anim Feed Sci Technol 2014; 198: 279-85.
[http://dx.doi.org/10.1016/j.anifeedsci.2014.10.012]

[123] Hu L, Shao Y, Jiang N, Gao X, Liu C, Lv X, *et al.* Effects of probiotic on the expression of IL-7 gene and immune response to Newcastle disease vaccine in broilers. Int J Health Sci Res 2016; 4: 140-8.

[124] Rewatkar H, Wankhede S, Agashe J, Padole R, jadhao A, Jadhao G. The Effect of Supplementation of Oregano Oil and Probiotic on Intestinal Microbes (E. coli spp, Salmonella spp, Clostridia spp.) of the Broiler Chicken. Int J Livest Res 2019; 1(0): 1.
[http://dx.doi.org/10.5455/ijlr.20190504063600]

[125] Lee KW, Lillehoj HS, Jang SI, *et al.* Effect of Bacillus-based direct-fed microbials on Eimeria maxima infection in broiler chickens. Comp Immunol Microbiol Infect Dis 2010; 33(6): e105-10.

[http://dx.doi.org/10.1016/j.cimid.2010.06.001] [PMID: 20621358]

[126] Rajput IR, Li LY, Xin X, *et al.* Effect of Saccharomyces boulardii and Bacillus subtilis B10 on intestinal ultrastructure modulation and mucosal immunity development mechanism in broiler chickens. Poult Sci 2013; 92(4): 956-65.
[http://dx.doi.org/10.3382/ps.2012-02845] [PMID: 23472019]

[127] Aalaei M, Khatibjoo A, Zaghari M, Taherpou K, Akbari-Gharaei M, Soltani M. Effect of single- and multi-strain probiotics on broiler breeder performance, immunity and intestinal toll-like receptors expression. J Appl Anim Res 2019; 47(1): 236-42.
[http://dx.doi.org/10.1080/09712119.2019.1618311]

Phytogenic Substances: A Promising Approach Towards Sustainable Aquaculture Industry

Abdelrazeq M. Shehata[1,2]**, Abdel-Moneim E. Abdel-Moneim**[3]**, Ahmed G. A. Gewida**[1]**, Mohamed E. Abd El-Hack**[4]**, Mahmoud Alagawany**[4] **and Mohammed A. E. Naiel**[5,*]

[1] *Department of Animal Production, Faculty of Agriculture, Al-Azhar University, Cairo 11651, Egypt*

[2] *Department of Dairy Science & Food Technology, Institute of Agricultural Sciences, Banaras Hindu University, Varanasi 221005, India*

[3] *Biological Applications Department, Nuclear Research Center, Egyptian Atomic Energy Authority, 13759, Egypt*

[4] *Department of Poultry, Faculty of Agriculture, Zagazig University, Zagazig 44519, Egypt*

[5] *Department of Animal Production, Faculty of Agriculture, Zagazig University, Zagazig 44519, Egypt*

Abstract: The aquaculture industry has shown rapid growth over the last three decades, especially with improving the farming systems. However, the rapid expansion and intensification practices in the aquaculture sector have been marred by increased stress levels and disease outbreaks, and subsequently, high fish mortality. Excessive use of veterinary drugs and antibiotics in aquaculture poses a great threat to human and aquatic animals' health, as well as to the biosystem. Furthermore, exposure to various pollutants such as industrial effluents and agricultural pesticides may cause devastating toxicological aspects of fish and adversely affect their health and growth. Besides, with a growing world population, there is a growing interest in intensifying aquaculture production to meet the global demand for nutritional security needs. Uncontrolled intensification of aquaculture production makes aquatic animals both vulnerable to, and potential sources of a wide range of hazards include pathogen transmission, disease outbreak, immunosuppression, impaired growth performance, malnutrition, foodborne illness, and high mortality. Plant-derived compounds are generally recognized as safe for fish, humans, and the environment and possess great potential as functional ingredients to be applied in aquaculture for several purposes. Phytogenic additives comprise a wide variety of medicinal plants and their bioactive compounds with multiple biological functions. The use of phytogenic compounds can open a promising approach towards enhancing the health status of aquatic animals. However, further *in-vivo* trials are necessary under favorable conditions with controlled amounts of identi-

* **Corresponding author Mohammed A. E. Naiel:** Department of Animal Production, Faculty of Agriculture, Zagazig University, Zagazig 44519, Egypt; E-mail: mohammednaiel.1984@gmail.com

Mohamed E. Abd El-Hack & Mahmoud Alagawany (Eds.)

fied bioactive compounds along with toxicity testing for fish safety towards a realistic evaluation of the tested substance efficacy.

Keywords: Antibiotic alternative, Antimicrobial activity, Antioxidant activity, Aquaculture, Gut microbiota, Immunostimulants, Phytogenic.

INTRODUCTION

Fish supply worldwide mainly depends on aquaculture, which presents about two-thirds of the total fish market. The growth of this sector has been predicted to increase by more than 60% between 2010 and 2030 [1]. However, the demand for animal proteins, particularly from fish sources, is forecast to increase with world population growth. Consequently, intensification of aquaculture operations has become necessary to meet the market demands of fish products. However, the application of this practice may entail some challenges, such as crowding, stress, and the rapid spread of diseases, which make aquatic animals more vulnerable to different stressors, resulting in immunosuppression and high mortality, and consequently significant economic losses.

To avoid economic losses caused by different factors in aquaculture, producers tend to use several veterinary medicinal products to control various diseases and improve the health status of fish. The regular administration of these chemical products to the ponds through feed additives or other routes may affect the ecosystem and fish and human health *via* their undegradable residual fractions in the biological system. In these circumstances, there was an urgent need to improve the health status of fish by stimulating growth, the immune system, and resistance to various diseases and environmental stressors with natural and inexpensive alternatives. For decreasing the global concern about this health and environmental safety risk, the use of phytochemicals in aquaculture presents an environmental-friendly substitute. It contributes to more sustainable aquaculture production systems due to the diversity of biological functions of these compounds and their ability to biodegradation in the ecosystem. Plant-derived compounds are generally recognized as safe for fish, humans, and the environment and possess great potential as functional ingredients to be applied in aquaculture for several purposes [2].

Due to the presence of various molecules in plant products with different chemical structures and the possibility of synergetic effect among these molecules, plant extracts and other phytochemicals can effectively improve fish health with various mechanisms at the same time [3]. The use of phytogenic feed additives contribute to modulating gut microbiota [4], improving gut functions [5], boosting the immune system [6], triggering disease resistance [7 - 9] and alleviating the inflammation and oxidative stress [10, 11], leading to enhancement of growth performance [7, 12, 13]. Furthermore, the use of plant-derived compounds can be used for the treatment and prevention of parasitic and fungal diseases in aquatic animals *via* their application in therapeutic baths [14, 15].

With banning antibiotic growth promoters' usage and growing concerns about superbugs, the search for novel alternatives to reduce antibiotic use in the aqua feed will significantly increase in the coming years. In this chapter, we reviewed scientific evidence that phytogenic compounds promote growth performance, boost immunity functions, mitigate oxidative stress and inflammation, maintain gut health, enhance beneficial intestinal bacteria, and reduce the adverse effects of pathogens in aquaculture (Table **1**).

Table 1. Potential of phytogenic feed additives in aquaculture.

Fish Species	Phytogenic Additives		Form/Part Utilized	Dietary Supplementation and Dose	Duration	Growth-promoting Activity	Immunomodulatory Activity	Antioxidant Activity	Antimicrobial Activity	Challenge with Pathogen	Ref.
	Scientific name	Common/ Commercial name									
Rainbow trout [*Oncorhynchus mykiss*]	*Origanum onites.*	Cretan oregano	Essential oil	0.125, 1.5, 2.5 and 3.0 mL/kg	90 days	FCR in fish-fed diets containing 1.5 and 3.0 mL/kg essential oil of *O. onites* was lower than other treatments.	Plasma lysozyme activity was significantly higher in fish-fed diet containing 3.0 mL/kg essential oil of *O. onites.*	Treatment diets did not improve the activity of antioxidant enzymes	Treatment diets improved disease resistance and survival rate.	*L. garvieae*	[21]
	Curcma longa	Curcumin	A yellow pigment derived from the plant *Curcuma longa*	1, 2 and 3%	56 days	The maximal weight gain and specific growth rate occurred at fish fed the diet containing 2% curcumin and feed conversion rate was improved in all treatments than control.	Treatment diets enhanced all the chosen immune parameters.	The highest values for antioxidant parameters were found in the group fed 2% curcumin.	Dietary supplements improved disease resistance and the relative percentage of survivals.	*A. salmonicida* sub sp. *Achromogenes*	[22]
	Zataria multiflora	A thyme-like plant	Hydroalcoholic extract	1, 2 and 3 g/ kg	56 days	Treatment diets did not affect growth; however, the survival rate was significantly improved in fish fed on 2 and 3 g Z. *multiflora*	Humoral immune responses and immune-related gene expressions were significantly increased in fish fed on 2 and 3 g Z. multiflora compared to the control group. Treatment 2g exhibited significantly high mucosal bactericidal activity and mucosal lysozyme activity.	Fish fed diet supplemented with 2g Z. *multiflora* had significantly higher plasma SOD and CAT and lower MDA levels compared to other groups	-	-	[23]
Channel catfish [*Ictalurus punctatus*]	-	Digestarom® P.E.P. MGE.	Essential oils that include carvacrol, thymol, anethol, and limonene.	200 g/ton	42 days	Dietary treatments did not affect weight gain and feed conversion ratio.	In the EO fish, mannose-binding lectin [MBL] levels were similar to non-challenged fish but significantly higher than non-treated fed fish.	The results demonstrate that essential oils improved antioxidant status.	The essential oils improved the survival of channel catfish challenged with *E. ictaluri.*	*Edwardsiellaictaluri*	[24]
African sharptooth catfish [*Clariasgariepinus*]	*Asimina triloba* and *Allium cepa* L.	Pawpaw–onion powder [POP] mixture	Pawpaw seed and onion peel powder	2.5, 5 and 10 g/kg.	60 days	The supplementation exerted no effect on the growth.	A significantly higher lymphocyte count was observed in treatment 10g/kg, whereas the 5.0g/ kg group recorded the highest hepatosomatic value.	Diet supplements have direct effects on the antioxidant response of *Clariasgariepinus*	-	-	[25]

(Table 1) cont.....

Fish Species	Phytogenic Additives		Form/Part Utilized	Dietary Supplementation and Dose	Duration	Growth-promoting Activity	Immunomodulatory Activity	Antioxidant Activity	Antimicrobial Activity	Challenge with Pathogen	Ref.
	Scientific name	Common/ Commercial name									
Common carp [*Cyprinus carpio* L.]	*Origanum onites* L.	Cretan oregano	Essential oil	0, 5, 10, 15, and 20 g/kg	60 Days	The treatments had significantly improved the growth parameters in a dose-dependent regime; meanwhile, the feed conversion ratio was not affected concerning the control group.	The serum protein profile, activities of liver function enzymes, and renal markers were not significantly altered by supplemental diets.	The treatments have improved the antioxidant status.	-	-	[20]
	Psidium guajava	lemon guava	Leaf powder	0, 0.25, 0.5 and 1%	56 days	Growth performance revealed improvement in guava leaf powder [GLP] treated groups.	All levels of GLP increased the total Ig in the serum, while fish fed 0.25% had noticeably higher serum lysozyme activity.	The level of expression of IL-8 was strongly upregulated in fish fed 0.5 and 1% of GLP.	-	-	[26]
	Olea europaea	European olive extract	Leaf	0, 0.1, 0.25, 0.50 and 1%	60 days	There was a significant increase in growth performance, digestive enzyme activities, and growth hormone expression in the 0.1 and OLE 0.25 groups.				-	[27]
Grass carp [*Ctenopharyngodonidella*]	*Curcuma longa*	Turmeric [Curcumin]	Rhizome	0, 196.11, 393.67, 591.46 and 788.52 mg/kg	60 days	The results showed that optimal dietary curcumin [393.67 mg/kg diet] improved weight gain, specific growth rate and reduced feed conversion ratio.	Optimal dietary curcumin increased the activities of lysozyme and acid phosphatase, decreased alanine amino transferase, and aspartate amino transferase activities in serum after injection.	Diet supplementation with 393.67 mg/kgcurcumin enhanced the antioxidant capacity of fish	Dietary curcumin up-regulated the mRNA levels of LYZ, C3 and antimicrobial peptides, and anti-inflammatory cytokines of interleukin-10 [IL-10].	*A. hydrophila*	[28]
Nile tilapia [*Oreochromis niloticus*]		Limonene [L] and thymol [T]	Powder	400 ppm [L], 500 ppm [T] & 400 +500 ppm [LT]	63 days	Fish weight was significantly improved to similar extents by diet [LT] and [L]. Dietary thymol had shown a strong tendency to improve somatic growth.				-	[29]
	Psidium guajava	lemon guava	Leaf	0.25, 0.50, 0.75, and 1.00%	84 days	Supplementation with *P. guajava* leaf extract resulted in significantly higher weight gain and feed intake.	Immune response and antioxidants were improved with PGE inclusion in fish diets as total protein, SOD, glutathione peroxidase and glutathione S-transferase significantly increased.		The challenge test revealed that the highest mortality was recorded in the control.	*A. hydrophila*	[3]
	Cocos nucifera and *Zea mays*	Coconut palm and corn	Virgin coconut oil [VCO] and corn oil [CO]	60 g/kg [VCO] 60g/ kg [CO] 30g/kg [FO]+ 30g/kg VCO[3FVCO] 30g/kg FO + 30g/kg CO [3FCO] 30g/kg VCO + 30g/kg CO [VO]	56 days	Fish fed 3FCO showed higher final weight, percentage weight gain and specific growth rate but not significantly higher than all other groups.	Lysozyme activity was significantly higher in fish fed diet CO while groups FO, 3FCO and VCO recorded the least activities. Although alternative complement activity [ACH50], complement proteins [C3 and C4], was not influenced, antibody titre production was significantly higher in fish fed diet 3FCO and lower in group CO.		Lower mortalities of fish were recorded in groups fed 3FCO and VO after 14 days postchallenge with *A. hydrophila* disease, indicating the enhancing effects of vegetable oils to boost immune response and resistance to disease.	*A. hydrophila*	[30]
	Castanea sativa	Spanish chestnut	Powder	0, 1, 2, 4, and 8 g/kg	56 days	Fish fed with *C. sativa* polyphenols [CSP] enriched diets significantly improved growth and feed conversion ratio.	The effects were already evident four weeks after the CSP administration.	CSP had a strong antioxidant activity, and they may represent a valid and environmentally safe alternative to antibiotics in aquaculture.	The disease protection test displayed that the fish's survival rate was significantly higher.	*Streptococcus agalactiae*	[31]
	Zingiber officinale and *Glycyrrhiza glabra*	Ginger and licorice	Aqueous extract.	5 ml ginger/ kg, 4 ml liquorice/kg and 2.5 ml ginger plus 2 ml liquorice /kg.	120 days	The ginger-licorice mix supply improved the growth performance and feed efficiency.	The mix group increased the hematocrit and hemoglobin, leucocytes, neutrophils, serum total protein, albumin and globulin, but decreased the blood urea nitrogen and creatinine.	There was an increase in hemoglobin and hematocrit values in experimental groups, which may be attributed to the antioxidant properties of both herbal plants.			[32]

(Table 1) cont.....

Fish Species	Phytogenic Additives		Form/Part Utilized	Dietary Supplementation and Dose	Duration	Growth-promoting Activity	Immunomodulatory Activity	Antioxidant Activity	Antimicrobial Activity	Challenge with Pathogen	Ref.
	Scientific name	Common/ Commercial name									
Zebra fish [*Danio rerio*]	*Zingiber officinale*	Ginger	Rhizomes	0, 1, 2 and 3%	56 days	The ginger dietary supplementation showed no significant effect on growth performance in treatment group compared to the control group.	Treatments 2% and 3% ginger increased total protein level, immunoglobulin level and alternative complement activity. Alkaline phosphatase activity increased in all treatment groups, whereas lysozyme activity was higher in fish that received 2% and 3% ginger compared to the control group.	Dietary ginger, especially at 3%, might augment some biochemical responses and expression of genes relevant to antioxidants.	-	-	[33]
European sea bass [*Dicentrarchuslabrax*]	*Origanum onites, Pimpinella anisum* and Citrus L.	[Digestarom PEP M.G.E 150]	Blend of anise, citrus, and oregano essential oils.	10%	60 days	Supplementation of the phytogenic product demonstrated improved performance and nutrient utilization together with increased protein and energy retention.	Supplementation with Digestarom PEP fully compensated for the negative intestinal changes observed in sea bass fed a low-FM diet. Still, it showed little improvement in fish immunological response, except for the 30% increase in lysozyme activity observed in fish fed the low FM-supplemented diet compared to control.	-	-	[34]
Silver catfish [*Rhamdiaquelen*]	Citrus × aurantium	Bitter orange	Citrus aurantium essential oil [EOCA]	0.25, 0.5, 1.0, and 2.0 ml/ kg	60 days	Fish fed with 2.0 ml EOCA per kg exhibited significantly higher growth performance than those fed the control diet.	EOCA improved immune responses.	Dietary addition of EOCA improved antioxidant parameters in silver catfish and could be useful as a dietary supplement.	-	-	[35]

POTENTIAL OF PHYTOGENIC FEED ADDITIVES IN AQUACULTURE

Phytogenic Feed Additives as Appetite Stimulators and Growth Promoters

The excessive use of antibiotics at sub-therapeutic doses in aquafeed as growth promoters leads to the development of antibiotic-resistant bacteria. Moreover, the accumulation of non-biodegradable antibiotics residuals in aquaculture environment also contribute to developing antibiotic resistance and can be associated with a potential threat to the ecosystem and human health [12, 16, 17]. Hence, a growing public interest has been given to develop strategies to replace antibiotics and identify natural alternatives. The use of phytogenic feed additives in aqua feed is a promising and safe alternative for promoting the growth performance and improving the productivity of aquatic organisms [12, 13, 17].

The essential oil obtained from *Citrus sinensis* improved growth performance and reduced tilapia's mortality rate [12]. Supplementation with *Citrus limon* peels essential oil at different concentrations [0.5%, 0.75%, and 1%] in tilapia diet for two months also improved growth performance and decreased mortality rate [7]. The high bioavailability of citrus essential oils and the beneficial activities of their volatile components [geraniol, terpinolene, and γ-terpinene] can explain part of these results. *Oregano* essential oil significantly increased growth performance and weight gain and decreased the feed conversion ratio when supplemented into channel catfish and common carp diets [18, 19]. These beneficial effects of oregano essential oil may be attributed to its high content of polyphenols [*e.g.*, caffeic acid and its esters and other hydroxycinnamic acids and their derivatives] (Fig. **1**). Besides, this growth-promoter activity of oregano essential oil may be accredited to the beneficial effects of carvacrol and thymol, the major components of oregano essential oil, as it has been proven that they can improve intestinal histomorphology and consequently effective feed utilization [20].

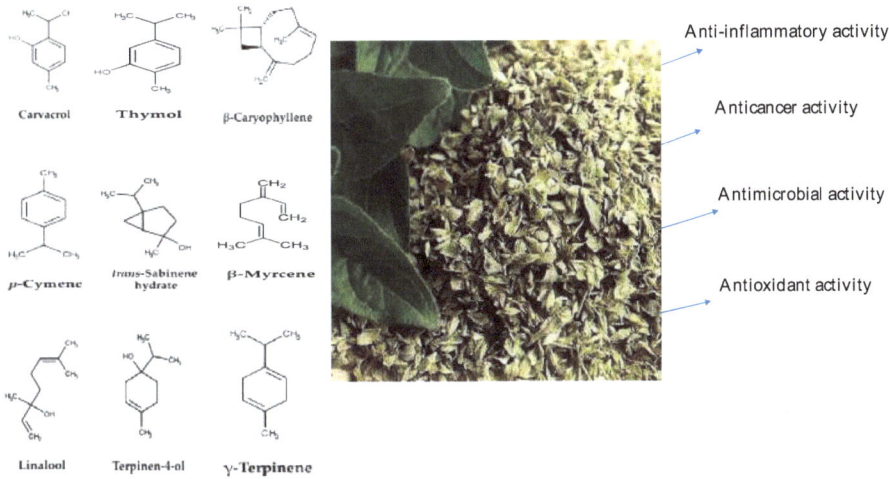

Fig. (1). Essential oils of oregano and their biological activity.

Supplementation of dietary peppermint extract at different levels [2% and 3%] for two months significantly improved weight gain and feed conversion rats of Caspian brown trout, rainbow trout and fry Caspian white fish in a dose-dependent manner [36 - 38]. Diet supplementation with different doses [20 mg/kg and 40 mg/kg] of *Gelsemium elegans* alkaloids for 84 days showed significant improvement in the growth performance of Wuchang bream. In addition, dietary supplementation with 10 gm/kg Thai ginseng powder for 56 days significantly improved the growth performance and survival rate of Nile tilapia [39]. These results demonstrated that fish fed *Gelsemium elegans* supplemented diets had increased lipids and amino acids in both muscle and whole body [40]. In the same context, Adeshina *et al.* [41], reported that dietary supplementation with *Eugenia caryophyllata*, buds extracted twice daily for 84 days, showed a significant increase in feed intake and growth performance of African catfish in a dose-dependent manner. Whereas, Tannins obtained from chestnut wood were observed to significantly improve the weight gain and feed conversion rate of juvenile beluga sturgeon when supplemented in their diet at 0.1% for 42 days [42]. Tannins can act as a growth promoter *via* their ability to upregulate the expression of growth-related genes such as GH and IGF1 [42]. Diet supplementation with a mixture of polyphenols for 56 days showed significant improvement in the growth performance of common carp fish *via* increasing weight gain, final weight, specific growth rate, and survival rate, and decreasing the feed conversion ratio [43]. These results may be attributed to several mechanisms of action, such as the ability of plant extracts to improve the digestibility and nutrient availability *via* improving gut morphology and microbiota or to their capability to enhance the appetite [44].

Moreover, intestinal amylase activity was found to be increased in fish fed dietary peppermint compared to control fish in a dose-dependent manner [36 - 38]. This capability of phytogenic feed additives to stimulate intestinal enzymes secretion can improve digestive and gut microbiota functions and consequently improve body weight and growth performance. Diet supplementation with 1 and 2% Ferula powder for eight weeks significantly upregulated the expression level of appetite and growth genes [Ghrl, GH, and IGF1] of common carp in a dose-dependent manner [11]. In fish fed different levels of dietary Ferula powder up to 2.5% for 63 days, significant improvement in body weight, specific growth rate, survival rate, while the feed conversion rate value was reduced with an increasing dietary Ferula level [45].

Phytogenic Feed Additives as Immunostimulants

Teleost fish have well-developed immune responses. However, studies on the immunomodulatory activity of phytogenic feed additives have been mostly focused on their effects on innate immune responses. Non-specific immune system responses include physical and chemical barriers and cellular and humoral components [46, 47]. The use of dietary phytogenic supplements to enhance fish immunity and disease resistance has gained more interest over the last few years.

Short-term dietary supplementation with thymol essential oils for 15 days enhanced the innate immune responses of Nile tilapia [48]. In the tilapia diet, supplementation with *C. sinensis* essential oil for two months improved lysozyme and myeloperoxidase activities and hemato-immunological parameters [12]. Dietary inclusion of essential oil obtained from *C. limon* peels at different concentrations [0.5%, 0.75%, and 1%] were found to promote the innate immunity and resistance to disease in tilapia fish [7]. In addition, Ngugi *et al.* [6], reported that diet supplemented with C. limon essential oil up to 5% stimulated the immunological responses in juvenile fish.

The complement system has a crucial role in non-specific immunity and is a major contributor to the specific immune responses in aquaculture organisms; therefore, its stimulation by phytogenic feed additives is a novel approach in aqua feed [49]. Moreover, the complement system also can directly kill pathogens *via* pore-forming on their surface membranes or *via* activating phagocytosis [50]. The activation of the complement system was associated with the dietary supplementation with medicinal plants and their derivatives. The complement system's rapid and efficient activation was shown in Nile tilapia supplemented with dietary peppermint and tea tree essential oils for two months, indicating the immunomodulatory effect of phytogenic feed additives [5]. Lysozyme is a lytic enzyme and a non-specific immune element that plays an essential role in the defense against pathogens. Phytogenic feed additives showed enhancement in the complement system and lysozyme activity in golden pompano, rainbow trout, Caspian white fish [36, 51 - 53]. Tan *et al.* [52], displayed that dietary dandelion significantly modulated the expression of selected genes related to lysozyme activity. These results also indicate that fish fed diets supplemented with phytogenic feed additives may show a better response during the exposure to different pathogens in the aquaculture environment or exhibit a better response to immunization with vaccines *via* activation of the complement immune system.

One of the most important targets of dietary immunostimulants is to provide resistance to pathogenic microorganisms by stimulating the immune responses, particularly in the mucosal immune system that directly is in contact with the external environment [46, 47]. Dietary supplementation with dandelion extracts stimulated intestinal immune responses *via* increasing goblet cell numbers and modulating mRNA levels of immune-related genes of juvenile golden pompano [52]. It has also been reported that *G. elegans* alkaloids can regulate mRNA levels of cytokine-related genes in the intestine, liver, spleen, and head kidney of fish fed supplemented diets for 12 weeks [40]. These results suggest the potential of the plant extract to enhance fish immune systems. In a six-week feeding trial, chestnut tannins significantly stimulated the immune system of juvenile beluga *via* enhancing the activity of skin and mucus lysozyme [42]. Inclusion of peppermint extracts [1%, 2%, and 3%] for 60 days in the diet of Caspian brown trout fish also triggered local [skin mucus] and systemic immune responses [38]. A significant increase in mRNA levels of mucosal immune genes of common carp [TNF2-α, IL1B, IL8, and LYZ] was found following dietary supplementation with 0.5%, 1%, or 2% Ferula for 56 days [11]. In another study, hemato-immunological analysis also revealed a significant increase in total protein levels, total immunoglobulins concentrations, lysozyme activity, and alternative hemolytic complement activity [ACH50] in the blood of koi carp fed diets supplemented with Ferula for 63 days in a dose-dependent manner [45]. These results indicate that the inclusion of dietary Ferula in aqua feed can enhance the humoral innate immune responses in supplemented fish. Moreover, Ferula is considered an important source of organo sulfur compounds, and these compounds' immunomodulatory and anti-inflammatory activities have been observed [54]. Diet supplementation with a combination of two medicinal plants

[*Astragalus* 0.5% and *Ganoderma* 0.5%] for 35 days boosted respiratory burst activity, phagocytosis, lysozyme activity, and serum antibody titers level in carp fish vaccinated with *A. hydrophila* [55]. Stimulation of lysozyme activity can enhances disease resistance *via* bonding to bacterial cell wall peptidoglycans, leading to damage of bacterial cytoplasmic membrane and cell lysis [56]. Serum and mucosal immunity and disease resistance were increased in tilapia and carp after feeding *Boesenbergia rotunda* or *Psidium guajava* powder for 8 weeks [26, 39]. Immune responses and disease resistance were also enhanced in Barramundi fish and fried Caspian white fish when peppermint was included in their diets [36, 57]. Eugenol extracted from clove buds was demonstrated to enhance resistance against pathogens and immune responses in African catfish after feeding the supplemented diet for 12 weeks [41].

These results strongly indicate the immunomodulatory role of dietary phytogenic in aquaculture and its contribution to improving fish health and disease resistance [40]. Besides, they are widely available and not expensive, and the by-products are also available. Furthermore, the excellent biodegradability of phytochemicals makes them friendlier to the aquaculture environment than chemotherapeutic agents.

Phytogenic Feed Additives as Natural Antioxidant Agents

Reactive oxygen/nitrogen species are important in the physiological control of cell function. Regulated generation of free-radical species and the balance between ROS production and neutralization by endogenous cellular defensive machinery are necessary for an organism's health [58]. The accumulation of the different pollutants in aquatic ecosystems such as toxic metals, agricultural pesticides, and other chemicals added to ponds for management purposes presents a major public health concern [59 - 61]. Exposure of aquatic organisms to these pollutants induces the generation of reactive oxygen species [ROS] in increased levels, resulting in oxidative damage in the different tissues, including DNA damage, lipid peroxidation, and protein degradation [58]. Moreover, with the growth of intensification practices in aquaculture, an increase in stress incidence has been detected in aquatic animals [62].

Triggering redox status *via* phytogenic feed additives with antioxidant activity can control the excessive production of ROS and repair the oxidative damage in different tissues of fish. Antioxidants are arguably a group of bioactive substances of great interest among plant bioactive compounds. The antioxidant activity of plant-derived molecules seems to be associated with their bioactive constituents, such as polyphenols and flavonoids [10]. The antioxidant activity *in-vitro* of plant-derived compounds has been reported [63 - 65]. Furthermore, the use of medicinal plants and their extracts in human and animal diets as a natural antioxidant has been long known [66 - 69]. Evidence from recent studies demonstrated the antioxidant activity of these natural additives in the aquaculture industry on a worldwide basis [42, 43, 70].

Antioxidant effects of dandelion *in-vitro* and *in-vivo* have been demonstrated [71 - 73]. Inclusion of dandelion in the diet of golden pompano for eight weeks showed significant improvement in intestinal antioxidants by increasing the activity of antioxidant enzymes and modulating mRNA expression of antioxidant-related genes [52]. A 12-weeks trial carried out by Ye *et al.* [40], has also reported that diet supplementation with *G. elegans* alkaloids [20 and 40 mg/kg] improved the antioxidant status in plasma, liver, and intestine of Wuchang bream *via* elevating the activity of antioxidant enzymes and mRNA levels of related genes. Inclusion of 1 and 2% Ferula powder in carp diet for 60 days showed a significant upregulation of antioxidant enzyme [GSR and GSTA] genes expression in a dependent-dose manner [11]. The antioxidant activity of Ferula has also been reported in several *in-vitro* studies on rats [53, 74, 75]. The antioxidant activity of Ferula seems to be attributed to its content of polyphenols. The antioxidant activity of phenolic compounds obtained from different plant sources has been demonstrated in several studies [76]. There are several mechanisms of phenols to act as antioxidants, such as hydrogen donor capacity, metal chelating, or their capability to scavenge free radicals [10, 77]. Dietary supplementation with tannins at a concentration of 0.05% and 0.1% for six weeks significantly increased antioxidant enzymes activity, catalase [CAT], superoxide dismutase [SOD], and polyphenol oxidase [PPO] in Beluga juveniles [42]. Recent evidence suggests that serum albumin is among the most critical antioxidant defences that organism has developed to prevent oxidative stress [78]. This antioxidant activity of serum albumin is fundamentally related to its potential to chelate metal ions and scavenge free radicals. Elevated serum albumin levels in fish fed diet supplemented with chestnut tannins [42] may explain part of the antioxidant activity of dietary tannins. The activity of CAT and peroxidase was also significantly elevated in common carp juveniles fed diets supplemented with a mixture of polyphenols for 56 days [43]. In the same context, the essential oil obtained from *Origanum heracleoticum* L. was found to act as an effective antioxidant when administered in channel catfish [19] and common carp diets [18] *via* increasing the activity of antioxidant enzymes.

Phytogenic Feed Additives as Modulators of Gut Health and Microbiota

It is now well established from various studies over the past decades that the gut microbiome plays a central role in host health. Gut microbiota and its metabolites contribute to animal metabolic capabilities, immune responses, and growth performance [79]. Fish possess high diversity and a dynamic community of gut microbiota, consisting of various bacterial genera profiles. Proteobacteria is the core bacterial community in all fish species, followed by Firmicutes, Bacteroidetes, and Actinobacteria [79]. Gut microbiota can modulate the expression of genes involved in the development of the gastrointestinal tract, nutrient digestion and uptake, host metabolic pathway, and innate immunity [80]. Promoting the relative abundance of these beneficial bacteria and reducing the pathogens enhances fish health and is associated with gut homeostasis [9]. While gut dysbiosis, an imbalance in the intestinal bacteria, was associated with the incidence of several diseases and metabolic disorders [80]. Shaping the composition of the gut microbiota of fish is influenced by dietary, genetic, and environmental factors. Also, the interaction of gut microbiota with the host immune system affects both the composition of gut microbiota and the immune system's development.

In recent years, phytogenic feed additives have been gained considerable interest in the aquaculture industry due to their capacity to improve fish performance by maintaining gut homeostasis [81]. Generally, phytogenic feed additives have been found to exert antimicrobial activity towards pathogens than beneficial bacteria [2, 82, 83]. Thus, the antimicrobial activity of phytogenic feed additives seems to be one of the critical mechanisms of action to modulate the gut microbial ecology [84]. Several studies have demonstrated the positive effects of phytochemicals obtained from different plant sources on gut microbiota. However, there are a few studies on the gut composition of the beneficial bacteria of fish [2].

Supplementation of macro-algae in shrimp diet improved gut microbiota by promoting the relative abundance of Bacteroidetes, Firmicutes, and Bacillaceae and reducing the abundance of harmful bacteria such as bacteria Gammaproteo bacteria and Vibrionaceae that associated with microbial dysbiosis [4]. Among the different microbial families, Rhodobacteraceae had a higher abundance in shrimp gut, followed by Flavobacteriaceae, when shrimp fed diet supplemented with different species of seaweed [4, 9]. These bacterial families are a dominant group in the shrimp gut during all life stages and play an important role in eliminating harmful bacterial species. However, the abundance of their members was affected by diet, environment, and diseases [9]. These results indicate that the inclusion of seaweeds in shrimp diet can effectively maintain gut homeostasis *via* enhancing the abundances of beneficial bacteria and decreasing those detrimental species.

Furthermore, it has previously been observed that plant-derived compounds can alter the diversity and abundance of dominant bacterial communities in the fish gut [9, 40]. The relative abundances of Proteobacteria, Bacteroidetes, and Verrucomicrobia were significantly increased. The relative abundance of

Firmicutes was significantly reduced in the gut of Wuchang bream after feeding on diets supplemented with different levels [5, 10, 20, and 40 mg/kg] of *G. elegans* alkaloids for 12 weeks compared to those fed control diet [40]. These elevations in the diversity and richness of Wuchang bream gut bacterial communities due to supplementation with *G. elegans* alkaloid can effectively maintain gut microbiota homeostasis. Lactic acid bacteria in fish gut as a vital component of the gut microbiota could play an essential role in gut homeostasis. They can prevent pathogenic bacteria [85]. The anti-pathogenic effect of lactic acid bacteria is due to the linear combination of different action mechanisms [86, 87]. Hence, enhancing the count of lactic acid bacteria in the fish gut can effectively improve the microbial balance in the gastrointestinal tract and improve gut health. The ratio of lactic acid bacteria to total viable counts was significantly increased [1.4, 1.86, 2.24, 2.46, and 2.59] times in the intestinal of koi carp with feeding on different levels of dietary Ferula [0.5, 1, 1.5, 2, and 2.5%] for 63 days in a dose-dependent manner [45]. This stimulatory effect on the population of lactic acid bacteria may be attributed to the crude fiber content of Ferula.

The gut is an essential functional organ, the primary function of which is to digest and absorb food-derived nutrients. In addition to this primary function, the gut acts as a chemical and physical barrier to spatially segregate exogenous microorganisms and host immune cells [88]. Intestinal barriers contribute to maintaining gut homeostasis and the symbiotic relationship between the intestinal microbiota and the host immunity [89, 90]. The dysfunction of mucosal barriers allows pathogens to invade the intestine epithelium, which induces immune responses, resulting in the development of inflammatory. The mucus layer is also involved in maintaining gut haemostasis and is dynamically affected by diet and environmental perturbations. In animals, a healthier intestinal tract is usually associated with improved intestinal histomorphology, such as high intestinal villus and shallow crypts, which effectively increase nutrient absorption capacity and promote growth performance. Phytogenic feed additives can improve gut health by altering intestinal morphology. Common carp fed diet incorporated with different levels of oregano essential oil [5, 10, 15, and 20 g/kg diet] had significantly increased villus height and villus width in a concentration-dependent pattern compared to those fed un-supplemented diet [18]. Also, a dramatic increase in villus length was reported when tilapia fed diet supplemented with thymol [91].

Intestinal histomorphometry analysis of a recent study conducted by Adeshina *et al.* [41], revealed that diet supplementation with *Eugenia caryophyllata*, buds extract significantly improved the intestinal villi length/width and absorption surface of African catfish in a dose-dependent manner [41]. Diet supplementation with different levels of dandelion extract [0.5, 1, 2, 4, and 10 g/ kg] for 56 days improved intestinal histo-morphology of juvenile golden pompano *via* increasing villus height, villus width, and villus number [52]. Moreover, dietary dandelion was found to improve intestinal barrier function and tight junction integrity *via* modulating mRNA expression of tight junction proteins [52]. A recent study conducted by Ye *et al.* [92], for 12 weeks has demonstrated that dietary alkaloids from *G. elegans* [20 and 40 mg/kg] exhibited positive effects on intestinal health of Wuchang bream *via* improving intestine morphometric parameters. While, Nile tilapia fed a diet containing 250 mg/kg tea tree essential oil showed higher villus and larger intestinal surface than those fed control diet [5].

PHYTOTHERAPY AS AN ALTERNATIVE FOR TREATING FISH DISEASE

Over the last three decades, the aquaculture industry showed rapid growth, especially with improving the farming systems. However, the aquaculture sector's rapid expansion and intensification practices have been marred by the increase of stress levels, disease outbreaks, and, subsequently, high fish mortality. Furthermore, unlike terrestrial ecosystems, the spread of diseases in aquaculture ecosystems seems more rapid due to intensification practices and other factors. To maintain fish health in these conditions, plant-derived molecules with antimicrobial properties have been widely used in aqua feed to improve defense systems and disease resistance in aquatic organisms.

Medicinal plants include several bioactive substances that simultaneously possess various biological effects, such as antibacterial, antifungal, and antiviral activities. Unlike chemotherapy, phytotherapy is less likely to develop microbial resistance because the mechanism of action of phytotherapeutic agents affects several targets simultaneously. Moreover, the broad-spectrum antimicrobial activities of phytotherapeutic agents in conjunction with their immunostimulatory activity may also prevent microbial resistance development [93].

Anti-bacterial Activity

Bacterial diseases still present major challenges for rainbow trout, carp, tilapia, and catfish aquaculture [94, 95], particularly under intensive farming systems [96, 97]. Excessive use of antibiotics to promote growth performance and/ or treat diseases in aquaculture has contributed to the global threat of antibiotic resistance in fish pathogens [98]. This global threat cannot be prevented but can nevertheless be controlled and addressed in the most effective way [99]. Traditional approaches to antibiotic discovery have failed to keep pace with the development and widespread antibiotic resistance in bacteria [98]. Recently, the investigation of novel non-antibiotic alternatives for treating and controlling infectious diseases has gained great importance. These alternatives include natural plant extracts and their derivatives [81].

Phytogenic feed additives are generally considered safe and effective against different bacterial strains. The antibacterial activity of phytogenic feed additives has been recognized for many years [84]. Plant extracts include several bioactive compounds that appear to be responsible for their antibacterial activity [100]. For example, phytogenic feed additives containing phenols and phenolic acids were an excellent alternative to antibiotics [84]. Phenolic compounds are secondary metabolites that occur in many plants and exert various biological activities, including antibacterial activity [61, 101]. The use of phenolic compounds as a promising alternative to antibiotics depends on the difficulty of developing bacterial [cross]-resistance to phytotherapeutic agents [102]. The hexane extract of *Hesperozygis ringens* showed a weak antibacterial activity *in-vitro* against *Aeromonas hydrophila*. However, it could increase the survival rate [93·33%] of silver catfish experimentally infected with *A. hydrophila*. This antibacterial effect appears to be related to the phytochemicals content in *H. ringens* extract, such as pulegone, the major compound in this extract [103]. To search for alternative antibiotic agents, Nile tilapia fingerlings fed different levels [0, 1, 2, 4, and 8 g/kg] of polyphenols extracted from *Castanea sativa* for eight weeks. The results showed that aqueous extract of *C. sativa* exhibits antibacterial activity against *Streptococcus agalactiae* with significant declines in mortality, especially in fish fed diets containing a level of 2 g/kg [31]. These results may be attributed to the immunomodulatory activity of dietary polyphenols, including the enhancement of phagocytosis, respiratory burst, lysozyme activity, and complement activity [43, 104 - 106]. Moreover, bioactive plant components have been reported to interact with bacterial cell membrane causing alteration of the membrane proteins functions and consequently increasing membrane permeability and leading to cytoplasmic leakage and cell death [107]. Cysteine-rich peptides isolated from *Sambucus nigra* exhibited antibacterial action against various Gram-negative bacteria *via* disrupting bacterial membrane [108]. Lipophilic substances such as essential oils can easily alter the structure of the phospholipid's membrane of the bacterial cells *via* altering the percentage of unsaturated fatty acids. Penetration of essential oils into Gram-positive bacterial cells contributes to their effect on the bacterial cell membrane and its cytoplasmic granules. One of the most important factors that can potentially affect the antibacterial activity of the bioactive plant components is the structure of the bacterial cell wall. For example, Gram-negative bacteria have a somewhat complex cell compared to Gram-positive bacteria, making it difficult for essential oils to penetrate the bacterial cell and, consequently, increasing its resistance to phytochemicals, including essential oils and phenolic compounds [109]. However, the cell wall of Gram-negative bacteria does not wholly prevent the penetration of lipophilic substances [110]. Essential oils of cinnamon, oregano, clove, and thyme, and their significant constitutes showed *in-vitro* strong antibacterial effect against four *Aeromonas salmonicida* subsp. *salmonicida* bacterial strains. Cinnamaldehyde had the highest efficacy with a minimum inhibitory concentration [MIC] of 62 to 125 μg/ ml [111]. The authors attributed the antibacterial activity of these phytochemicals to their ability to disrupt the bacterial cell membrane by increasing its permeability and alternating the structure of its fatty acids [111]. Furthermore, due to the great variety of bioactive compounds in plant extracts, their antibacterial properties seem to be attributed to multiple mechanisms simultaneously at various sites.

Quorum sensing-related processes mainly mediate the successful establishment of pathogenic bacterial infection. In addition, virulence factors in pathogenic bacteria such as the production of lytic enzymes, biofilm formation, and flagellar motility-dependent swarming are regulated by the expression of quorum sensing-related genes [112, 113]. Biofilm formation plays a critical role in accomplishing drug resistance for several pathogens, such as *Vibrio harveyi*, the most common and well-known aquatic pathogens [114]. Moreover, antibiotic resistance is widespread in biofilm-forming pathogens [112, 115]. Therefore, inhibition of bacterial quorum sensing system and related processes presents a promising target in identifying novel therapeutic agents. Phytochemicals from plants were found to possess an inhibitory effect on quorum sensing in human and animal pathogens [8, 116]. The ability of plant extracts to reduce the swarming and pathogenicity at sub-lethal concentrations also have been reported [113]. Essential oils isolated from aromatic plants had strong antibiofilm and antivirulence activity against clinical and drug-resistant *Staphylococcus aureus* strains [117]. Dietary supplementation with thymol [100 mg/ kg diet] showed a significant decrease in mortality rate grass carp infected by *Aeromonas hydrophila via* prevention of *A. hydrophila*-induced branchial bioenergetics [13]. While supplementation of 1g thymol/ kg diet for two months decreased *Aeromonas* spp. in the gut of rainbow trout [118]. Carvacrol and thymol also showed *in-vitro* antibacterial and antibiofilm effects against *A. hydrophila* isolated from silver catfish [119]. A recent *in-vivo* study showed that 30 min daily baths administration of carvacrol [5 mg/L], terpinen-4-of [5, 10, 15, and 25 mg/L], and thymol [5, 10, 15, and 25 mg/L] in their nano-encapsulated forms for six days prevented mortality in silver catfish infected with *A. hydrophila* [120]. Also, the treatment of infected fish with the pure forms of these compounds showed a significant increase in survival rate [120].

Aerolysin, a pore-forming toxin, plays a central role in the pathogenicity of *A. hydrophila* as an essential virulence factor controlled by quorum sensing related gens [121]. Magnolol, a natural compound, could significantly improve the survival rate in channel catfish exposed to *A. hydrophila* infection as a result of reducing the production of aerolysinvia, inhibiting the transcription of the aerA gene but without significant impact on biofilm formation or quorum sensing signalling pathway [122]. In this sense, thymol showed a significant antibiofilm activity. It significantly reduced the aerolysin production by inhibiting the transcription of genes aerA, ahyI, and ahy R, indicating the ability to inhibit the quorum sensing system at sub-inhibitory concentrations [8]. *in-vivo* study also showed the potential of thymol to reduce the mortality rate of channel catfish exposed to *A. hydrophila* infection [8]. A similar recent study reported that rosmarinic acid possesses an antipathogenic effect against *A. hydrophila* and can reduce the mortality rate of challenged zebra fish *via* down-regulating virulence genes expression and inhibiting quorum sensing system [123].

Safari *et al.* [45], displayed that the skin mucus of fish fed dietary Ferula showed strong resistance to several pathogenic bacteria [45]. Some researchers attributed these results to nitrogenous compounds in skin mucus [*e.g.*, proteases, lysozyme, immunoglobulins, and mucin] [124]. Indeed, as an immunological site, the mucosal immune system possesses the ability to raise immune responses and confer effective resistance to the host against pathogens. Also, it can be attributed to the positive effect of dietary Ferula on the population of beneficial bacteria, particularly lactic acid bacteria [45] that possess a strong antibacterial effect against intestinal pathogenic microbiota [85 - 87].

Anti-viral Activity

The high mortality and morbidity rates due to the infection with viral diseases result in critical economic losses in the aquaculture industry. Vaccination against viral diseases in aquaculture is the most effective strategy for preventing and controlling the various viral diseases [125]. The majority of commercial vaccines in aquaculture are killed whole-cell vaccines. Their common administration route is the intraperitoneal injection, resulting in increased handling stress on fish and high labor cost [126]. However, effective vaccines against viral diseases in aquaculture are still less than required, and this condition poses severe problems in some fish species [127]. Consequently, alternative biocontrol agents with antiviral activity against fish viral diseases are urgently needed as a novel approach. Natural substances possess a variety of structures that enable them to exert antiviral activity *via* different targets and with various mechanisms. The polyphenolic complex of ellagitannins extracted from sea buckthorn [*Hippophae rhamnoides* L.] has been demonstrated to possess antiviral activity against some human and animal pathogenic viruses. The major bioactive substances in sea buckthorn are the phenolic compounds and flavonoids, which appear to play a central role in its antiviral activity [128]. Extracts of sea buckthorn were reported to have high concentrations of flavonol glycosides and their derivatives. Accordingly, extreme antiviral activity has been demonstrated with these extracts

[129]. In addition, ethyl acetate and butanol extracts of the sea buckthorn leaf could be used to develop novel and potent antiviral agents. Among the different antiviral mechanisms of sea buckthorn extracts, their efficacy to inhibit viral neuraminidase [93].

Many published studies reported the antiviral activity of natural flavonoids [130]. The antiviral action of polyphenolic plant products may be attributed to direct interactions with the viral particle [131] or *via* suppressing the initial steps of virus replication without direct interaction with virus particles [132]. Tea polyphenols were found to interact directly with viral particles [133]. Catechins from green tea, including [-]-epigallocatechin gallate [EGCG], [-]-epicatechin gallate [ECG], and [-]-epigallocatechin [EGC], were observed to exert inhibitory activity on hemagglutination, and viral neuraminidase activity and suppress viral RNA synthesis at high concentrations [133].Western blot analysis obtained from an *in-vitro* study showed that ECG inhibits grass carp reovirus, the most pathogenic pathogen of the aquareo viruses, in a concentration-dependent pattern. ECG at a concentration above 20 µg/ml exhibited significant inhibition in the activity of grass carp reovirus [134]. Catechins possessing 3-galloyl group in their skeleton have been reported to play a vital role in their antiviral activity [133].

On the molecular basis, the extract of Cistus plant, a polyphenol-rich plant extract, seems to interact with the virus surface proteins and inhibit its binding to cellular receptors [130]. It has been established that polyphenols possess protein-binding capacity, indicating that polyphenol-rich plants may directly interfere with pathogens *via* such a mechanism.

White Spot Syndrome Virus [WSSV] has been associated with high mortality and economic damage in shrimp-producing countries [9, 135]. The ethanolic extract of *Cynodon dactylon* has been reported to prevent WSSV infection with 100% survival in black tiger shrimp *Penaeus monodon* [136, 137].VP28, a dominant protein in WSSV envelope proteins, plays a key role in shrimp's systemic infection, which serves as a viral-binding protein to the host cell [138, 139]. In this study, the extract of *C. dactylon* was found to be inhibitor of VP28 gene expression, indicating that the extract of *C. dactylon* exerts its antiviral activity against WSSV through direct interaction with the envelope of the virus [140].

The antiviral activity of several synthesized coumarins was evaluated against spring viremia of carp virus [SVCV] in EPC cells. Among the evaluated coumarins, two imidazole coumarins derivative [7-[3-benzimidazole propoxy] coumarin [B4] and 7-[4-[4-methyl-imidazole]]-coumarin [C2]] exhibited inhibitory effect against SVCV replication in fish cells by more than 90% *via* activating the Nrf-2-ARE pathway [141], inhibiting the SVCV-induced mitochondria injury and activating antioxidant enzymes to maintain intracellular redox homeostasis [142]. Based on the findings of this study, a new imidazole coumarin derivative, 7-[4-benzimidazole-butoxy]-coumarin, was synthesized, and its antiviral effect was also examined against SVCV. The new compound showed anti-SVCV greater than that obtained in the previous study. Moreover, it showed significant inhibition of cell death induced by SVCV [125]. Most recently, the authors examined the antiviral activity of imidazole coumarins derivatives C2 and B4 *in-vivo* and found that these groups can positively reduce horizontal transmission of SVCV and both coumarin derivatives showed stability in water for a prolonged inhibitory half-life [3.5 and 4 days, respectively] [143, 144]. A novel synthesized coumarin derivative [2-amino-4-[4-nitrophenyl]-5-oxo-4H, 5H-pyrano[3,2-c]chromene-3-carbonitrile] also significantly inhibited SVCV replication *in-vitro* and *in-vivo* [145]. Liu *et al.* [146], evaluated the efficacy of Lonicera japonica natural components against grouper iridovirus *in-vitro* and *in-vivo*. They reported that these natural components have excellent antiviral effects against grouper iridovirus infection in a dose-dependent manner.

To find safe and efficient natural antiviral products, *Olea europaea* extracts were evaluated for their activity against viral haemorrhagic septicaemia virus. Preincubation of the virus with *O. europaea* extracts before the infection and administration after infection showed *in-vitro* potency to inhibit viral infectivity in a dose-dependent manner [147]. The authors suggested that *O. europaea* extracts can directly interact with the viral envelope by interacting with the surface of a phospholipid bilayer [148], enhancing drastic alterations on the membrane surface and, consequently, inhibiting the early fusion steps [147]. These results indicate that polyphenols and other bioactive compounds isolated from medicinal plants can be used as a promising source of antiviral agents for aquaculture (Fig. **2**).

Anthelminthic [Monogeneans] Activity

For decades, parasites outbreak was one of the significant health obstacles due to the intensification of aquaculture [149]. Monogeneans are parasitic flatworms commonly found on the skin, gills, and eventually the eyes of fish. Some monogeneans are highly pathogenic to cultured finfish, resulting in high host mortality great economic losses in aquaculture [150]. Indeed, control and prevention of monogenean and other parasitic infections in the cultured fish present a significant challenge due to the lack of effective methods [81, 151]. Numerous chemotherapeutic compounds are used in aquaculture to control and prevent monogenean infections, but most of these compounds are toxic to fish and pollutants of the environment [152].

Phenolic acids (hydroxy-benzoic and cinnamic acids)   **Phenolic acids**

Flavonoids   **Stilbenes**   **Anthocyanins**

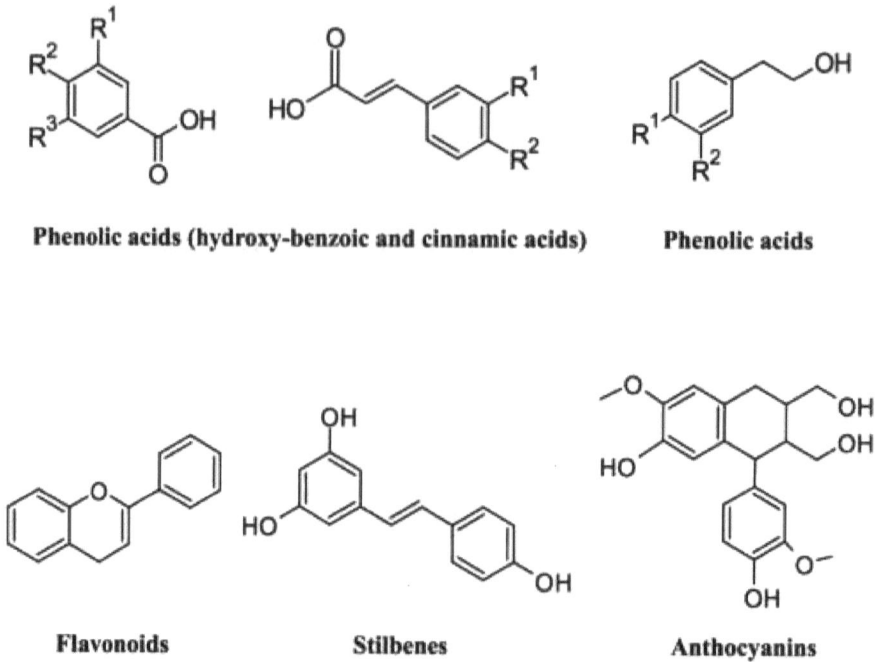

Fig. (2). Chemical structure of different classes of polyphenols.

Consequently, research on the effectiveness and ability of phytogenic feed additives to control monogenean infections has greatly increased in aquaculture. Methanol extract of *Macleaya cordata* showed strong anthelmintic activity against monogenean in goldfish with EC50 and EC90 values of 8.6 and 25.5 mg/L, respectively, after 48-h exposure [152]. However, the toxic concentration [LC50] of methanol extract of *M. cordata* after 48 h of exposure was 3.2 times higher than the effective one [EC90] [150]. Also in goldfish, the chloroform extract of *Fructus Arctii* showed 100% anthelmintic efficacy against *Dactylogyrus intermedius* [Monogenea] at a level of 240 mg/L after 48 h of exposure [153]. Methanolic and aqueous extracts of *Semen aesculi* exhibited great anthelmintic activity against *D. intermedius* [Monogenea] with EC50 values of 5.23 and 6.48 mg/L, respectively, after 48 h of exposure in goldfish [154]. However, a very weak anthelmintic effect of *S. aesculi* extract against *G. kobayashii* was recently observed [152]. These differences may indicate the species-specific anthelmintic activity of *S. aesculi* or maybe due to the different conditions of extraction and exposure.

The anthelminthic activity of essential oils against monogeneans has also been reported [14, 151, 155 - 157]. *Cymbopogon citratus* essential oil showed *in-vitro*

100% parasite mortality against the monogeneans at different levels ranging from 100-500 mg/L depending on the exposure time. However, *in-vivo* assays showed a low efficiency due to the low dose [60 mg/L in therapeutic baths] of *C. citratus* essential oil tolerated by tambaqui fish [14]. de Oliveira Hashimoto *et al.* [156], have evaluated the antiparasitic effect of essential oils of pepper rosemary and peppermint in tilapia parasitized by the monogeneans. The results showed that baths containing 20 mg/L essential oil of pepper rosemary or 40mg/L essential oil of peppermint showed an efficient antiparasitic effect against monogenean parasites. However, hematological alterations were observed in fish treated with pepper rosemary [156]. The antihelminthic activity of essential oil of peppermint also was evaluated *in-vitro* and *in-vivo*. The results of *in-vitro* test indicated the antihelminthic effect of peppermint essential oil in a dose-dependent manner. However, treatment with 40 mg/L peppermint essential oil in a water bath did not show an effective antihelminthic effect against monogeneans and caused tissue alterations in the gills of *Arapaima gigas* [151].

Compared to the essential oils of tea tree and peppermint, *Copaifera oleoresin* possessed stronger efficacy *in-vitro* against monogenean parasites. Tea tree and peppermint essential oils showed 100% anthelmintic efficacy against monogeneans at 400 mg/L, while *C. duckei* showed the same effect only at 100mg/L. In addition, *in-vivo* parasitological analysis showed that 50mg/L in therapeutic bath reduced the parasite loads by 45% in parasitized pacu fish without hematological or histological alterations [155]. The mechanism of action of *Copaiferaoleoresins* [*C. duckei*] seems to be *via* causing swelling and lysis of monogenean parasites [158]. Treatment of naturally parasitized tambaqui juveniles with eugenol at 10 mg/L in a water bath showed a strong antiparasitic effect against monogenean parasites and reduced the parasite loads by 81% [159].

Anti-fungal Activity

Aquaculture systems can be influenced by fungal pathogens threatening fish and human health and causing economic losses. Indeed, a few fungal diseases are found in fish, such as saprolegniasis andicthyophonus disease [102, 160]. Saprolegniasis, caused by *Saprolegnia Parasitica*, is the most common fungal disease in freshwater fish and results in high fish mortality and huge economic losses [160]. Several studies have demonstrated the antifungal activity of plant-derived compounds [15, 160, 161]. Ethanolic extracts of pomegranate and thyme exhibited antifungal activity against *S. diclinain-vitro* at 0.5 mg/ml with 200 and 400 ppm MIC, respectively [15]. Methanolic extracts and essential oils of some plants also showed promising *in-vitro* inhibitory activity against Saprolegnia [160]. Ameen *et al.* [162], have evaluated *in-vitro* fungicidal activity of different levels of the aqueous extracts of various plants against *Ichthyophonus* sp. The results showed that extracts of linseed, henna, and turmeric had antifungal activity against *Ichthyophonus* sp. at the highest levels [20%], whereas the ginger had the inhibitory effect at the lowest concentration [2%] [162].

CONCLUDING REMARKS

Phytogenic molecules can affect the physiological functions of aquatic animals, indicating the potential to be utilized as growth-promoters and immunostimulants in aquaculture. Phytogenic feed additives can modulate gut microbiota and maintain gut homeostasis. Besides, they have the potency to stimulate immune responses and antioxidant enzymes *via* the upregulation of mRNA expression of related gens. In addition, phytogenic compounds can effectively enhance fish disease resistance *via* their antimicrobial activity. The pathogenesis of several diseases in aquatic organisms is rather multi-factorial, not a result of an individual cause; however, the mechanism of action of veterinary drugs, to be more effective in a short time, mainly targets a single cause.

On the contrary, phytogenic compounds can act on different targets simultaneously due to the diversity of their constituents with different mechanisms of action. However, the biological activity of plant-derived molecules is greatly influenced by several factors that should be considered, such as extraction method, synergistic or antagonistic interaction between several used molecules, dose and administration route, and so on. Despite phytogenic feed additives being groups of natural compounds, studies on their mechanisms of action, compatibility with diet, and safety assessment need to be accomplished before they can be used more widely in aquaculture.

CONSENT FOR PUBLICATION

Not applicable.

CONFLICT OF INTEREST

The author declares no conflict of interest, financial or otherwise.

ACKNOWLEDGMENT

All the authors of the manuscript thank and acknowledge their respective Institutes, Universities and Organizations for necessary help in the compilation of this manuscript.

REFERENCES

[1]   The state of world fisheries and aquaculture, Opportunities and challenges. Food and Agriculture Organization of the United Nations Rome. American Fisheries Society 2012.

[2]   Sutili FJ, Gatlin DM III, Heinzmann BM, Baldisserotto B. Plant essential oils as fish diet additives: benefits on fish health and stability in feed. Rev Aquacult 2018; 10(3): 716-26.
[http://dx.doi.org/10.1111/raq.12197]

[3]   Omitoyin BO, Ajani EK, Orisasona O, Bassey HE, Kareem KO, Osho FE. Effect of guava *Psidium guajava* (L.) aqueous extract diet on growth performance, intestinal morphology, immune response and survival of *Oreochromis niloticus* challenged with *Aeromonas hydrophila*. Aquacult Res 2019; 50(7): 1851-61.
[http://dx.doi.org/10.1111/are.14068]

[4]   Niu J, Xie J-J, Guo T-Y, Fang H-H, Zhang Y-M, Liao S-Y, *et al.* Comparison and Evaluation of Four Species of Macro-Algaes as Dietary Ingredients in *Litopenaeus vannamei* Under Normal Rearing and WSSV Challenge Conditions: Effect on Growth, Immune Response, and Intestinal Microbiota. Front Physiol 2019.
[http://dx.doi.org/10.3389/fphys.2018.01880]

[5]   Valladão GMR, Gallani SU, Pala G, *et al.* Practical diets with essential oils of plants activate the complement system and alter the intestinal morphology of Nile tilapia. Aquacult Res 2017; 48(11): 5640-9.
[http://dx.doi.org/10.1111/are.13386]

[6]   Ngugi CC, Oyoo-Okoth E, Muchiri M. Effects of dietary levels of essential oil (EO) extract from bitter lemon (*C itrus limon*) fruit peels on growth, biochemical, haemato-immunological parameters and disease resistance in Juvenile *L abeo victorianus* fingerlings challenged with *A eromonas hydrophila*. Aquacult Res 2017; 48(5): 2253-65.
[http://dx.doi.org/10.1111/are.13062]

[7]   Baba E, Acar Ü, Öntaş C, Kesbiç OS, Yılmaz S. Evaluation of *Citrus limon* peels essential oil on growth performance, immune response of Mozambique tilapia *Oreochromis mossambicus* challenged with *Edwardsiella tarda*. Aquaculture 2016; 465: 13-8.
[http://dx.doi.org/10.1016/j.aquaculture.2016.08.023]

[8]   Dong J, Zhang L, Liu Y, *et al.* Thymol Protects Channel Catfish from *Aeromonas hydrophila* Infection by Inhibiting Aerolysin Expression and Biofilm Formation. Microorganisms 2020; 8(5): 636.
[http://dx.doi.org/10.3390/microorganisms8050636] [PMID: 32349419]

[9]   Schleder DD, Blank M, Peruch LGB, *et al.* Impact of combinations of brown seaweeds on shrimp gut microbiota and response to thermal shock and white spot disease. Aquaculture 2020; 519: 734779.
[http://dx.doi.org/10.1016/j.aquaculture.2019.734779]

[10]   Rice-Evans CA, Miller NJ, Paganga G. Structure-antioxidant activity relationships of flavonoids and phenolic acids. Free Radic Biol Med 1996; 20(7): 933-56.
[http://dx.doi.org/10.1016/0891-5849(95)02227-9] [PMID: 8743980]

[11]   Safari R, Hoseinifar SH, Nejadmoghadam S, Jafar A. Transciptomic study of mucosal immune, antioxidant and growth related genes and non-specific immune response of common carp (Cyprinus carpio) fed dietary Ferula (Ferula assafoetida). Fish Shellfish Immunol 2016; 55: 242-8.
[http://dx.doi.org/10.1016/j.fsi.2016.05.038] [PMID: 27241284]

[12]   Acar Ü, Kesbiç OS, Yılmaz S, Gültepe N, Türker A. Evaluation of the effects of essential oil extracted from sweet orange peel (*Citrus sinensis*) on growth rate of tilapia (*Oreochromis mossambicus*) and possible disease resistance against Streptococcus iniae. Aquaculture 2015; 437: 282-6.
[http://dx.doi.org/10.1016/j.aquaculture.2014.12.015]

[13]   Morselli MB, Baldissera MD, Souza CF, *et al.* Effects of thymol supplementation on performance, mortality and branchial energetic metabolism in grass carp experimentally infected by *Aeromonas hydrophila*. Microb Pathog 2020; 139: 103915.
[http://dx.doi.org/10.1016/j.micpath.2019.103915] [PMID: 31809794]

[14]   Gonzales APPF, Yoshioka ETO, Mathews PD, *et al.* Anthelminthic efficacy of *Cymbopogon citratus* essential oil (Poaceae) against monogenean parasites of *Colossoma macropomum* (Serrasalmidae), and

blood and histopathological effects. Aquaculture 2020; 528: 735500.
[http://dx.doi.org/10.1016/j.aquaculture.2020.735500]

[15]   Mostafa AAF, Al-Askar AA, Taha Yassin M. Anti-saprolegnia potency of some plant extracts against *Saprolegnia diclina,* the causative agent of saprolengiasis. Saudi J Biol Sci 2020; 27(6): 1482-7.
[http://dx.doi.org/10.1016/j.sjbs.2020.04.008] [PMID: 32489284]

[16]   Böger B, Surek M, Vilhena RO, *et al.* Occurrence of antibiotics and antibiotic resistant bacteria in subtropical urban rivers in Brazil. J Hazard Mater 2021; 402: 123448.
[http://dx.doi.org/10.1016/j.jhazmat.2020.123448] [PMID: 32688189]

[17]   Dawood MAO, Koshio S, Esteban MÁ. Beneficial roles of feed additives as immunostimulants in aquaculture: a review. Rev Aquacult 2018; 10(4): 950-74.
[http://dx.doi.org/10.1111/raq.12209]

[18]   Abdel-Latif HMR, Abdel-Tawwab M, Khafaga AF, Dawood MAO. Dietary origanum essential oil improved antioxidative status, immune-related genes, and resistance of common carp (*Cyprinus carpio* L.) to Aeromonas hydrophila infection. Fish Shellfish Immunol 2020; 104: 1-7.
[http://dx.doi.org/10.1016/j.fsi.2020.05.056] [PMID: 32474085]

[19]   Zheng ZL, Tan JYW, Liu HY, Zhou XH, Xiang X, Wang KY. Evaluation of oregano essential oil (*Origanum heracleoticum* L.) on growth, antioxidant effect and resistance against Aeromonas hydrophila in channel catfish (Ictalurus punctatus). Aquaculture 2009; 292(3-4): 214-8. [Ictalurus punctatus].
[http://dx.doi.org/10.1016/j.aquaculture.2009.04.025]

[20]   Abdel-Latif HMR, Abdel-Tawwab M, Khafaga AF, Dawood MAO. Dietary *oregano* essential oil improved the growth performance *via* enhancing the intestinal morphometry and hepato-renal functions of common carp (*Cyprinus carpio* L.) fingerlings. Aquaculture 2020; 526: 735432.
[http://dx.doi.org/10.1016/j.aquaculture.2020.735432]

[21]   Diler O, Gormez O, Diler I, Metin S. Effect of oregano (*Origanum onites* L.) essential oil on growth, lysozyme and antioxidant activity and resistance against *Lactococcus garvieae* in rainbow trout, *Oncorhynchus mykiss* (Walbaum). Aquacult Nutr 2017; 23(4): 844-51. [Walbaum].
[http://dx.doi.org/10.1111/anu.12451]

[22]   Yonar ME, Mişe Yonar S, İspir Ü, Ural MŞ. Effects of curcumin on haematological values, immunity, antioxidant status and resistance of rainbow trout (Oncorhynchus mykiss) against Aeromonas salmonicida subsp. achromogenes. Fish Shellfish Immunol 2019; 89: 83-90.
[http://dx.doi.org/10.1016/j.fsi.2019.03.038] [PMID: 30898618]

[23]   Taheri Mirghaed A, Hoseini SM, Hoseinifar SH, Van Doan H. Effects of dietary thyme (Zataria multiflora) extract on antioxidant and immunological responses and immune-related gene expression of rainbow trout (Oncorhynchus mykiss) juveniles. Fish Shellfish Immunol 2020; 106: 502-9.
[http://dx.doi.org/10.1016/j.fsi.2020.08.002] [PMID: 32810529]

[24]   Peterson BC, Peatman E, Ourth DD, Waldbieser GC. Effects of a phytogenic feed additive on growth performance, susceptibility of channel catfish to Edwardsiella ictaluri and levels of mannose binding lectin. Fish Shellfish Immunol 2015; 44(1): 21-5.
[http://dx.doi.org/10.1016/j.fsi.2015.01.027] [PMID: 25659231]

[25]   Fawole FJ, Adeoye AA, Tiamiyu LO, Samuel FC, Omosuyi OM, Amusa MT. Dietary combination of pawpaw seed and onion peel powder: Impact on growth, haematology, biochemical and antioxidant status of *Clarias gariepinus.* Aquacult Res 2020; 51(7): 2903-12.
[http://dx.doi.org/10.1111/are.14629]

[26]   Hoseinifar SH, Sohrabi A, Paknejad H, Jafari V, Paolucci M, Van Doan H. Enrichment of common carp (*Cyprinus carpio*) fingerlings diet with Psidium guajava: The effects on cutaneous mucosal and serum immune parameters and immune related genes expression. Fish Shellfish Immunol 2019; 86: 688-94.
[http://dx.doi.org/10.1016/j.fsi.2018.12.001] [PMID: 30521968]

[27]   Zemheri-Navruz F, Acar Ü, Yılmaz S. Dietary supplementation of olive leaf extract enhances growth performance, digestive enzyme activity and growth related genes expression in common carp *Cyprinus carpio*. Gen Comp Endocrinol 2020; 296: 113541.
[http://dx.doi.org/10.1016/j.ygcen.2020.113541] [PMID: 32585215]

[28]   Ming J, Ye J, Zhang Y, *et al.* Optimal dietary curcumin improved growth performance, and modulated innate immunity, antioxidant capacity and related genes expression of NF-κB and Nrf2 signaling pathways in grass carp (Ctenopharyngodon idella) after infection with Aeromonas hydrophila. Fish Shellfish Immunol 2020; 97: 540-53.
[http://dx.doi.org/10.1016/j.fsi.2019.12.074] [PMID: 31881329]

[29]   Aanyu M, Betancor MB, Monroig Ó. The effects of combined phytogenics on growth and nutritional physiology of Nile tilapia *Oreochromis niloticus*. Aquaculture 2020; 519: 734867.
[http://dx.doi.org/10.1016/j.aquaculture.2019.734867]

[30]   Apraku A, Huang X, Ayisi CL. Effects of alternative dietary oils on immune response, expression of immune-related genes and disease resistance in juvenile Nile tilapia, *Oreochromis niloticus*. Aquacult Nutr 2019; 25(3): 597-608.
[http://dx.doi.org/10.1111/anu.12882]

[31]   Van Doan H, Hoseinifar SH, Hung TQ, *et al.* Dietary inclusion of chestnut (Castanea sativa) polyphenols to Nile tilapia reared in biofloc technology: Impacts on growth, immunity, and disease resistance against Streptococcus agalactiae. Fish Shellfish Immunol 2020; 105: 319-26.
[http://dx.doi.org/10.1016/j.fsi.2020.07.010] [PMID: 32702475]

[32]   Mohammed E, Kamel M, El Iraqi K, Tawfik AM, Khattab MS, Elsabagh M. *Zingiber officinale* and *Glycyrrhiza glabra*, individually or in combination, reduce heavy metal accumulation and improve growth performance and immune status in Nile tilapia, *Oreochromis niloticus*. Aquacult Res 2020; 51(5): 1933-41.
[http://dx.doi.org/10.1111/are.14544]

[33]   Ahmadifar E, Sheikhzadeh N, Roshanaei K, Dargahi N, Faggio C. Can dietary ginger (Zingiber officinale) alter biochemical and immunological parameters and gene expression related to growth, immunity and antioxidant system in zebrafish (Danio rerio)? Aquaculture 2019; 507: 341-8. [Danio rerio].
[http://dx.doi.org/10.1016/j.aquaculture.2019.04.049]

[34]   Gonçalves RA, Serradeiro R, Machado M, Costas B, Hunger C, Dias J. Interactive effects of dietary fishmeal level and plant essential oils supplementation on European sea bass, *Dicentrarchus labrax* : Growth performance, nutrient utilization, and immunological response. J World Aquacult Soc 2019; 50(6): 1078-92.
[http://dx.doi.org/10.1111/jwas.12616]

[35]   Lopes JM, de Freitas Souza C, Saccol EMH, *et al. Citrus x aurantium* essential oil as feed additive improved growth performance, survival, metabolic, and oxidative parameters of silver catfish (*Rhamdia quelen*). Aquacult Nutr 2019; 25(2): 310-8. [Rhamdia quelen].
[http://dx.doi.org/10.1111/anu.12854]

[36]   Adel M, Abedian Amiri A, Zorriehzahra J, Nematolahi A, Esteban MÁ. Effects of dietary peppermint (Mentha piperita) on growth performance, chemical body composition and hematological and immune parameters of fry Caspian white fish (Rutilus frisii kutum). Fish Shellfish Immunol 2015; 45(2): 841-7.
[http://dx.doi.org/10.1016/j.fsi.2015.06.010] [PMID: 26067171]

[37]   Adel M, Pourgholam R, Zorriehzahra J, Ghiasi M. Hemato – Immunological and biochemical parameters, skin antibacterial activity, and survival in rainbow trout (Oncorhynchus mykiss) following the diet supplemented with Mentha piperita against Yersinia ruckeri. Fish Shellfish Immunol 2016; 55: 267-73.
[http://dx.doi.org/10.1016/j.fsi.2016.05.040] [PMID: 27245867]

[38] Adel M, Safari R, Pourgholam R, Zorriehzahra J, Esteban MÁ. Dietary peppermint (Mentha piperita) extracts promote growth performance and increase the main humoral immune parameters (both at mucosal and systemic level) of Caspian brown trout (Salmo trutta caspius Kessler, 1877). Fish Shellfish Immunol 2015; 47(1): 623-9.
[http://dx.doi.org/10.1016/j.fsi.2015.10.005] [PMID: 26455650]

[39] Van Doan H, Hoseinifar SH, Chitmanat C, *et al.* The effects of Thai ginseng, *Boesenbergia rotunda* powder on mucosal and serum immunity, disease resistance, and growth performance of Nile tilapia (*Oreochromis niloticus*) fingerlings. Aquaculture 2019; 513: 734388.
[http://dx.doi.org/10.1016/j.aquaculture.2019.734388]

[40] Ye Q, Feng Y, Wang Z, *et al.* Effects of dietary Gelsemium elegans alkaloids on intestinal morphology, antioxidant status, immune responses and microbiota of Megalobrama amblycephala. Fish Shellfish Immunol 2019; 94: 464-78.
[http://dx.doi.org/10.1016/j.fsi.2019.09.048] [PMID: 31546035]

[41] Adeshina I, Jenyo-Oni A, Emikpe BO, Ajani EK, Abdel-Tawwab M. Stimulatory effect of dietary clove, *Eugenia caryophyllata*, bud extract on growth performance, nutrient utilization, antioxidant capacity, and tolerance of African catfish, *Clarias gariepinus* (B.), to *Aeromonas hydrophila* infection. J World Aquacult Soc 2019; 50(2): 390-405.
[http://dx.doi.org/10.1111/jwas.12565]

[42] Safari R, Hoseinifar SH, Imanpour MR, Mazandarani M, Sanchouli H, Paolucci M. Effects of dietary polyphenols on mucosal and humoral immune responses, antioxidant defense and growth gene expression in beluga sturgeon (Huso huso). Aquaculture 2020; 528: 735494. [Huso huso].
[http://dx.doi.org/10.1016/j.aquaculture.2020.735494]

[43] Jahazi MA, Hoseinifar SH, Jafari V, Hajimoradloo A, Van Doan H, Paolucci M. Dietary supplementation of polyphenols positively affects the innate immune response, oxidative status, and growth performance of common carp, *Cyprinus carpio* L. Aquaculture 2020; 517: 734709.
[http://dx.doi.org/10.1016/j.aquaculture.2019.734709]

[44] Abdel-Moneim AME, Shehata AM, Alzahrani SO, *et al.* The role of polyphenols in poultry nutrition. J Anim Physiol Anim Nutr (Berl) 2020; 104(6): 1851-66.
[http://dx.doi.org/10.1111/jpn.13455] [PMID: 32969538]

[45] Safari O, Sarkheil M, Paolucci M. Dietary administration of ferula (*Ferula asafoetida*) powder as a feed additive in diet of koi carp, *Cyprinus carpio* koi: effects on hemato-immunological parameters, mucosal antibacterial activity, digestive enzymes, and growth performance. Fish Physiol Biochem 2019; 45(4): 1277-88.
[http://dx.doi.org/10.1007/s10695-019-00674-x] [PMID: 31256305]

[46] Ringø E, Zhou Z, Vecino JLG, *et al.* Effect of dietary components on the gut microbiota of aquatic animals. A never-ending story? Aquacult Nutr 2016; 22(2): 219-82.
[http://dx.doi.org/10.1111/anu.12346]

[47] Vallejos-Vidal E, Reyes-López F, Teles M, MacKenzie S. The response of fish to immunostimulant diets. Fish Shellfish Immunol 2016; 56: 34-69.
[http://dx.doi.org/10.1016/j.fsi.2016.06.028] [PMID: 27389620]

[48] Valladão GMR, Gallani SU, Kotzent S, Assane IM, Pilarski F. Effects of dietary thyme essential oil on hemato-immunological indices, intestinal morphology, and microbiota of Nile tilapia. Aquacult Int 2019; 27(2): 399-411.
[http://dx.doi.org/10.1007/s10499-018-0332-5]

[49] Boshra H, Li J, Sunyer JO. Recent advances on the complement system of teleost fish. Fish Shellfish Immunol 2006; 20(2): 239-62.
[http://dx.doi.org/10.1016/j.fsi.2005.04.004] [PMID: 15950490]

[50] Morgan BP, Boyd C, Bubeck D. Molecular cell biology of complement membrane attack. Semin Cell Dev Biol 2017; 72: 124-32.

[http://dx.doi.org/10.1016/j.semcdb.2017.06.009] [PMID: 28647534]

[51] Baba E, Uluköy G, Öntaş C. Effects of feed supplemented with Lentinula edodes mushroom extract on the immune response of rainbow trout, *Oncorhynchus mykiss*, and disease resistance against Lactococcus garvieae. Aquaculture 2015; 448: 476-82.
[http://dx.doi.org/10.1016/j.aquaculture.2015.04.031]

[52] Tan X, Sun Z, Zhou C, *et al*. Effects of dietary dandelion extract on intestinal morphology, antioxidant status, immune function and physical barrier function of juvenile golden pompano Trachinotus ovatus. Fish Shellfish Immunol 2018; 73: 197-206.
[http://dx.doi.org/10.1016/j.fsi.2017.12.020] [PMID: 29258755]

[53] Zhou C, Lin H, Ge X, *et al*. The Effects of dietary soybean isoflavones on growth, innate immune responses, hepatic antioxidant abilities and disease resistance of juvenile golden pompano Trachinotus ovatus. Fish Shellfish Immunol 2015; 43(1): 158-66.
[http://dx.doi.org/10.1016/j.fsi.2014.12.014] [PMID: 25541076]

[54] Schepetkin IA, Kirpotina LN, Khlebnikov AI, Balasubramanian N, Quinn MT. Neutrophil Immunomodulatory Activity of Natural Organosulfur Compounds. Molecules 2019; 24(9): 1809.
[http://dx.doi.org/10.3390/molecules24091809] [PMID: 31083328]

[55] Yin G, Ardó L, Thompson KD, Adams A, Jeney Z, Jeney G. Chinese herbs (Astragalus radix and Ganoderma lucidum) enhance immune response of carp, Cyprinus carpio, and protection against Aeromonas hydrophila. Fish Shellfish Immunol 2009; 26(1): 140-5.
[http://dx.doi.org/10.1016/j.fsi.2008.08.015] [PMID: 18817878]

[56] Magnadóttir B. Innate immunity of fish (overview). Fish Shellfish Immunol 2006; 20(2): 137-51.
[http://dx.doi.org/10.1016/j.fsi.2004.09.006] [PMID: 15950491]

[57] Talpur AD. *Mentha piperita* (Peppermint) as feed additive enhanced growth performance, survival, immune response and disease resistance of Asian seabass, Lates calcarifer (Bloch) against Vibrio harveyi infection. Aquaculture 2014; 420-421: 71-8.
[http://dx.doi.org/10.1016/j.aquaculture.2013.10.039]

[58] Valavanidis A, Vlahogianni T, Dassenakis M, Scoullos M. Molecular biomarkers of oxidative stress in aquatic organisms in relation to toxic environmental pollutants. Ecotoxicol Environ Saf 2006; 64(2): 178-89.
[http://dx.doi.org/10.1016/j.ecoenv.2005.03.013] [PMID: 16406578]

[59] Naiel MAE, Ismael NEM, Abd El-hameed SAA, Amer MS. The antioxidative and immunity roles of chitosan nanoparticle and vitamin C-supplemented diets against imidacloprid toxicity on Oreochromis niloticus. Aquaculture 2020; 523: 735219.
[http://dx.doi.org/10.1016/j.aquaculture.2020.735219]

[60] Naiel MAE, Ismael NEM, Negm SS, Ayyat MS, Al-Sagheer AA. Rosemary leaf powder–supplemented diet enhances performance, antioxidant properties, immune status, and resistance against bacterial diseases in Nile Tilapia (Oreochromis niloticus). Aquaculture 2020; 526: 735370. [Oreochromis niloticus].
[http://dx.doi.org/10.1016/j.aquaculture.2020.735370]

[61] Naiel MAE, Shehata AM, Negm SS, *et al*. The new aspects of using some safe feed additives on alleviated imidacloprid toxicity in farmed fish: a review. Rev Aquacult 2020; 12(4): 2250-67.
[http://dx.doi.org/10.1111/raq.12432]

[62] Hanke I, Hassenrück C, Ampe B, Kunzmann A, Gärdes A, Aerts J. Chronic stress under commercial aquaculture conditions: Scale cortisol to identify and quantify potential stressors in milkfish (*Chanos chanos*) mariculture. Aquaculture 2020; 526: 735352.
[http://dx.doi.org/10.1016/j.aquaculture.2020.735352]

[63] Guo R, Guo X, Li T, Fu X, Liu RH. Comparative assessment of phytochemical profiles, antioxidant and antiproliferative activities of Sea buckthorn (*Hippophaë rhamnoides* L.) berries. Food Chem 2017; 221: 997-1003.

[http://dx.doi.org/10.1016/j.foodchem.2016.11.063] [PMID: 27979305]

[64]   Hamed YS, Abdin M, Akhtar HMS, *et al.* Extraction, purification by macrospores resin and *in vitro* antioxidant activity of flavonoids from Moringa oliefera leaves. S Afr J Bot 2019; 124: 270-9.
[http://dx.doi.org/10.1016/j.sajb.2019.05.006]

[65]   Trampetti F, Pereira C, Rodrigues MJ, *et al.* Exploring the halophyte *Cistanche phelypaea* (L.) Cout as a source of health promoting products: *in vitro* antioxidant and enzyme inhibitory properties, metabolomic profile and computational studies. J Pharm Biomed Anal 2019; 165: 119-28.
[http://dx.doi.org/10.1016/j.jpba.2018.11.053] [PMID: 30529825]

[66]   Azab SS, Abdel-Daim M, Eldahshan OA. Phytochemical, cytotoxic, hepatoprotective and antioxidant properties of *Delonix regia* leaves extract. Med Chem Res 2013; 22(9): 4269-77.
[http://dx.doi.org/10.1007/s00044-012-0420-4]

[67]   EL-Sabagh MR, Abd Eldaim MA, Mahboub DH, Abdel-Daim M. Effects of *Spirulina Platensis* Algae on Growth Performance, Antioxidative Status and Blood Metabolites in Fattening Lambs. J Agric Sci (Toronto) 2014; 6(3).
[http://dx.doi.org/10.5539/jas.v6n3p92]

[68]   Katalinic V, Milos M, Kulisic T, Jukic M. Screening of 70 medicinal plant extracts for antioxidant capacity and total phenols. Food Chem 2006; 94(4): 550-7.
[http://dx.doi.org/10.1016/j.foodchem.2004.12.004]

[69]   Yeung AWK, Tzvetkov NT, El-Tawil OS, Bungău SG, Abdel-Daim MM, Atanasov AG. Antioxidants: Scientific Literature Landscape Analysis. Oxid Med Cell Longev. 2019.
[http://dx.doi.org/10.1155/2019/8278454]

[70]   García-Beltrán JM, Mansour AT, Alsaqufi AS, Ali HM, Esteban MÁ. Effects of aqueous and ethanolic leaf extracts from drumstick tree (Moringa oleifera) on gilthead seabream (*Sparus aurata* L.) leucocytes, and their cytotoxic, antitumor, bactericidal and antioxidant activities. Fish Shellfish Immunol 2020; 106: 44-55.
[http://dx.doi.org/10.1016/j.fsi.2020.06.054] [PMID: 32739532]

[71]   Dias MI, Barros L, Alves RC, Oliveira MBPP, Santos-Buelga C, Ferreira ICFR. Nutritional composition, antioxidant activity and phenolic compounds of wild Taraxacum sect. Ruderalia. Food Res Int 2014; 56: 266-71.
[http://dx.doi.org/10.1016/j.foodres.2014.01.003]

[72]   Hu C, Kitts DD. Dandelion (*Taraxacum officinale*) flower extract suppresses both reactive oxygen species and nitric oxide and prevents lipid oxidation *in vitro*. Phytomedicine 2005; 12(8): 588-97.
[http://dx.doi.org/10.1016/j.phymed.2003.12.012] [PMID: 16121519]

[73]   Kenny O, J Smyth T, M Hewage C, P Brunton N. Antioxidant properties and quantitative UPLC-MS/MS analysis of phenolic compounds in dandelion (*Taraxacum officinale*) root extracts. Free Radic Antioxid 2014; 4(1): 55-61.
[http://dx.doi.org/10.5530/fra.2014.1.9]

[74]   Ebrahimzadeh MA, Fazel Nabavi S, Mohammad Nabavi S, Eslami B. Antioxidant activity of flower, stem and leaf extracts of <i>Ferula gummosa</i> Boiss. Grasas Aceites 2010; 61(3): 244-50.
[http://dx.doi.org/10.3989/gya.110809]

[75]   Mallikarjuna GU, Dhanalakshmi S, Raisuddin S, Ramesha Rao A. Chemomodulatory influence of Ferula asafoetida on mammary epithelial differentiation, hepatic drug metabolizing enzymes, antioxidant profiles and N-methyl-N-nitrosourea-induced mammary carcinogenesis in rats. Breast Cancer Res Treat 2003; 81(1): 1-10.
[http://dx.doi.org/10.1023/A:1025448620558] [PMID: 14531492]

[76]   Balasundram N, Sundram K, Samman S. Phenolic compounds in plants and agri-industrial by-products: Antioxidant activity, occurrence, and potential uses. Food Chem 2006; 99(1): 191-203.
[http://dx.doi.org/10.1016/j.foodchem.2005.07.042]

[77]   Amarowicz R, Pegg RB, Rahimi-Moghaddam P, Barl B, Weil JA. Free-radical scavenging capacity and antioxidant activity of selected plant species from the Canadian prairies. Food Chem 2004; 84(4): 551-62.
[http://dx.doi.org/10.1016/S0308-8146(03)00278-4]

[78]   Bourdon E, Loreau N, Lagrost L, Blache D. Differential effects of cysteine and methionine residues in the antioxidant activity of human serum albumin. Free Radic Res 2005; 39(1): 15-20.
[http://dx.doi.org/10.1080/10715760400024935] [PMID: 15875807]

[79]   Dulski T, Kozłowski K, Ciesielski S. Habitat and seasonality shape the structure of tench [*Tinca tinca* L.] gut microbiome. Sci Rep 2020; 10(1): 4460-10.
[http://dx.doi.org/10.1038/s41598-020-61351-1]

[80]   Nicholson JK, Holmes E, Kinross J, *et al*. Host-gut microbiota metabolic interactions. Science 2012; 336(6086): 1262-7.
[http://dx.doi.org/10.1126/science.1223813] [PMID: 22674330]

[81]   Reverter M, Bontemps N, Lecchini D, Banaigs B, Sasal P. Use of plant extracts in fish aquaculture as an alternative to chemotherapy: Current status and future perspectives. Aquaculture 2014; 433: 50-61.
[http://dx.doi.org/10.1016/j.aquaculture.2014.05.048]

[82]   Abd El-Hack ME, El-Saadony MT, Shehata AM, *et al*. Approaches to prevent and control Campylobacter spp. colonization in broiler chickens: a review. Environ Sci Pollut Res Int 2021; 28(5): 4989-5004.
[http://dx.doi.org/10.1007/s11356-020-11747-3] [PMID: 33242194]

[83]   Abdel-Moneim AME, El-Saadony MT, Shehata AM, *et al*. Antioxidant and antimicrobial activities of *Spirulina platensis* extracts and biogenic selenium nanoparticles against selected pathogenic bacteria and fungi. Saudi J Biol Sci 2022; 29(2): 1197-209.
[http://dx.doi.org/10.1016/j.sjbs.2021.09.046] [PMID: 35197787]

[84]   Si W, Gong J, Tsao R, *et al*. Antimicrobial activity of essential oils and structurally related synthetic food additives towards selected pathogenic and beneficial gut bacteria. J Appl Microbiol 2006; 100(2): 296-305.
[http://dx.doi.org/10.1111/j.1365-2672.2005.02789.x] [PMID: 16430506]

[85]   Merrifield DL, Balcázar JL, Daniels C, Zhou Z, Carnevali O, Sun Y-Z, *et al*. Indigenous Lactic Acid Bacteria in Fish and Crustaceans Aquaculture Nutrition. John Wiley & Sons, Ltd 2014; pp. 128-68.

[86]   Brandi J, Cheri S, Manfredi M, Di Carlo C, Vita Vanella V, Federici F, *et al*. Exploring the wound healing, anti-inflammatory, anti-pathogenic and proteomic effects of lactic acid bacteria on keratinocytes. Sci Rep 2020; 10(1): 11572-0.
[http://dx.doi.org/10.1038/s41598-020-68483-4]

[87]   Vázquez JA, González MP, Murado MA. Effects of lactic acid bacteria cultures on pathogenic microbiota from fish. Aquaculture 2005; 245(1-4): 149-61.
[http://dx.doi.org/10.1016/j.aquaculture.2004.12.008]

[88]   Magnotti LJ, Deitch EA. Burns, bacterial translocation, gut barrier function, and failure. J Burn Care Rehabil 2005; 26(5): 383-91.
[http://dx.doi.org/10.1097/01.bcr.0000176878.79267.e8] [PMID: 16151282]

[89]   Okumura R, Takeda K. Roles of intestinal epithelial cells in the maintenance of gut homeostasis. Exp Mol Med 2017; 49(5): e338-10.
[http://dx.doi.org/10.1038/emm.2017.20]

[90]   Okumura R, Takeda K. Maintenance of intestinal homeostasis by mucosal barriers. Inflamm Regen 2018; 38: 5-10.
[http://dx.doi.org/10.1186/s41232-018-0063-z]

[91]   Abd El-Naby AS, Al-Sagheer AA, Negm SS, Naiel MAE. Dietary combination of chitosan nanoparticle and thymol affects feed utilization, digestive enzymes, antioxidant status, and intestinal

morphology of *Oreochromis niloticus*. Aquaculture 2020; 515: 734577.
[http://dx.doi.org/10.1016/j.aquaculture.2019.734577]

[92]   Ye Q, Feng Y, Wang Z, *et al*. Effects of dietary Gelsemium elegans alkaloids on growth performance, immune responses and disease resistance of Megalobrama amblycephala. Fish Shellfish Immunol 2019; 91: 29-39.
[http://dx.doi.org/10.1016/j.fsi.2019.05.026] [PMID: 31100439]

[93]   Enioutina EY, Teng L, Fateeva TV, *et al*. Phytotherapy as an alternative to conventional antimicrobials: combating microbial resistance. Expert Rev Clin Pharmacol 2017; 10(11): 1203-14.
[http://dx.doi.org/10.1080/17512433.2017.1371591] [PMID: 28836870]

[94]   Dong HT, Nguyen VV, Le HD, *et al*. Naturally concurrent infections of bacterial and viral pathogens in disease outbreaks in cultured Nile tilapia (*Oreochromis niloticus*) farms. Aquaculture 2015; 448: 427-35.
[http://dx.doi.org/10.1016/j.aquaculture.2015.06.027]

[95]   Moriarty DJW. The role of microorganisms in aquaculture ponds. Aquaculture 1997; 151(1-4): 333-49.
[http://dx.doi.org/10.1016/S0044-8486(96)01487-1]

[96]   Flores-Kossack C, Montero R, Köllner B, Maisey K. Chilean aquaculture and the new challenges: Pathogens, immune response, vaccination and fish diversification. Fish Shellfish Immunol 2020; 98: 52-67.
[http://dx.doi.org/10.1016/j.fsi.2019.12.093] [PMID: 31899356]

[97]   Pulkkinen K, Suomalainen LR, Read AF, Ebert D, Rintamäki P, Valtonen ET. Intensive fish farming and the evolution of pathogen virulence: the case of columnaris disease in Finland. Proc Biol Sci 2010; 277(1681): 593-600.
[http://dx.doi.org/10.1098/rspb.2009.1659]

[98]   Abreu AC, McBain AJ, Simões M. Plants as sources of new antimicrobials and resistance-modifying agents. Nat Prod Rep 2012; 29(9): 1007-21.
[http://dx.doi.org/10.1039/c2np20035j] [PMID: 22786554]

[99]   Bush K, Courvalin P, Dantas G, *et al*. Tackling antibiotic resistance. Nat Rev Microbiol 2011; 9(12): 894-6.
[http://dx.doi.org/10.1038/nrmicro2693] [PMID: 22048738]

[100] Hammer KA, Carson CF, Riley TV. Antimicrobial activity of essential oils and other plant extracts. J Appl Microbiol 1999; 86(6): 985-90.
[http://dx.doi.org/10.1046/j.1365-2672.1999.00780.x] [PMID: 10438227]

[101] Williamson G. The role of polyphenols in modern nutrition. Nutr Bull 2017; 42(3): 226-35.
[http://dx.doi.org/10.1111/nbu.12278] [PMID: 28983192]

[102] Simões M, Lemos M, Simões LC. Phytochemicals Against Drug-Resistant Microbes Dietary Phytochemicals and Microbes. Springer Netherlands 2012; pp. 185-205.
[http://dx.doi.org/10.1007/978-94-007-3926-0_6]

[103] Rosa IA, Rodrigues P, Bianchini AE, *et al*. Extracts of *Hesperozygis ringens* (Benth.) Epling: *in vitro* and *in vivo* antibacterial activity against fish pathogenic bacteria. J Appl Microbiol 2019; 126(5): 1353-61.
[http://dx.doi.org/10.1111/jam.14219] [PMID: 30735293]

[104] Hoseinifar SH, Jahazi MA, Nikdehghan N, Van Doan H, Volpe MG, Paolucci M. Effects of dietary polyphenols from agricultural by-products on mucosal and humoral immune and antioxidant responses of convict cichlid (Amatitlania nigrofasciata). Aquaculture 2020; 517: 734790. [Amatitlania nigrofasciata].
[http://dx.doi.org/10.1016/j.aquaculture.2019.734790]

[105] Salomón R, Firmino JP, Reyes-López FE, *et al*. The growth promoting and immunomodulatory effects

of a medicinal plant leaf extract obtained from Salvia officinalis and Lippia citriodora in gilthead seabream (Sparus aurata). Aquaculture 2020; 524: 735291. [Sparus aurata].
[http://dx.doi.org/10.1016/j.aquaculture.2020.735291]

[106]   Wu Y, Gong Q, Fang H, Liang W, Chen M, He R. Effect of Sophora flavescens on non-specific immune response of tilapia (GIFT Oreochromis niloticus) and disease resistance against Streptococcus agalactiae. Fish Shellfish Immunol 2013; 34(1): 220-7.
[http://dx.doi.org/10.1016/j.fsi.2012.10.020] [PMID: 23092731]

[107]   Tsuchiya H. Membrane Interactions of Phytochemicals as Their Molecular Mechanism Applicable to the Discovery of Drug Leads from Plants. Molecules 2015; 20(10): 18923-66.
[http://dx.doi.org/10.3390/molecules201018923] [PMID: 26501254]

[108]   Álvarez C, Barriga A, Albericio F, Romero M, Guzmán F. Identification of Peptides in Flowers of *Sambucus nigra* with Antimicrobial Activity against Aquaculture Pathogens. Molecules 2018; 23(5): 1033.
[http://dx.doi.org/10.3390/molecules23051033] [PMID: 29702623]

[109]   Sakkas H, Gousia P, Economou V, Sakkas V, Petsios S, Papadopoulou C. *In vitro* antimicrobial activity of five essential oils on multi-drug resistant Gram-negative clinical isolates. J Intercult Ethnopharmacol 2016; 5(3): 212-8.
[http://dx.doi.org/10.5455/jice.20160331064446] [PMID: 27366345]

[110]   Plésiat P, Nikaido H. Outer membranes of Gram-negative bacteria are permeable to steroid probes. Mol Microbiol 1992; 6(10): 1323-33.
[http://dx.doi.org/10.1111/j.1365-2958.1992.tb00853.x] [PMID: 1640833]

[111]   Hayatgheib N, Fournel C, Calvez S, Pouliquen H, Moreau E. *In vitro* antimicrobial effect of various commercial essential oils and their chemical constituents on *Aeromonas salmonicida* subsp. *salmonicida.* J Appl Microbiol 2020; 129(1): 137-45.
[http://dx.doi.org/10.1111/jam.14622] [PMID: 32119179]

[112]   Salini R, Santhakumari S, Veera Ravi A, Karutha Pandian S. Synergistic antibiofilm efficacy of undecanoic acid and auxins against quorum sensing mediated biofilm formation of luminescent Vibrio harveyi. Aquaculture 2019; 498: 162-70.
[http://dx.doi.org/10.1016/j.aquaculture.2018.08.038]

[113]   Vattem DA, Mihalik K, Crixell SH, McLean RJC. Dietary phytochemicals as quorum sensing inhibitors. Fitoterapia 2007; 78(4): 302-10.
[http://dx.doi.org/10.1016/j.fitote.2007.03.009] [PMID: 17499938]

[114]   Arunkumar M, LewisOscar F, Thajuddin N, Pugazhendhi A, Nithya C. *In vitro* and *in vivo* biofilm forming Vibrio spp: A significant threat in aquaculture. Process Biochem 2020; 94: 213-23.
[http://dx.doi.org/10.1016/j.procbio.2020.04.029]

[115]   Borges A, Abreu A, Dias C, Saavedra M, Borges F, Simões M. New Perspectives on the Use of Phytochemicals as an Emergent Strategy to Control Bacterial Infections Including Biofilms. Molecules 2016; 21(7): 877.
[http://dx.doi.org/10.3390/molecules21070877] [PMID: 27399652]

[116]   Nazzaro F, Fratianni F, Coppola R. Quorum sensing and phytochemicals. Int J Mol Sci 2013; 14(6): 12607-19.
[http://dx.doi.org/10.3390/ijms140612607] [PMID: 23774835]

[117]   Rubini D, Banu SF, Nisha P, *et al.* Essential oils from unexplored aromatic plants quench biofilm formation and virulence of Methicillin resistant *Staphylococcus aureus.* Microb Pathog 2018; 122: 162-73.
[http://dx.doi.org/10.1016/j.micpath.2018.06.028] [PMID: 29920307]

[118]   Giannenas I, Triantafillou E, Stavrakakis S, *et al.* Assessment of dietary supplementation with carvacrol or thymol containing feed additives on performance, intestinal microbiota and antioxidant status of rainbow trout (Oncorhynchus mykiss). Aquaculture 2012; 350-353: 26-32. [Oncorhynchus

mykiss].
[http://dx.doi.org/10.1016/j.aquaculture.2012.04.027]

[119] Bandeira Junior G, Sutili FJ, Gressler LT, *et al.* Antibacterial potential of phytochemicals alone or in combination with antimicrobials against fish pathogenic bacteria. J Appl Microbiol 2018; 125(3): 655-65.
[http://dx.doi.org/10.1111/jam.13906] [PMID: 29741243]

[120] da Cunha JA, Bandeira Junior G, da Silva EG, *et al.* The survival and hepatic and muscle glucose and lactate levels of Rhamdia quelen inoculated with *Aeromonas hydrophila* and treated with terpinen--ol, carvacrol or thymol. Microb Pathog 2019; 127: 220-4.
[http://dx.doi.org/10.1016/j.micpath.2018.12.005] [PMID: 30529428]

[121] Bücker R, Krug SM, Rosenthal R, *et al.* Aerolysin from *Aeromonas hydrophila* perturbs tight junction integrity and cell lesion repair in intestinal epithelial HT-29/B6 cells. J Infect Dis 2011; 204(8): 1283-92.
[http://dx.doi.org/10.1093/infdis/jir504] [PMID: 21917902]

[122] Dong J, Ding H, Liu Y, *et al.* Magnolol protects channel catfish from *Aeromonas hydrophila* infection *via* inhibiting the expression of aerolysin. Vet Microbiol 2017; 211: 119-23.
[http://dx.doi.org/10.1016/j.vetmic.2017.10.005] [PMID: 29102106]

[123] Rama Devi K, Srinivasan R, Kannappan A, *et al.* *In vitro* and *in vivo* efficacy of rosmarinic acid on quorum sensing mediated biofilm formation and virulence factor production in *Aeromonas hydrophila*. Biofouling 2016; 32(10): 1171-83.
[http://dx.doi.org/10.1080/08927014.2016.1237220] [PMID: 27739324]

[124] Ángeles Esteban M. An Overview of the Immunological Defenses in Fish Skin. ISRN Immunology 2012; 2012: 1-29.
[http://dx.doi.org/10.5402/2012/853470]

[125] Shen YF, Liu L, Feng CZ, *et al.* Synthesis and antiviral activity of a new coumarin derivative against spring viraemia of carp virus. Fish Shellfish Immunol 2018; 81: 57-66.
[http://dx.doi.org/10.1016/j.fsi.2018.07.005] [PMID: 29981474]

[126] Adams A. Progress, challenges and opportunities in fish vaccine development. Fish Shellfish Immunol 2019; 90: 210-4.
[http://dx.doi.org/10.1016/j.fsi.2019.04.066] [PMID: 31039441]

[127] Rodger HD. Fish Disease Causing Economic Impact in Global Aquaculture. Fish Vaccines: Springer Basel 2016; pp. 1-34.
[http://dx.doi.org/10.1007/978-3-0348-0980-1_1]

[128] Krejcarová J, Straková E, Suchý P, Herzig I, Karásková K. Sea buckthorn (*Hippophae rhamnoides* L.) as a potential source of nutraceutics and its therapeutic possibilities - a review. Acta Vet Brno 2015; 84(3): 257-68.
[http://dx.doi.org/10.2754/avb201584030257]

[129] Enkhtaivan G, Maria John KM, Pandurangan M, Hur JH, Leutou AS, Kim DH. Extreme effects of Seabuckthorn extracts on influenza viruses and human cancer cells and correlation between flavonol glycosides and biological activities of extracts. Saudi J Biol Sci 2017; 24(7): 1646-56.
[http://dx.doi.org/10.1016/j.sjbs.2016.01.004] [PMID: 30294231]

[130] Ehrhardt C, Hrincius E, Korte V, *et al.* A polyphenol rich plant extract, CYSTUS052, exerts anti influenza virus activity in cell culture without toxic side effects or the tendency to induce viral resistance. Antiviral Res 2007; 76(1): 38-47.
[http://dx.doi.org/10.1016/j.antiviral.2007.05.002] [PMID: 17572513]

[131] Haslam E. Natural polyphenols (vegetable tannins) as drugs: possible modes of action. J Nat Prod 1996; 59(2): 205-15.
[http://dx.doi.org/10.1021/np960040+] [PMID: 8991956]

[132] Choi HJ, Song JH, Kwon DH. Quercetin 3-rhamnoside exerts antiinfluenza A virus activity in mice. Phytother Res 2012; 26(3): 462-4.
[http://dx.doi.org/10.1002/ptr.3529] [PMID: 21728202]

[133] Song JM, Lee KH, Seong BL. Antiviral effect of catechins in green tea on influenza virus. Antiviral Res 2005; 68(2): 66-74.
[http://dx.doi.org/10.1016/j.antiviral.2005.06.010] [PMID: 16137775]

[134] Zhang Y, Wang H, Su M, Lu L. (-)-Epicatechin gallate, a metabolite of (-)-epigallocatechin gallate in grass carp, exhibits antiviral activity *in vitro* against grass carp reovirus. Aquacult Res 2020; 51(4): 1673-80.
[http://dx.doi.org/10.1111/are.14513]

[135] Gallaga-Maldonado EP, Montaldo HH, Castillo-Juárez H, *et al.* Crossbreeding effects for White Spot Disease resistance in challenge tests and field pond performance in Pacific white shrimp *Litopenaeus vannamei* involving susceptible and resistance lines. Aquaculture 2020; 516: 734527.
[http://dx.doi.org/10.1016/j.aquaculture.2019.734527]

[136] Balasubramanian G, Sarathi M, Venkatesan C, Thomas J, Sahul Hameed AS. Oral administration of antiviral plant extract of *Cynodon dactylon* on a large scale production against White spot syndrome virus (WSSV) in *Penaeus monodon.* Aquaculture 2008; 279(1-4): 2-5.
[http://dx.doi.org/10.1016/j.aquaculture.2008.03.052]

[137] Howlader P, Ghosh AK, Islam SS, Bir J, Banu GR. Antiviral activity of Cynodon dactylon on white spot syndrome virus (WSSV)-infected shrimp: an attempt to mitigate risk in shrimp farming. Aquacult Int 2020; 28(4): 1725-38.
[http://dx.doi.org/10.1007/s10499-020-00553-w]

[138] Ma Y, Liu Y, Wu Y, *et al.* An attenuated Vibrio harveyi surface display of envelope protein VP28 to be protective against WSSV and vibriosis as an immunoactivator for Litopenaeus vannamei. Fish Shellfish Immunol 2019; 95: 195-202.
[http://dx.doi.org/10.1016/j.fsi.2019.10.016] [PMID: 31604149]

[139] Zhai YF, Shi DJ, He PM, Cai CE, Yin R, Jia R. Effect of trans-vp28 gene Synechocystis sp. PCC6803 on growth and immunity of Litopenaeus vannamei and defense against white spot syndrome virus (WSSV). Aquaculture 2019; 512: 734306. [WSSV].
[http://dx.doi.org/10.1016/j.aquaculture.2019.734306]

[140] Kiataramgul A, Maneenin S, Purton S, *et al.* An oral delivery system for controlling white spot syndrome virus infection in shrimp using transgenic microalgae. Aquaculture 2020; 521: 735022.
[http://dx.doi.org/10.1016/j.aquaculture.2020.735022]

[141] Liu L, Hu Y, Shen YF, Wang GX, Zhu B. Evaluation on antiviral activity of coumarin derivatives against spring viraemia of carp virus in epithelioma papulosum cyprini cells. Antiviral Res 2017; 144: 173-85.
[http://dx.doi.org/10.1016/j.antiviral.2017.06.007] [PMID: 28624462]

[142] Song DW, Liu L, Shan LP, Qiu TX, Chen J, Chen JP. Therapeutic potential of phenylpropanoid-based small molecules as anti-SCVC agents in aquaculture. Aquaculture 2020; 526: 735349.
[http://dx.doi.org/10.1016/j.aquaculture.2020.735349]

[143] Liu G, Wang C, Wang H, *et al.* Antiviral efficiency of a coumarin derivative on spring viremia of carp virus *in vivo.* Virus Res 2019; 268: 11-7.
[http://dx.doi.org/10.1016/j.virusres.2019.05.007] [PMID: 31095989]

[144] Qiu TX, Song DW, Shan LP, Liu GL, Liu L. Potential prospect of a therapeutic agent against spring viraemia of carp virus in aquaculture. Aquaculture 2020; 515: 734558.
[http://dx.doi.org/10.1016/j.aquaculture.2019.734558]

[145] Liu L, Qiu TX, Song DW, Shan LP, Chen J. Inhibition of a novel coumarin on an aquatic rhabdovirus by targeting the early stage of viral infection demonstrates potential application in aquaculture.

Antiviral Res 2020; 174: 104672.
[http://dx.doi.org/10.1016/j.antiviral.2019.104672] [PMID: 31825851]

[146] Liu M, Yu Q, Yi Y, *et al.* Antiviral activities of Lonicera japonica Thunb. Components against grouper iridovirus *in vitro* and *in vivo*. Aquaculture 2020; 519: 734882.
[http://dx.doi.org/10.1016/j.aquaculture.2019.734882]

[147] Micol V, Caturla N, Pérezfons L, Más V, Pérez L, Estepa A. The olive leaf extract exhibits antiviral activity against viral haemorrhagic septicaemia rhabdovirus (VHSV). Antiviral Res 2005; 66(2-3): 129-36.
[http://dx.doi.org/10.1016/j.antiviral.2005.02.005] [PMID: 15869811]

[148] Leri M, Oropesa-Nuñez R, Canale C, *et al.* Oleuropein aglycone: A polyphenol with different targets against amyloid toxicity. Biochim Biophys Acta, Gen Subj 2018; 1862(6): 1432-42.
[http://dx.doi.org/10.1016/j.bbagen.2018.03.023] [PMID: 29571746]

[149] Valladão GMR, Gallani SU, Pilarski F. Phytotherapy as an alternative for treating fish disease. J Vet Pharmacol Ther 2015; 38(5): 417-28.
[http://dx.doi.org/10.1111/jvp.12202] [PMID: 25620601]

[150] Doan HV, Soltani E, Ingelbrecht J, Soltani M. Medicinal Herbs and Plants: Potential Treatment of Monogenean Infections in Fish. Rev Fish Sci Aquacult 2020; 28(2): 260-82.
[http://dx.doi.org/10.1080/23308249.2020.1712325]

[151] Malheiros DF, Maciel PO, Videira MN, Tavares-Dias M. Toxicity of the essential oil of *Mentha piperita* in *Arapaima gigas* (pirarucu) and antiparasitic effects on Dawestrema spp. (Monogenea). Aquaculture 2016; 455: 81-6. [Monogenea].
[http://dx.doi.org/10.1016/j.aquaculture.2016.01.018]

[152] Zhou S, Li WX, Wang YQ, Zou H, Wu SG, Wang GT. Anthelmintic efficacies of three common disinfectants and extracts of four traditional Chinese medicinal plants against Gyrodactylus kobayashii (Monogenea) in goldfish (Carassius auratus). Aquaculture 2017; 466: 72-7. [Carassius auratus].
[http://dx.doi.org/10.1016/j.aquaculture.2016.09.048]

[153] Wang G, Han J, Feng T, Li F, Zhu B. Bioassay-guided isolation and identification of active compounds from Fructus Arctii against Dactylogyrus intermedius (Monogenea) in goldfish (Carassius auratus). Parasitol Res 2009; 106(1): 247-55.
[http://dx.doi.org/10.1007/s00436-009-1659-7] [PMID: 19859737]

[154] Liu YT, Wang F, Wang GX, Han J, Wang Y, Wang YH. *in vivo* anthelmintic activity of crude extracts of Radix angelicae pubescentis, Fructus bruceae, Caulis spatholobi, Semen aesculi, and Semen pharbitidis against Dactylogyrus intermedius (Monogenea) in goldfish (*Carassius auratus*). Parasitol Res 2010; 106(5): 1233-9.
[http://dx.doi.org/10.1007/s00436-010-1799-9] [PMID: 20191290]

[155] da Costa JC, Valladão GMR, Pala G, *et al.* Copaifera duckei oleoresin as a novel alternative for treatment of monogenean infections in pacu *Piaractus mesopotamicus*. Aquaculture 2017; 471: 72-9.
[http://dx.doi.org/10.1016/j.aquaculture.2016.11.041]

[156] de Oliveira Hashimoto GS, Neto FM, Ruiz ML, *et al.* Essential oils of *Lippia sidoides* and *Mentha piperita* against monogenean parasites and their influence on the hematology of Nile tilapia. Aquaculture 2016; 450: 182-6.
[http://dx.doi.org/10.1016/j.aquaculture.2015.07.029]

[157] Valentim DSS, Duarte JL, Oliveira AEM F M, *et al.* Nanoemulsion from essential oil of *Pterodon emarginatus* (Fabaceae) shows *in vitro* efficacy against monogeneans of *Colossoma macropomum* (Pisces: Serrasalmidae). J Fish Dis 2018; 41(3): 443-9.
[http://dx.doi.org/10.1111/jfd.12739] [PMID: 29194663]

[158] Valladão GMR, Gallani SU, Ikefuti CV, *et al.* Essential oils to control ichthyophthiriasis in pacu, *Piaractus mesopotamicus* (Holmberg): special emphasis on treatment with *Melaleuca alternifolia*. J Fish Dis 2016; 39(10): 1143-52.

[http://dx.doi.org/10.1111/jfd.12447] [PMID: 26776242]

[159] de Lima Boijink C, Miranda WSC, Chagas EC, Dairiki JK, Inoue LAKA. Anthelmintic activity of eugenol in tambaquis with monogenean gill infection. Aquaculture 2015; 438: 138-40.
[http://dx.doi.org/10.1016/j.aquaculture.2015.01.014]

[160] Emara EKM, Gaafar AY, Shetaia YM. *In vitro* screening for the antifungal activity of some Egyptian plant extracts against the fish pathogen *Saprolegnia parasitica*. Aquacult Res 2020; 51(11): 4461-70.
[http://dx.doi.org/10.1111/are.14791]

[161] Khosravi AR, Shokri H, Sharifrohani M, Mousavi HE, Moosavi Z. Evaluation of the antifungal activity of Zataria multiflora, Geranium herbarium, and Eucalyptus camaldolensis essential oils on Saprolegnia parasitica-infected rainbow trout (*Oncorhynchus mykiss*) eggs. Foodborne Pathog Dis 2012; 9(7): 674-9.
[http://dx.doi.org/10.1089/fpd.2011.1086] [PMID: 22690761]

[162] Ameen F, Al-Niaeem K, Taher MM, Sultan FAH. Potential of Plant Extracts to Inhibit the Ichthyophonus sp. Infection in Blue Tilapia: A Preliminary Study *in vitro*. Natl Acad Sci Lett 2018; 41(2): 129-32.
[http://dx.doi.org/10.1007/s40009-018-0617-2]

The Beneficial Impacts of Essential Oils Application against Parasitic Infestation in Fish Farm

Samar S. Negm[1], Mohamed E. Abd El-Hack[2,*], Mahmoud Alagawany[2], Amlan Kumar Patra[3] and Mohammed A. E. Naiel[4,*]

[1] *Fish Biology and Ecology Department, Central Lab for Aquaculture Research Abbassa, Agriculture Research Centre, Giza, Egypt*

[2] *Department of Poultry, Faculty of Agriculture, Zagazig University, Zagazig 44519, Egypt*

[3] *Department of Animal Nutrition, West Bengal University of Animal and Fishery Sciences, Kolkata 700037, India*

[4] *Department of Animal Production, Faculty of Agriculture, Zagazig University, Zagazig 44519, Egypt*

Abstract: Aquaculture is a growing sector due to the high rising demand for fish, shrimp, oysters, and other products, which is partially conflicted by various infectious diseases. The infectious diseases affecting the production and inducing high mortalities cause substantial economic losses in this sector. Also, parasitic infections may induce severe mortality and morbidity in fish farms. Therefore, most farmers apply several kinds of antibiotics to control the problems induced by bacterial diseases and, to some extent, parasitic infections. The extensive usage of antibiotics to control or prevent pathogens may lead to the development of pathogenic resistant strains that might cause hazards to human health. Besides, there is a global trend toward reducing the application of antibiotics in aquaculture farms. Thus, there is a great effort to discover new natural and safe products with pharmaceutical properties, such as natural essential oils (EO). Essential oils are secondary metabolites of many plants (roots, flowers, seeds, leaves, fruits and peels) and their molecular structures provide a high antimicrobial and antiparasitic efficiency against pathogens. Consequently, it is essential to provide sufficient knowledge about the mode of action of EO against fish parasites and its future applications and directions in aquaculture.

Keywords: Antibiotics, Antiparasitic, Diseases, Essential oils, Parasites.

* **Corresponding author Mohamed E. Abd El-Hack and Mohammed A. E. Naiel:** Department of Poultry, Faculty of Agriculture, Zagazig University, Zagazig 44519, Egypt and Department of Animal Production, Faculty of Agriculture, Zagazig University, Zagazig 44519, Egypt; E-mails: dr.mohamed.e.abdalhaq@gmail.com and mohammednaiel1984@gmail.com

INTRODUCTION

Worldwide, aquaculture is a fast-rising practice due to high demands for aquaculture-derived foods and other products for human uses. It is a vital source of nutrition, food, livelihood and economic income for millions of people worldwide [1]. In 2018, the world aquaculture production was around 114.5 million tons from aquatic animals, with a total farm gate sale value of US$ 263.6 billion [2].

Several factors affect fish culture production, such as inadequate diet, high stocking density, low oxygen water content, pollution, and infectious diseases, including parasitosis. The intensive production from fish farms leads to a high incidence of parasitic diseases that threaten the sustainability of fish farming [3]. Many parasite infections may be responsible for huge economic losses; thereby, the application of sufficient controlling procedures and treatments is a vital challenge for fish producers [4]. Given the secondary adverse effects of synthetic treatments on antimicrobial resistance and residue in foods, phytotherapy has become a leading alternative strategy in controlling parasitic infections. Herbs may include one or more bioactive constituents with the potency for therapeutic effects [5].

Essential oils (EO) are naturally aromatic and volatile liquids extracted from various parts of medicinal herbs [6]. Several recent studies have proven the efficiency of some EO against bacterial [7], fungal and protozoal diseases [8, 9]. Even though there are more than 3000 EOs identified, less than 0.4% of them have been verified on fish parasites [1]. So, there is a need to conduct more experiments on the therapeutic potential of these products to determine the antiparasitic effects of other EOs for preventing fish parasites to maximize their benefits to hosts.

This chapter focuses on recent knowledge about *in-vivo* studies on the EOs against fish parasites. Also, we have highlighted the antiparasitic mode of action of EOs, along with their immunomodulatory role and potential use against both fish ectoparasites and endoparasites. There is an urgent need to explore EOs and their effective, safe dose levels to control endoparasite infections in fish. Therefore, the development of new products may be attained by considering the beneficial features of EOs. Moreover, EOs could be used as a practical alternative strategy for therapeutic compounds against several infections.

ESSENTIAL OIL RESOURCES, STRUCTURE AND BIOACTIVE MOLECULES

EOs are volatile compounds often present as a complex mixture of several constituents that are extracted from herbs [1]. The EOs are classified according to their chemical structures, mainly terpenes (hydrocarbons) like terpinene, myrcene, limonene, pinene, *p*-cymene, *a*- and *b*-phellandrene; and terpenoids like geraniol, linalool, menthol, 4-carvomenthenol, terpineol, carveol, borneol, citral, citronellal, perillaldehyde, carvacrol, thymol, safrole, eugenol, verbenol, menthone, pulegone, carvone, thujone, verbenone, fenchone, citronellic acid, cinnamic acid and linalyl acetate. Also, some EOs may contain the structure of oxides [1,8-cineole), methyl anthranilate, coumarins and sesquiterpenes such as zingiberene, curcumin, farnesol, sesquiphellandrene, turmerone, and nerolidol [10, 11]. Many of these secondary constituents extracted from herbs showed varied antiparasitic activity [5, 12].

Many herbal plants such as *Thymus hyemalis, Thymus glandulosus, Thymus zygis, Thymus vulgaris, Origanum dictamnus, Monarda fistulosa, Origanum vulgare, Origanum onites* and *Origanum compactum* contain a natural monoterpene compound called thymol and 2-isopropyl-5-methylphenol [13]. The ginger EO is extracted from ginger (*Zingiber officinale* Rosc, Zingiberaceae). The main bioactive molecule is citral, which has high antibacterial activities and could be used as a dietary supplement to treat pathogens [15]. In addition, *Hedeoma patens, Lippia graveolens, Lippia palmeri, Lippia alba, Origanum dictamnus, Origanum hirtum, Origanumonites*, and *Origanum vulgare* are common oregano plant species belonging to the Verbenaceae family, which are used for extraction of *Origanum* EO [16]. Table **1** illustrates many studies showing the effectiveness of various EO in reducing parasitic infection in several fish species.

Table **1.** *In-vivo* and in *in vitro* influence of different essential oils and their derivative molecules for different fish species infected with parasites.

References	Solvent	Frequently	Dose	Fish Species	Essential Oil	Parasite
Firmino, Vallejos-Vidal [86]	water	104 days	0.5%	*Sparusaurata*	microencapsulated blend of garlic, carvacrol and thymol essential oils	*Sparicotylechrysophrii*
Pereira, Oliveira [102]	ethyl alcohol (70%).	24 hours	LC$_{50}$ 67.97 µg/L and 59.55 µg/L.	In *in vitro*	*Cymbopogon citratus* (Poaceae)	*Argulussp and Dolops discoidalis*
Barriga, Gonzales [103]	ethyl alcohol (70%).	After 9 hours	700 mg/L	*Colossoma. macropomum* fingerlings (30 ± 5 g) and a large Serrasalmidae	*Lippia grata* (Verbenaceae)	monogeneans

(Table 1) cont.....

References	Solvent	Frequently	Dose	Fish Species	Essential Oil	Parasite
Barriga, Gonzales [103]	ethyl alcohol (70%).	After 9 hours	700 mg/L	*Colossoma. macropomum* fingerlings (30 ± 5 g) and a large Serrasalmidae	*Lippia grata* (Verbenaceae)	monogeneans
de Souza Costa, da Cruz [104]	3% Tween 80	2 hs	540 mg L^{-1}	*In-in vitro*	*Mentha piperita*	acanthocephalan *Neoechinorhynchus buttnerae*
	water	30 days	0.54 and 2.88 g kg^{-1} Treat 85.46% and 70.03%	*Colossoma. macropomum*	*Mentha piperita*	acanthocephalan *Neoechinorhynchus buttnerae*
de Souza, Baldisserotto [105]	water	30 days	1.44 and 2.88 g kg^{-1} treatment reduced parasite with 67.10% and 56.93% respectively	*Colossoma. macropomum*	*Lippia alba*	acanthocephalan *Neoechinorhynchus buttnerae*
	water	30 days	1.44 and 2.88 g kg^{-1} treatment reduced parasite with11.70% and 5.19%, respectively	*Colossoma. macropomum*	*Zingiber officinale*	acanthocephalan *Neoechinorhynchus buttnerae*
Gonzales, Yoshioka [106]	70% alcohol.	5 min, 10min, 30min and 47.1% after therapeutic bath for 3 days 30 min and 100 min	500 mg L^{-1}, 400 mg L^{-1}, 300 mg L^{-1} and 60 mg L^{-1} 200 mg L^{-1} and 100 mg L^{-1}	*Colossoma. macropomum* (Serrasalmidae) *In-in vitro*	*Cymbopogon citratus*	*monogeneans Anacanthorus spathulatus*
dos Anjos and Isaac [107]	35 mL/L of absolute ethyl alcohol—99.8%	ineffectiveness after 1 hours of therapeutic bath	35 mg/L	*Oreochromis niloticus*	*Mentha piperita*	gill monogeneans
Soares, Cardoso [4]	water + alcohol	100% effective within 20 and 60 min 80% effective with 3 h and 100% with 6 h 6 h 8 h	320 and 160 mg·L^{-1} 80 mg·L^{-1} 40 mg·L^{-1} 80 mg·L^{-1}	*In-in vitro Colossoma macropomum*	*Lippia origanoides*	monogenoideans (*Anacanthorus spathulatus*)

(Table 1) cont.....

References	Solvent	Frequently	Dose	Fish Species	Essential Oil	Parasite
de Castro Nizio, Fujimoto [108]	water	100% mortality at1 h	0.5 and 2.0 mg/L.	*In-vivo* (fresh water fish)	*Varronia curassavica* especially VCUR-202 accession,	*Ichthyophthirius multifiliis*
Soares, Neves [5]	water	20 min 2–3 h 30 min reduce parasite 40.7% and 50.3%	1280 mg/L and 2560 mg/L 160 mg/L 100 and 150 mg/L	*In-in vitro Colossoma macropomum* (tambaqui)	*Lippia alba*	Monogenoideans protozoon I. multifiliis.

THE IMMUNOSTIMULATORY ROLE OF EOS

Essential oils could promote improvement of fish health by modulating several immune and other physiological responses [17]. Understanding fish health status is associated with determining numerous hematological and metabolic indices [18]. For instance, low hematocrit levels indicate chronic anemic conditions [19]. Also, several biochemical indices could be related to fish's entire metabolic state, such as plasma protein, glucose, and triglycerides[20]. Zargar, Rahimi-Afzal [21] reported a significant reduction in alkaline phosphatase and aspartate transaminase (AST) concentration in the rainbow trout fed thymol oil supplemented diets. Also, Valladão, Gallani [22] stated that fortified Nile tilapia diets with 1% thyme oil significantly increased lymphocytes and leucocytes counts, but other blood biochemical parameters were not influenced due to feeding supplements. In addition, thyme extract showed higher efficiency in reducing blood stress indications (glucose and cortisol) in fish treated with a high level of antibiotics as an environmental stressor [23]. Additionally, supplemented tilapia diets with 1 or 2 mL kg^{-1} thyme EO enhanced the immune response *via* increased lysozyme, immunoglobulin M (IgM) and immunoglobulin G (IgG) activity [24]. In the same context, EOs (thymol and carvacrol) at 200 mg kg^{-1} promoted tilapia plasma lysozyme and phagocytosis activity compared with the control, showing an immunostimulatory impact [25]. Moreover, Ran, Hu [26] investigated that using EOs in germ-free zebrafish at levels 2 to 20 mg L^{-1} for one day upregulated the expressions of interleukin-1b (*IL1b*) and claudin 1 genes. While, Zargar, Rahimi-Afzal [21] reported significant upregulation of immune-related genes expressions such as complement component 3 and cluster of differentiation 4, lysozyme and IL1B when thymol oil was blended with the rainbow trout diets. Supplemented fish diets with 1% thyme for 45 days were found to promote lysozyme and phagocytic activities as well as immunocompetent cell population [27].

The enhanced fish immunity after dietary thymol oil administration may be ascribed to the richness of thyme oil with several putative bioactive compounds (quercetin, rutin and naringenin flavonoids), as well as caffeic and rosmarinic acids which could promote immune response *via* the improvement of the innate and adaptive immune responses [28]. The thymol showed a higher ability to promote the proliferation of lymphocyte and phagocytic levels and increase the immunoglobulins such as IgM and IgA[29]. In addition, fortified feed diets with thyme led to higher upregulation of encoding immune gene expression after fish were exposed to contaminated water with pesticides [28]. The caffeic acid present in thyme oil (Fig. **1**) activated the upregulation of IgM, *HSP-70*, complement component chemokine, tumor necrosis factor, *IL1b*, *TLR-7*, interleukin 8 and *IFN-c* genes in Nile tilapia [30].

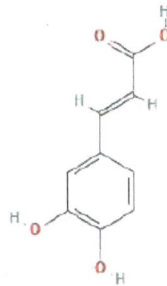

Fig. (1). The basic structural formula of caffeic acid.

At the same trend, Yilmaz and Ergün [31], Haghighi, Pourmoghim [32] and Abdel-Latif, Abdel-Tawwab [33] observed that supplemented carp diets with oregano EO significantly improved the total serum protein, albumin, and globulin levels, kidney (creatinine and BUN) and liver function enzymes (ALT, AST, and ALP) compared with the control group. Similarly, Rafieepour, Hajirezaee [34] illustrated that fortified rainbow trout diets with oregano extract resulted in non-significant changes in the serum enzymes (AST, ALT, and lactate dehydrogenase (LDH)) in comparison to the control group. The improvement in serum protein content and liver and kidney enzymes activity might be due to the role of oregano EO supplementation in improving the activity levels of digestive enzymes (protease, lipase, and amylase) [35]. Additionally, dietary supplementation with oregano EO stimulates appetite in fish [36], which contributes to the enhanced growth rate and general health status.

Moreover, Dawood, Metwally [37] observed that the increased red blood cell (RBCs), and white blood cell (WBCs) counts and promoted immunity in fish fed menthol EO may be due to higher levels of polyphenol in menthol EO, which increased the entire cell's content from iron [38]. Also, reduction in liver and kidney enzyme levels (AST, ALT, ALP, urea, uric acid and creatinine) and increased total protein and globulin levels in fish fed supplemented diets with menthol EO have been reported [37]. In addition, it was elucidated that the immunity indices such as phagocytic index, phagocytic activity, and lysozyme activity were linked with menthol EO *via* increasing the cell ascorbic acid tocopherol and mineral contents [39].

Blood serum protein is a responsive biochemical system whose alteration reflects an organism's response to assorted internal and external conditions. Hence, the highest levels of serum protein, albumin and globulin in fish could be correlated with a high innate immune activity (high lymphocyte count) due to enhanced protein production from liver tissue. Thus, supplemented young beluga fish diets with 0.01, 0.1 to 1% rosemary oil (RO) significantly increased total serum protein contents and enhanced immunity [40, 41]. Also, Naiel, Ismael [42] also illustrated an increment in serum total protein and globulin levels in Nile tilapia after using aflatoxin B1 contaminated diet with 0.5% rosemary, leading to activated immunity and improved immunity fish health. Ebrahimi, Haghjou [41] investigated that the immunity role of RO may be due to its efficiency in producing the peripheral blood neutrophil cells. In the same context, Brum, Pereira [43] found that fortified Nile tilapia (*Oreochromis niloticus*) diets with 1.0% clove basil oil reduced hematocrit and increased neutrophil levels. The hematocrit reduction may be due to energy set off to stimulate the innate immune (the neutrophil production process). In the same study, Nile tilapia diets supplemented with 0.5% and 1% ginger oil significantly increased the total leukocyte, lymphocyte and neutrophil blood counts and promoted phagocytic activity.

In addition, Geay, Mellery [44] investigated that replacement fish oil with linseed oil (LO) significantly increased the plasma lysozyme activity in Eurasian perch fish challenged with *A. salmonicida*. Nguyen, Mandiki [45] observed that the innate immune status of common carp fed the plant oil-based diets linseed oil (LO), sesame oil (SO) and their combination was similar to that of fish fed the fish oil-based diet, except for a decrease in complement activity in fish fed SO diet. Likewise, Amer, Metwally [24] demonstrated that all immunity parameters (IgG, IgM and lysozyme) were promoted in the fish groups fed diets supplemented with cinnamaldehyde or thymol. Also, Acar, Kesbiç [46] investigated that supplemented *O. mossambicus* diets with citrus EO promoted immunity and resistance against pathogens. In a similar study, Dorucu, Ispir [47] reported improvement in the total immunoglobulin level in *O. mykiss* fed diets supplemented with black cumin seed extract. The reported improvement of immune activity correlated with the biological effects of carvacrol [31]. Besides, Giannenas, Triantafillou [48] reported higher catalase activities and lysozyme levels in fish-fed diets supplemented with carvacrol compared with control. Similarly, Jamroz, Wertelecki [49] mentioned that the continuous administration of leaf extracts with higher antioxidant activities sustained immune response and protective stimulation in tilapia against several environmental stressors.

Directly, there are several investigations on the ability of herbal EO to treat infectious diseases *via* promoting the immune status [50]. Mu, Wei [51] showed that the total replacement of dietary fish oil with rapeseed oil significantly increased the serum *TNF-2α* levels in large yellow croakers. The immune response of EO may be due to its ability to promote the tumor necrosis factor (*TNF-2α*) cell production. *TNF-2α* is a main factor responsible for inflammation and regulation of immune function as well as increased non-esterified fatty acid (NEFA) content in serum and lipid levels in the liver of rainbow trouts [52]. Incorporating immunostimulants in diets supplemented with the oregano EO could be related to the expression of both *IL-1β* and *IL-10* genes in common carps (*Cyprinus carpio* L.) [53]. Also, it was detected that there was an increase in plasma lysozyme levels in fish-fed diets fortified with oregano EO in common carp (*Cyprinus carpio* L.)[53], and rainbow trouts [54]. The interleukin 12 (*IL-12*) is mainly produced by monocytes and macrophages, which is responsible for the pro-inflammatory cytokine (*IFN-γ* and *IL-8*) production to promote the activity of phagocyte to control the pathogenic agents [55]. Similarly, Dawood, Metwally [37] confirmed that fish fed diets supplemented with menthol EO significantly increased mRNA levels of *IFN-γ*, *IL-1β*, and *IL-8*, *HSP70*, and caspase-3 (*CASP3*) genes, and phagocytic activity and this may be due to the modulation of the cytokines secretions by T lymphocytes [56]. Also, de Souza Silva, de Pádua Pereira [56] observed that Nile tilapia diets enriched with 0.25% *Mentha* significantly increased the total leukocytes and circulating thrombocytes numbers resulting in a higher cellular innate response.

THE ANTIOXIDANT AND PROTECTIVE ROLE OF EO

The rise of parasitic infestation in wild and farmed fish species indicates the unsuitable living environmental conditions, which led to reduced fish resistance [57]. Several factors (endogenous or exogenous) induce oxidative stress in aquatic organisms. Many kinds of biotic and abiotic agents could be used to protect fish against oxidative stress induced by the parasitic infection. Fish have improved cellular defensive mechanisms against pathogens by stimulation of the antioxidant enzymes (superoxide dismutase, SOD; catalase CAT; peroxidase PER and glutathione) and related enzymes (GR, glutathione peroxidase (GP) and GST) [58]. According to a prior investigation, the parasitic infestation has been shown to modulate the entire biotransformation system, immune activity and antioxidant defensive mechanisms [57, 59 - 61]. Fortified *Nile tilapia* diets with 1 and 2 mL of thymol EO kg $^{-1}$ significantly improved the antioxidant activity by reducing the malondialdehyde (MDA) formation and promoting the glutathione reductases (GR) production and catalase activity in the tissues and blood, respectively [24]. Likewise, Giannenas, Triantafillou [48] found that the dietary administration of thymol and carvacrol significantly decreased the MDA formation in *Oncorhynchus mykiss* fish tissue. Besides, El Euony, Elblehi [62] concluded that African catfish fed diets supplemented with 500 ppm of thymol oil increased the proliferating cell nuclear antigen (PCNA) and caspase-3 splenocytes in response to their potent antioxidant and anti-apoptotic influences. Moreover, the antioxidative capability of tilapia fish (*Oreochromis niloticus*) was enhanced to resist the increased oxidative damage induced by the exposure to lambda-cyhalothrin [28]. In rainbow trout, thyme oil or its components (thymol and carvacrol) promoted glutathione peroxidase (GPx) and SOD activities. Still, they reduced the glutathione S-transferase, catalase and glutathione reductase activities [48]. This antioxidant activity may be due to the stimulatory impact of phytochemicals (flavonoids and phenolic molecules), and minerals such as manganese (Mn) that acts as a cofactor for the superoxide dismutase (SOD) enzyme [63]. Some common phytobiotic compounds, such as oregano EO, have mainly antioxidant properties [64], and this is due to their content of total phenolic molecules (thymol > carvacrol > γ-terpinene > myrcene > linalool > p-cymene > limonene > 1,8-cineole > α-pinene) [65]. Moreover, Hirano, Sasamoto [66] and Embuscado [67] stated that phytogenic constituents found in several EO could scavenge free radicals, chelate transition-metal ions, and decay peroxides by their antioxidant capacity through single-electron transfer process.

Abdel-Latif, Abdel-Tawwab [53] reported that dietary oregano EO could boost the antioxidative status of common carp fingerlings *via* modulating the SOD and CAT activities cooperated with a significant reduction of blood MDA levels. These results may be related to the high potency of oregano EO as natural antioxidants, such as carvacrol and thymol [68], which have a strong ability for scavenging free radicals, chelation of transition-metal ions, and transformation of peroxides [67], similarly occurring in channel catfish [69], rainbow trout [48], and *Nile tilapia* [36]. Moreover, Diler, Gormez [54] detected a significant increase in the hepatocytes CAT activity in rainbow trout fed on diets supplemented with oregano EO isolated from *O. onites* L. Also, Rafieepour, Hajirezaee [34] noted high hepatic SOD and CAT levels in rainbow trout fed on diets supplemented with 6 and 10 g of *O. vulgare* extract per kg diet. Newly, Zhang, Wang [35] reported high serum SOD and GSH-Px with reductions in MDA levels in koi carps fed diets supplemented with oregano EO.

The reactive oxygen species (ROS) production in animals is upregulated by modulating the anti-oxidative defense activities by increasing the SOD, CAT,

GPx levels. It was found that the total replacement of fish oil (FO) by rapeseed oil (RO) significantly reduced the total antioxidant (T-AOC) and SOD activities in tilapia serum [70]. However, the fish fed rapeseed oil (RO) had significantly lower T-AOC activity and expression levels of the anti-oxidation-related genes (SOD1, SOD2 and GPx) in large yellow croaker hepatocytes [51]. The α-linolenic acid found in essential oil could be responsible for the anti-oxidative activity by modulating the Nrf2 signaling pathway, which stimulates the type II detoxifying enzyme genes transcriptions (SOD, CAT and GPx) in fish [71]. Also, menthol oil was found to have antiradical components such as monoterpene ketones, menthone, and iso menthone that could display higher mRNA levels of CAT and GPX genes [72].

The Nrf2 is a redox-sensitive transcription factor which plays a vital role in cellular defense mechanisms against oxidative damage [74]. It was found that the Nrf2 mRNA expression was decreased with the increase of dietary PO levels in large yellow croakers (*Larimichthyscrocea*) [73]. Similarly, the higher levels of dietary vegetable oil reduce the antioxidant capacity through altering the Nrf2 pathway in Japanese sea bass [75]. Also, increasing dietary PO levels cause increases in the *TLR2, TLR3, TLR9, TLR22, MyD88* mRNA expression and pro-inflammatory gene expression [73]. The activation of TLR-NF-κB signaling pathway might be correlated with the induction *IL-1β* and TNFα mRNA expression activity [76]. Furthermore, studies have reported that activated NF-κB could suppress the expression or activity of anti-inflammatory cytokines like IL-10 and TGFβ [78]. Those results were in agreement with the findings of Tan, Peng [77] that high lipid diet activated the TLR-NF-κB signaling pathways and promoted the cytokine gene expression in trout.

THE ANTIPARASITIC EFFECTS OF EOS

Nutritional therapies are a useful strategy for controlling and preventing infectious diseases [79]. Several feed additives such as prebiotics, probiotics, organic acids, phytogenic plants and essential oils have gained huge interest in animal feeds [80]. Phytogenic molecules are plant-based natural extracts or compounds derived from herbs and spices, known for their beneficial characteristics and efficiency on production and health in aquatic animals [81]. In aquafeeds, using EOs as a feed additive has been investigated to promote appetite, enhance feed efficiency and production, and improve innate immune activity [82]. Thymol, carvacrol, cinnamaldehyde, and other EOs isolated from clove, coriander, star anise, ginger, garlic, rosemary, and mint have been blended with diets individually or as a mixture in fish, poultry and animal nutrition [83, 84]. Oregano is the most widespread used phytogenic plant because of its high contents of carvacrol and thymol as bioactive molecules (Fig. **2**) [85, 86]. These molecules have a wide spectrum of antibacterial, immunostimulatory and anti-oxidative efficiency, promoting the intestinal microbiota, growth promoters and decreasing diseases and mortality [48, 87 - 90]. The efficiency of garlic (*Allium sativum*) derived molecules as an immunostimulant, antibacterial and antiparasitic agent has been investigated in numerous fish species [91 - 93]; as well as fortified fish diets with garlic extract showing high effectiveness against monogenean parasite infestations [94, 95].

Several studies investigated the efficiency of several EOs against parasitic infestation in fish. Steverding, Morgan [96] investigated that the concentrations between 3 to 30 ppm from Australian tea tree (*Melaleuca alternifolia*) oil significantly lowered the parasitic infestation (*Gyrodactylus turnbulli*). The higher anthelminthic properties of cajuput and West Indian bay combined oils may be due to the high terpenes and phenol propanoids contents on guppies *Poecilia reticulata*. Also, Malheiros, Maciel [97] found that the *M. piperita* EO had high anthelminthic activity against *Dawestrema* spp. on the giant Amazon fish. The efficiency of *M. piperita* essential oil against *Dawestrema* spp. was dose-dependent. The *Lippia alba* EO exhibited potent antiparasitic efficiency at high concentration against monogenoideans *in-in vitro* and at low levels (100 and 150 mg/L) against the protozoon *I. multifiliis in-vivo* (fish fries) [5]. Besides, Soares, Neves [98] investigated that the *L. sidoides* EO has *in-in vitro* antiparasitic activity against *Ichthyophthirius multifiliis*, and monogenoideans found in the gills of *C. macropomum* and this EO at low concentrations (10 and 20 mg/L) revealed toxic effects.

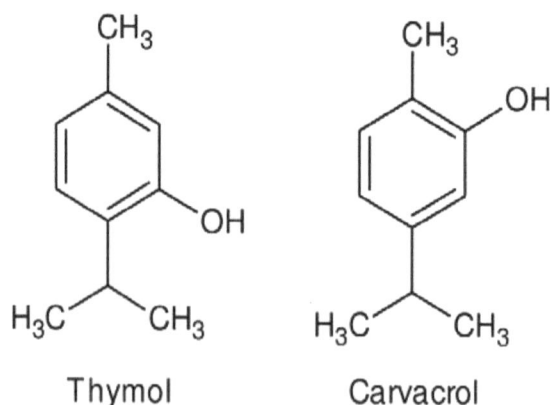

Fig. (2). The basic structural formula of carvacrol and thymol.

Soares, Cardoso [4] concluded that the *L. origanoides* EO has a higher *in-in vitro* effect against monogenoidean parasites of *C. macropomum*. Still, the low levels are safe for fish (20 and 40 mg·L^{-1}) showed no antiparasitic effectiveness. The *in-in vitro* results against *I. multifiliis* infections verified that all tested *Melaleuca alternifolia, Lavandula angustifolia* and *Mentha piperita* EOs [57, 114, 227 and 45 µL L^{-1}) showed a parasitic cytotoxic effect [99]. Moreover, Baldissera, Souza [100] proposed that the *Melaleuca alternifolia* EO treatment bath 1 h per day for 4 days could be an effective therapeutic protocol for treating infected silver catfish with *I. multifiliis*. Also, therapeutic baths containing EO of *Lippia sidoides* (pepper rosemary) and *Mentha piperita* (peppermint) in tilapia fish decreased the prevalence of monogeneans (*Cichlidogyrus tilapiae, Cichlidogyrus thurstonae, Cichlidogyrus halli,* and *Scutogyrus longicornis*) infection up to 70% [12]. In another study, Ling, Jiang [101] reported that cinnamaldehyde EO was effective against *Dactylogyrus intermedius*, and the cinnamaldehyde served as an alternative therapeutic agent in gold fish. Recently, Firmino, Vallejos-Vidal [86] observed a reduction of 78% prevalence of the total parasite count of *S. chrysophrii* and a reduction in the presence of most parasitic developmental stages detected in *Sparusaurata* fish fed the diet containing garlic, carvacrol and thymol EOs for 65 days.

CONCLUDING REMARKS

In cultured fish, parasitic infestations cause decreased fish production and huge economic losses. Thus, there is a need to develop new and safe alternative treatment protocols using bioactive molecules derived from herbs, such as EOs. Some EOs have shown a high ability to control or treat parasitic helminths in fish, particularly monogeneans and nematodes. Few EOs have shown anthelmintic activity in in *in vitro* or therapeutic baths. Because of many identified EOs, the issue of specificity remains unknown for most EOs. Also, according to the results from previous studies, we believe it would be functional to understand whether EOs may diminish the adverse effects of parasite infestations on fish health and production. Moreover, sustainability measures of anthelmintic behaviors of EOs should be verified before their use in fish parasite risk management.

Several previous studies demonstrated that antiparasitic drug resistance exists in animals if intensive treatment strategies are used. Thus, in farmed fish, these natural compounds and herbal products, including EOs, may be useful to diminish disease effects. Lastly, the low solubility of EOs on the water is the main problem that is still unresolved, which is considered the main disadvantage in wide application in fish farming. Consequently, EO in nanoforms or nano-capsulated forms can be applied to overcome such constraints. Still, there is a need to examine such EOs technologies and their effects on the treatment or control of fish parasite infections.

CONSENT FOR PUBLICATION

Not applicable.

CONFLICT OF INTEREST

The author declares no conflict of interest, financial or otherwise.

ACKNOWLEDGEMENTS

Declared none.

REFERENCES

[1]   Tavares-Dias M. Current knowledge on use of essential oils as alternative treatment against fish parasites. Aquat Living Resour 2018; 31: 13.
        [http://dx.doi.org/10.1051/alr/2018001]

[2]   The state of world fisheries and aquaculture 2020 Sustainability in action. Food and Agriculture Organization of the United Nations 2020.

[3]   Soler-Jiménez LC, Paredes-Trujillo AI, Vidal-Martínez VM. Helminth parasites of finfish commercial aquaculture in Latin America. J Helminthol 2017; 91(2): 110-36.
        [http://dx.doi.org/10.1017/S0022149X16000833] [PMID: 27976599]

[4]   Soares BV, Cardoso ACF, Campos RR, *et al.* Antiparasitic, physiological and histological effects of the essential oil of *Lippia origanoides* (Verbenaceae) in native freshwater fish Colossoma macropomum. Aquaculture 2017; 469: 72-8.
        [http://dx.doi.org/10.1016/j.aquaculture.2016.12.001]

[5]   Soares BV, Neves LR, Oliveira MSB, *et al.* Antiparasitic activity of the essential oil of Lippia alba on ectoparasites of *Colossoma macropomum* (tambaqui) and its physiological and histopathological effects. Aquaculture 2016; 452: 107-14.
        [http://dx.doi.org/10.1016/j.aquaculture.2015.10.029]

[6]   Ozogul Y, Kuley Boğa E, Akyol I, *et al.* Antimicrobial activity of thyme essential oil nanoemulsions on spoilage bacteria of fish and food-borne pathogens. Food Biosci 2020; 36100635.
        [http://dx.doi.org/10.1016/j.fbio.2020.100635]

[7]   Fabri RL, Nogueira MS, Moreira JR, Bouzada MLM, Scio E. Identification of antioxidant and antimicrobial compounds of Lippia species by bioautography. J Med Food 2011; 14(7-8): 840-6.
        [http://dx.doi.org/10.1089/jmf.2010.0141] [PMID: 21476886]

[8]   Escobar P, Milena Leal S, Herrera LV, Martinez JR, Stashenko E. Chemical composition and antiprotozoal activities of Colombian *Lippia* spp essential oils and their major components. Mem Inst Oswaldo Cruz 2010; 105(2): 184-90.
        [http://dx.doi.org/10.1590/S0074-02762010000200013] [PMID: 20428679]

[9]   Ocazionez RE, Meneses R, Torres FÁ, Stashenko E. Virucidal activity of Colombian Lippia essential oils on dengue virus replication *in vitro*. Mem Inst Oswaldo Cruz 2010; 105(3): 304-9.
        [http://dx.doi.org/10.1590/S0074-02762010000300010] [PMID: 20512244]

[10]   Sharifi-Rad J, Sureda A, Tenore G, *et al.* Biological activities of essential oils: From plant chemoecology to traditional healing systems. Molecules 2017; 22(1): 70.
        [http://dx.doi.org/10.3390/molecules22010070] [PMID: 28045446]

[11]   Santos Rd. Metabolismo básico e origem dos metabólitos secundários Simões, CMO; Schenkel, EP; Gosmann, G; Mello. JCP 2004; pp. 403-34.

[12]   de Oliveira Hashimoto GS, Neto FM, Ruiz ML, *et al.* Essential oils of *Lippia sidoides* and Mentha piperita against monogenean parasites and their influence on the hematology of Nile tilapia. Aquaculture 2016; 450: 182-6.
[http://dx.doi.org/10.1016/j.aquaculture.2015.07.029]

[13]   Figiel A, Szumny A, Gutiérrez-Ortíz A, Carbonell-Barrachina ÁA. Composition of oregano essential oil (*Origanum vulgare*) as affected by drying method. J Food Eng 2010; 98(2): 240-7.
[http://dx.doi.org/10.1016/j.jfoodeng.2010.01.002]

[14]   Orafidiya LO, Agbani EO, Iwalewa EO, Adelusola KA, Oyedapo OO. Studies on the acute and sub-chronic toxicity of the essential oil of *Ocimum gratissimum* L. leaf. Phytomedicine 2004; 11(1): 71-6.
[http://dx.doi.org/10.1078/0944-7113-00317] [PMID: 14971724]

[15]   Yang Y, Wang Q, Diarra MS, Yu H, Hua Y, Gong J. Functional assessment of encapsulated citral for controlling necrotic enteritis in broiler chickens. Poult Sci 2016; 95(4): 780-9.
[http://dx.doi.org/10.3382/ps/pev375] [PMID: 26740132]

[16]   Leyva-López N, Gutiérrez-Grijalva E, Vazquez-Olivo G, Heredia J. Essential oils of oregano: Biological activity beyond their antimicrobial properties. Molecules 2017; 22(6): 989.
[http://dx.doi.org/10.3390/molecules22060989] [PMID: 28613267]

[17]   Harikrishnan R, Balasundaram C, Heo MS. Impact of plant products on innate and adaptive immune system of cultured finfish and shellfish. Aquaculture 2011; 317(1-4): 1-15.
[http://dx.doi.org/10.1016/j.aquaculture.2011.03.039]

[18]   Hrubec TC, Smith SA, Robertson JL. Age-related changes in hematology and plasma chemistry values of hybrid striped bass (Morone chrysops X Morone saxatilis). Vet Clin Pathol 2001; 30(1): 8-15.
[http://dx.doi.org/10.1111/j.1939-165X.2001.tb00249.x] [PMID: 12024324]

[19]   Higuchi L, Feiden A, Maluf M, Dallagnol J, Zaminhan M, Boscolo W. Erythrocitary and biochemical evaluation of Rhamdia quelen submitted to diets with different proteic and energetic levels. Cienc Anim Bras 2011; 12(1): 70-5.

[20]   Cox RA, García-Palmieri MR. Cholesterol, triglycerides, and associated lipoproteins. Clinical methods: the history, physical, and laboratory examinations 1990.

[21]   Zargar A, Rahimi-Afzal Z, Soltani E, *et al.* Growth performance, immune response and disease resistance of rainbow trout (*Oncorhynchus mykiss*) fed *Thymus vulgaris* essential oils. Aquacult Res 2019; 50(11): 3097-106.
[http://dx.doi.org/10.1111/are.14243]

[22]   Valladão GMR, Gallani SU, Kotzent S, Assane IM, Pilarski F. Effects of dietary thyme essential oil on hemato-immunological indices, intestinal morphology, and microbiota of Nile tilapia. Aquacult Int 2019; 27(2): 399-411.
[http://dx.doi.org/10.1007/s10499-018-0332-5]

[23]   Hoseini SM, Yousefi M. Beneficial effects of thyme (*Thymus vulgaris*) extract on oxytetracycline-induced stress response, immunosuppression, oxidative stress and enzymatic changes in rainbow trout (*Oncorhynchus mykiss*). Aquacult Nutr 2019; 25(2): 298-309.
[http://dx.doi.org/10.1111/anu.12853]

[24]   Amer SA, Metwally AE, Ahmed SAA. The influence of dietary supplementation of cinnamaldehyde and thymol on the growth performance, immunity and antioxidant status of monosex Nile tilapia fingerlings (*Oreochromis niloticus*). Egypt J Aquat Res 2018; 44(3): 251-6.
[http://dx.doi.org/10.1016/j.ejar.2018.07.004]

[25]   Fachini-Queiroz FC, Kummer R, Estevao-Silva CF. Effects of thymol and carvacrol, constituents of *Thymus vulgaris* L. essential oil, on the inflammatory response. Evidence-Based Complementary and Alternative Medicine 2012.

[26]   Ran C, Hu J, Liu W, *et al.* Thymol and carvacrol affect hybrid tilapia through the combination of direct stimulation and an intestinal microbiota-mediated effect: insights from a germ-free zebrafish

model. J Nutr 2016; 146(5): 1132-40.
[http://dx.doi.org/10.3945/jn.115.229377] [PMID: 27075912]

[27] Gültepe N, Bilen S, Yılmaz S, Güroy D, Aydın S. Effects of herbs and spice on health status of tilapia (*Oreochromis mossambicus*) challenged with *Streptococcus iniae*. Acta Vet Brno 2014; 83(2): 125-31.
[http://dx.doi.org/10.2754/avb201483020125]

[28] Khalil SR, Elhakim YA, Abd El-fattah AH, Ragab Farag M, Abd El-Hameed NE, EL-Murr AE. Dual immunological and oxidative responses in *Oreochromis niloticus* fish exposed to lambda cyhalothrin and concurrently fed with Thyme powder (*Thymus vulgaris* L.): Stress and immune encoding gene expression. Fish Shellfish Immunol 2020; 100: 208-18.
[http://dx.doi.org/10.1016/j.fsi.2020.03.009]

[29] Abd El-Naby AS, Al-Sagheer AA, Negm SS, Naiel MAE. Dietary combination of chitosan nanoparticle and thymol affects feed utilization, digestive enzymes, antioxidant status, and intestinal morphology of *Oreochromis niloticus*. Aquaculture 2020; 515734577.
[http://dx.doi.org/10.1016/j.aquaculture.2019.734577]

[30] Yilmaz S. Effects of dietary caffeic acid supplement on antioxidant, immunological and liver gene expression responses, and resistance of Nile tilapia, *Oreochromis niloticus* to *Aeromonas veronii*. Fish Shellfish Immunol 2019; 86: 384-92.
[http://dx.doi.org/10.1016/j.fsi.2018.11.068] [PMID: 30502464]

[31] Yilmaz E, Ergün S, ilmaz S. Influence of carvacrol on the growth performance, hematological, non-specific immune and serum biochemistry parameters in rainbow trout (*Oncorhynchus mykiss*). Food Nutr Sci 2015; 6(5): 523-31.
[http://dx.doi.org/10.4236/fns.2015.65054]

[32] Haghighi M, Pourmoghim H, Sharif Rohani M. Effect of Origanum vulgare extract on immune responses and hematological parameters of rainbow trout (*Oncorhynchus mykiss*). Oceanography & Fisheries Open access Journal 2018; 6(3)555687.
[http://dx.doi.org/10.19080/OFOAJ.2018.06.555687]

[33] Abdel-Latif HMR, Abdel-Tawwab M, Khafaga AF, Dawood MAO. Dietary oregano essential oil improved the growth performance *via* enhancing the intestinal morphometry and hepato-renal functions of common carp (*Cyprinus carpio* L.) fingerlings. Aquaculture 2020; 526735432.
[http://dx.doi.org/10.1016/j.aquaculture.2020.735432]

[34] Rafieepour A, Hajirezaee S, Rahimi R. Dietary oregano extract (*Origanum vulgare* L.) enhances the antioxidant defence in rainbow trout, *Oncorhynchus mykiss* against toxicity induced by organophosphorus pesticide, diazinon. Toxin Rev 2018; 1-11.

[35] Zhang R, Wang XW, Liu LL, Cao YC, Zhu H. Dietary oregano essential oil improved the immune response, activity of digestive enzymes, and intestinal microbiota of the koi carp, *Cyprinus carpio*. Aquaculture 2020; 518734781.
[http://dx.doi.org/10.1016/j.aquaculture.2019.734781]

[36] Abdel-Latif HM, Khalil RH. Evaluation of two Phytobiotics, Spirulina platensis and Origanum vulgare extract on Growth, Serum antioxidant activities and Resistance of Nile tilapia (*Oreochromis niloticus*) to pathogenic Vibrio alginolyticus. Int J Fish Aquat Stud 2014; 1: 250-5.

[37] Dawood MAO, El-Salam Metwally A, Elkomy AH, *et al.* The impact of menthol essential oil against inflammation, immunosuppression, and histopathological alterations induced by chlorpyrifos in Nile tilapia. Fish Shellfish Immunol 2020; 102: 316-25.
[http://dx.doi.org/10.1016/j.fsi.2020.04.059] [PMID: 32371257]

[38] Layrisse M, García-Casal MN, Solano L, *et al.* New property of vitamin A and beta-carotene on human iron absorption: effect on phytate and polyphenols as inhibitors of iron absorption. Arch Latinoam Nutr 2000; 50(3): 243-8.
[PMID: 11347293]

[39] Abdel-Daim MM, Dawood MAO, Elbadawy M, Aleya L, Alkahtani S. Spirulina platensis reduced

oxidative damage induced by chlorpyrifos toxicity in Nile tilapia (*Oreochromis niloticus*). Animals (Basel) 2020; 10(3): 473.
[http://dx.doi.org/10.3390/ani10030473] [PMID: 32178251]

[40]   Akrami R, Gharaei A, Mansour MR, Galeshi A. Effects of dietary onion (*Allium cepa*) powder on growth, innate immune response and hemato–biochemical parameters of beluga (*Huso huso* Linnaeus, 1754) juvenile. Fish Shellfish Immunol 2015; 45(2): 828-34.
[http://dx.doi.org/10.1016/j.fsi.2015.06.005] [PMID: 26067169]

[41]   Ebrahimi E, Haghjou M, Nematollahi A, Goudarzian F. Effects of rosemary essential oil on growth performance and hematological parameters of young great sturgeon (Huso huso). Aquaculture 2020; 521734909.
[http://dx.doi.org/10.1016/j.aquaculture.2019.734909]

[42]   Naiel MAE, Ismael NEM, Shehata SA. Ameliorative effect of diets supplemented with rosemary (*Rosmarinus officinalis*) on aflatoxin B1 toxicity in terms of the performance, liver histopathology, immunity and antioxidant activity of Nile Tilapia (*Oreochromis niloticus*). Aquaculture 2019; 511734264.
[http://dx.doi.org/10.1016/j.aquaculture.2019.734264]

[43]   Brum A, Pereira SA, Owatari MS, *et al.* Effect of dietary essential oils of clove basil and ginger on Nile tilapia (*Oreochromis niloticus*) following challenge with *Streptococcus agalactiae*. Aquaculture 2017; 468: 235-43.
[http://dx.doi.org/10.1016/j.aquaculture.2016.10.020]

[44]   Geay F, Mellery J, Tinti E, *et al.* Effects of dietary linseed oil on innate immune system of Eurasian perch and disease resistance after exposure to *Aeromonas salmonicida* achromogen. Fish Shellfish Immunol 2015; 47(2): 782-96.
[http://dx.doi.org/10.1016/j.fsi.2015.10.021] [PMID: 26497094]

[45]   Nguyen TM, Mandiki SNM, Gense C, Tran TNT, Nguyen TH, Kestemont P. A combined *in vivo* and *in vitro* approach to evaluate the influence of linseed oil or sesame oil and their combination on innate immune competence and eicosanoid metabolism processes in common carp (*Cyprinus carpio*). Dev Comp Immunol 2020; 102103488.
[http://dx.doi.org/10.1016/j.dci.2019.103488] [PMID: 31476324]

[46]   Acar Ü, Kesbiç OS, Yılmaz S, Gültepe N, Türker A. Evaluation of the effects of essential oil extracted from sweet orange peel (*Citrus sinensis*) on growth rate of tilapia (*Oreochromis mossambicus*) and possible disease resistance against Streptococcus iniae. Aquaculture 2015; 437: 282-6.
[http://dx.doi.org/10.1016/j.aquaculture.2014.12.015]

[47]   Dorucu M, Ispir U, Colak S, Altinterim B, Celayir Y. The effect of black cumin seeds, Nigella sativa, onthe immune response of rainbow trout, *Oncorhynchus mykiss*. Mediterranean Aquaculture Journal 2009; 2(1): 27-33.
[http://dx.doi.org/10.21608/maj.2009.2667]

[48]   Giannenas I, Triantafillou E, Stavrakakis S, *et al.* Assessment of dietary supplementation with carvacrol or thymol containing feed additives on performance, intestinal microbiota and antioxidant status of rainbow trout (*Oncorhynchus mykiss*). Aquaculture 2012; 350-353: 26-32.
[http://dx.doi.org/10.1016/j.aquaculture.2012.04.027]

[49]   Jamroz D, Wertelecki T, Houszka M, Kamel C. Influence of diet type on the inclusion of plant origin active substances on morphological and histochemical characteristics of the stomach and jejunum walls in chicken. J Anim Physiol Anim Nutr (Berl) 2006; 90(5-6): 255-68.
[http://dx.doi.org/10.1111/j.1439-0396.2005.00603.x] [PMID: 16684147]

[50]   Mahima , Rahal A, Deb R, *et al.* Immunomodulatory and therapeutic potentials of herbal, traditional/indigenous and ethnoveterinary medicines. Pak J Biol Sci 2012; 15(16): 754-74.
[http://dx.doi.org/10.3923/pjbs.2012.754.774] [PMID: 24175417]

[51]   Mu H, Wei C, Xu W, Gao W, Zhang W, Mai K. Effects of replacement of dietary fish oil by rapeseed

oil on growth performance, anti-oxidative capacity and inflammatory response in large yellow croaker *Larimichthys crocea.* Aquacult Rep 2020; 16100251.
[http://dx.doi.org/10.1016/j.aqrep.2019.100251]

[52]   Albalat A, Liarte C, MacKenzie S, Tort L, Planas JV, Navarro I. Control of adipose tissue lipid metabolism by tumor necrosis factor-α in rainbow trout (Oncorhynchus mykiss). J Endocrinol 2005; 184(3): 527-34.
[http://dx.doi.org/10.1677/joe.1.05940] [PMID: 15749811]

[53]   Abdel-Latif HMR, Abdel-Tawwab M, Khafaga AF, Dawood MAO. Dietary origanum essential oil improved antioxidative status, immune-related genes, and resistance of common carp (*Cyprinus carpio* L.) to Aeromonas hydrophila infection. Fish Shellfish Immunol 2020; 104: 1-7.
[http://dx.doi.org/10.1016/j.fsi.2020.05.056] [PMID: 32474085]

[54]   Diler O, Gormez O, Diler I, Metin S. Effect of oregano (*Origanum onites* L.) essential oil on growth, lysozyme and antioxidant activity and resistance against *Lactococcus garvieae* in rainbow trout, *Oncorhynchus mykiss* (Walbaum). Aquacult Nutr 2017; 23(4): 844-51.
[http://dx.doi.org/10.1111/anu.12451]

[55]   Liu J, Cao S, Kim S, *et al.* Interleukin-12: an update on its immunological activities, signaling and regulation of gene expression. Curr Immunol Rev 2005; 1(2): 119-37.
[http://dx.doi.org/10.2174/1573395054065115] [PMID: 21037949]

[56]   de Souza Silva LT, de Pádua Pereira U, de Oliveira HM, *et al.* Hemato-immunological and zootechnical parameters of Nile tilapia fed essential oil of Mentha piperita after challenge with *Streptococcus agalactiae.* Aquaculture 2019; 506: 205-11.
[http://dx.doi.org/10.1016/j.aquaculture.2019.03.035]

[57]   Skuratovskaya E, Skuratovskaya E, Zav'yalov A, Rudneva I. Health parameters and antioxidant response in Black Sea whiting Merlangius merlangus euxinus (Nordmann, 1840) parasitized by nematode Hysterothylacium aduncum (Rud., 1802). Comun Sci 2019; 9(4): 700-9.
[http://dx.doi.org/10.14295/cs.v9i4.2441]

[58]   Livingstone DR. Contaminant-stimulated reactive oxygen species production and oxidative damage in aquatic organisms. Mar Pollut Bull 2001; 42(8): 656-66.
[http://dx.doi.org/10.1016/S0025-326X(01)00060-1] [PMID: 11525283]

[59]   Skuratovskaya E, Zav'yalov A, Rudneva I. Response of the antioxidant system of black sea whiting merlangus *merlangus euxinus* (Nordmann, 1840) to parasitic nematode *hysterothylacium aduncum* (Rudolphi, 1802) infection. Bull Eur Assoc Fish Pathol 2015; 35(5): 170-6.

[60]   Dautremepuits C, Betoulle S, Vernet G. Antioxidant response modulated by copper in healthy or parasitized carp (Cyprinus carpio L.) by Ptychobothrium sp. (Cestoda). Biochim Biophys Acta, Gen Subj 2002; 1573(1): 4-8.
[http://dx.doi.org/10.1016/S0304-4165(02)00328-8] [PMID: 12383935]

[61]   Ganjewala D. Cymbopogon essential oils: Chemical compositions and bioactivities. International journalof essential oil therapeutics 2009; 3(2-3): 56-65.

[62]   El Euony OI, Elblehi SS, Abdel-Latif HM, Abdel-Daim MM, El-Sayed YS. Modulatory role of dietary Thymus vulgaris essential oil and Bacillus subtilis against thiamethoxam-induced hepatorenal damage, oxidative stress, and immunotoxicity in African catfish (*Clarias garipenus*). Environ Sci Pollut Res Int 2020; 27(18): 23108-28.
[http://dx.doi.org/10.1007/s11356-020-08588-5] [PMID: 32333347]

[63]   Jabri-Karoui I, Bettaieb I, Msaada K, Hammami M, Marzouk B. Research on the phenolic compounds and antioxidant activities of Tunisian Thymus capitatus. J Funct Foods 2012; 4(3): 661-9.
[http://dx.doi.org/10.1016/j.jff.2012.04.007]

[64]   Lo KM, Cheung PCK. Antioxidant activity of extracts from the fruiting bodies of Agrocybe aegerita var. alba. Food Chem 2005; 89(4): 533-9.
[http://dx.doi.org/10.1016/j.foodchem.2004.03.006]

[65]   Youdim KA, Deans SG, Finlayson HJ. The antioxidant properties of thyme (*Thymus zygis* L.) essential oil: an inhibitor of lipid peroxidation and a free radical scavenger. J Essent Oil Res 2002; 14(3): 210-5. [http://dx.doi.org/10.1080/10412905.2002.9699825]

[66]   Hirano R, Sasamoto W, Matsumoto A, Itakura H, Igarashi O, Kondo K. Antioxidant ability of various flavonoids against DPPH radicals and LDL oxidation. J Nutr Sci Vitaminol (Tokyo) 2001; 47(5): 357-62. [http://dx.doi.org/10.3177/jnsv.47.357] [PMID: 11814152]

[67]   Embuscado ME. Spices and herbs: Natural sources of antioxidants – a mini review. J Funct Foods 2015; 18: 811-9. [http://dx.doi.org/10.1016/j.jff.2015.03.005]

[68]   García-Beltrán J, Esteban M. Properties and applications of plants of Origanum Sp. Genus SM J Biol 2016; 2: 1006-15.

[69]   Zheng ZL, Tan JYW, Liu HY, Zhou XH, Xiang X, Wang KY. Evaluation of oregano essential oil (*Origanum heracleoticum* L.) on growth, antioxidant effect and resistance against Aeromonas hydrophila in channel catfish (Ictalurus punctatus). Aquaculture 2009; 292(3-4): 214-8. [http://dx.doi.org/10.1016/j.aquaculture.2009.04.025]

[70]   Peng X, Li F, Lin S, Chen Y. Effects of total replacement of fish oil on growth performance, lipid metabolism and antioxidant capacity in tilapia (*Oreochromis niloticus*). Aquacult Int 2016; 24(1): 145-56. [http://dx.doi.org/10.1007/s10499-015-9914-7]

[71]   Jin M, Lu Y, Yuan Y, *et al.* Regulation of growth, antioxidant capacity, fatty acid profiles, hematological characteristics and expression of lipid related genes by different dietary n-3 highly unsaturated fatty acids in juvenile black seabream (*Acanthopagrus schlegelii*). Aquaculture 2017; 471: 55-65. [http://dx.doi.org/10.1016/j.aquaculture.2017.01.004]

[72]   Lalhminghlui K, Jagetia GC. Evaluation of the free-radical scavenging and antioxidant activities of Chilauni, *Schima wallichii* Korth *in vitro*. Future Sci OA 2018; 4(2)FSO272. [http://dx.doi.org/10.4155/fsoa-2017-0086] [PMID: 29379645]

[73]   Li X, Ji R, Cui K, *et al.* High percentage of dietary palm oil suppressed growth and antioxidant capacity and induced the inflammation by activation of TLR-NF-κB signaling pathway in large yellow croaker (Larimichthys crocea). Fish Shellfish Immunol 2019; 87: 600-8. [http://dx.doi.org/10.1016/j.fsi.2019.01.055] [PMID: 30738147]

[74]   Kim J, Cha YN, Surh YJ. A protective role of nuclear factor-erythroid 2-related factor-2 (Nrf2) in inflammatory disorders. Mutat Res 2010; 690(1-2): 12-23. [http://dx.doi.org/10.1016/j.mrfmmm.2009.09.007] [PMID: 19799917]

[75]   Tan P, Dong X, Xu H, Mai K, Ai Q. Dietary vegetable oil suppressed non-specific immunity and liver antioxidant capacity but induced inflammatory response in Japanese sea bass (Lateolabrax japonicus). Fish Shellfish Immunol 2017; 63: 139-46. [http://dx.doi.org/10.1016/j.fsi.2017.02.006] [PMID: 28189766]

[76]   Ajuwon KM, Spurlock ME. Palmitate activates the NF-κB transcription factor and induces IL-6 and TNFα expression in 3T3-L1 adipocytes. The Journal of nutrition 2005; 135(8): 1841-6.

[77]   Tan P, Peng M, Liu D, *et al.* Suppressor of cytokine signaling 3 (SOCS3) is related to pro-inflammatory cytokine production and triglyceride deposition in turbot (*Scophthalmus maximus*). Fish Shellfish Immunol 2017; 70: 381-90. [http://dx.doi.org/10.1016/j.fsi.2017.09.006] [PMID: 28882805]

[78]   Lawrence T. The nuclear factor NF-kappaB pathway in inflammation. Cold Spring Harb Perspect Biol 2009; 1(6)a001651. [http://dx.doi.org/10.1101/cshperspect.a001651] [PMID: 20457564]

[79]   Trichet VV. Nutrition and immunity: an update. Aquacult Res 2010; 41(3): 356-72.
        [http://dx.doi.org/10.1111/j.1365-2109.2009.02374.x]

[80]   Ghanbari M, Kneifel W, Domig KJ. A new view of the fish gut microbiome: Advances from next-generation sequencing. Aquaculture 2015; 448: 464-75.
        [http://dx.doi.org/10.1016/j.aquaculture.2015.06.033]

[81]   Encarnação P. Functional feed additives in aquaculture feeds Aquafeed formulation. Elsevier 2016; pp. 217-37.
        [http://dx.doi.org/10.1016/B978-0-12-800873-7.00005-1]

[82]   Sutili FJ, Gatlin DM III, Heinzmann BM, Baldisserotto B. Plant essential oils as fish diet additives: benefits on fish health and stability in feed. Rev Aquacult 2018; 10(3): 716-26.
        [http://dx.doi.org/10.1111/raq.12197]

[83]   Franz C, Baser KHC, Windisch W. Essential oils and aromatic plants in animal feeding - a European perspective. A review. Flavour Fragrance J 2010; 25(5): 327-40.
        [http://dx.doi.org/10.1002/ffj.1967]

[84]   Gadde U, Kim WH, Oh ST, Lillehoj HS. Alternatives to antibiotics for maximizing growth performance and feed efficiency in poultry: a review. Anim Health Res Rev 2017; 18(1): 26-45.
        [http://dx.doi.org/10.1017/S1466252316000207] [PMID: 28485263]

[85]   Zhou F, Ji B, Zhang H, *et al.* Synergistic effect of thymol and carvacrol combined with chelators and organic acids against *Salmonella Typhimurium.* J Food Prot 2007; 70(7): 1704-9.
        [http://dx.doi.org/10.4315/0362-028X-70.7.1704] [PMID: 17685346]

[86]   Firmino JP, Vallejos-Vidal E, Sarasquete C, *et al.* Unveiling the effect of dietary essential oils supplementation in *Sparus aurata* gills and its efficiency against the infestation by Sparicotyle chrysophrii. Sci Rep 2020; 10(1): 17764.
        [http://dx.doi.org/10.1038/s41598-020-74625-5] [PMID: 33082387]

[87]   Bandeira Junior G, Sutili FJ, Gressler LT, *et al.* Antibacterial potential of phytochemicals alone or in combination with antimicrobials against fish pathogenic bacteria. J Appl Microbiol 2018; 125(3): 655-65.
        [http://dx.doi.org/10.1111/jam.13906] [PMID: 29741243]

[88]   Volpatti D, Chiara B, Francesca T, Marco G. Growth parameters, innate immune response and resistance to *Listonella* (*Vibrio*) *anguillarum* of *Dicentrarchus labrax* fed carvacrol supplemented diets. Aquacult Res 2013; 45(1): 31-44.
        [http://dx.doi.org/10.1111/j.1365-2109.2012.03202.x]

[89]   Pérez-Sánchez J, Benedito-Palos L, Estensoro I, *et al.* Effects of dietary NEXT ENHANCE®150 on growth performance and expression of immune and intestinal integrity related genes in gilthead sea bream (Sparus aurata L.). Fish Shellfish Immunol 2015; 44(1): 117-28.
        [http://dx.doi.org/10.1016/j.fsi.2015.01.039] [PMID: 25681752]

[90]   Ahmadifar E, Falahatkar B, Akrami R. Effects of dietary thymol-carvacrol on growth performance, hematological parameters and tissue composition of juvenile rainbow trout, *Oncorhynchus mykiss.* J Appl Ichthyology 2011; 27(4): 1057-60.
        [http://dx.doi.org/10.1111/j.1439-0426.2011.01763.x]

[91]   Nya EJ, Dawood Z, Austin B. The garlic component, allicin, prevents disease caused by *Aeromonas hydrophila* in rainbow trout, *Oncorhynchus mykiss* (Walbaum). J Fish Dis 2010; 33(4): 293-300.
        [http://dx.doi.org/10.1111/j.1365-2761.2009.01121.x] [PMID: 20082660]

[92]   Militz TA, Southgate PC, Carton AG, Hutson KS. Efficacy of garlic (*Allium sativum*) extract applied as a therapeutic immersion treatment for *Neobenedenia* sp. management in aquaculture. J Fish Dis 2014; 37(5): 451-61.
        [http://dx.doi.org/10.1111/jfd.12129] [PMID: 23952605]

[93]   Foysal MJ, Alam M, Momtaz F, *et al.* Dietary supplementation of garlic (*Allium sativum*) modulates

gut microbiota and health status of tilapia (*Oreochromis niloticus*) against *Streptococcus iniae* infection. Aquacult Res 2019; 50(8): 2107-16.
[http://dx.doi.org/10.1111/are.14088]

[94]   Militz TA, Southgate PC, Carton AG, Hutson KS. Dietary supplementation of garlic (*Allium sativum*) to prevent monogenean infection in aquaculture. Aquaculture 2013; 408-409: 95-9.
[http://dx.doi.org/10.1016/j.aquaculture.2013.05.027]

[95]   Fridman S, Sinai T, Zilberg D. Efficacy of garlic based treatments against monogenean parasites infecting the guppy (Poecilia reticulata (Peters)). Vet Parasitol 2014; 203(1-2): 51-8.
[http://dx.doi.org/10.1016/j.vetpar.2014.02.002] [PMID: 24598083]

[96]   Steverding D, Morgan E, Tkaczynski P, Walder F, Tinsley R. Effect of Australian tea tree oil on Gyrodactylus spp. infection of the three-spined stickleback *Gasterosteus aculeatus*. Dis Aquat Organ 2005; 66(1): 29-32.
[http://dx.doi.org/10.3354/dao066029] [PMID: 16175965]

[97]   Malheiros DF, Maciel PO, Videira MN, Tavares-Dias M. Toxicity of the essential oil of Mentha piperita in Arapaima gigas (pirarucu) and antiparasitic effects on Dawestrema spp. (Monogenea). Aquaculture 2016; 455: 81-6.
[http://dx.doi.org/10.1016/j.aquaculture.2016.01.018]

[98]   Soares BV, Neves LR, Ferreira DO, *et al.* Antiparasitic activity, histopathology and physiology of Colossoma macropomum (tambaqui) exposed to the essential oil of Lippia sidoides (Verbenaceae). Vet Parasitol 2017; 234: 49-56.
[http://dx.doi.org/10.1016/j.vetpar.2016.12.012] [PMID: 28115182]

[99]   Valladão GMR, Gallani SU, Ikefuti CV, *et al.* Essential oils to control ichthyophthiriasis in pacu, *Piaractus mesopotamicus* (Holmberg): special emphasis on treatment with *Melaleuca alternifolia*. J Fish Dis 2016; 39(10): 1143-52.
[http://dx.doi.org/10.1111/jfd.12447] [PMID: 26776242]

[100] Baldissera MD, Souza CF, Moreira KLS, da Rocha MIUM, da Veiga ML, Baldisserotto B. Melaleuca alternifolia essential oil prevents oxidative stress and ameliorates the antioxidant system in the liver of silver catfish (*Rhamdia quelen*) naturally infected with Ichthyophthirius multifiliis. Aquaculture 2017; 480: 11-6.
[http://dx.doi.org/10.1016/j.aquaculture.2017.07.042]

[101] Ling F, Jiang C, Liu G, Li M, Wang G. Anthelmintic efficacy of cinnamaldehyde and cinnamic acid from cortex cinnamon essential oil against *Dactylogyrus intermedius*. Parasitology 2015; 142(14): 1744-50.
[http://dx.doi.org/10.1017/S0031182015001031] [PMID: 26442478]

[102] Pereira EC, Oliveira EC, Sousa EMO, *et al.* Lethal concentration of *Cymbopogon citratus* (Poaceae) essential oil for *Dolops discoidalis* and *Argulus* sp. (Crustacea: Argulidae). J Fish Dis 2020; 43(12): 1497-504.
[http://dx.doi.org/10.1111/jfd.13250] [PMID: 32924179]

[103] Barriga IB, Gonzales APPF, Brasiliense ARP, Castro KNC, Tavares-Dias M. Essential oil of *Lippia grata* (Verbenaceae) is effective in the control of monogenean infections in *Colossoma macropomum* gills, a large Serrasalmidae fish from Amazon. Aquacult Res 2020; 51(9): 3804-12.
[http://dx.doi.org/10.1111/are.14728]

[104] Costa CMS, da Cruz MG, Lima TBC, *et al.* Efficacy of the essential oils of *Mentha piperita, Lippia alba* and *Zingiber officinale* to control the acanthocephalan Neoechinorhynchus buttnerae in *Colossoma macropomum*. Aquacult Rep 2020; 18100414.
[http://dx.doi.org/10.1016/j.aqrep.2020.100414]

[105] de Souza RC, Baldisserotto B, Melo JFB, da Costa MM, de Souza EM, Copatti CE. Dietary *Aloysia triphylla* essential oil on growth performance and biochemical and haematological variables in Nile tilapia. Aquaculture 2020; 519734913.

[http://dx.doi.org/10.1016/j.aquaculture.2019.734913]

[106] Gonzales APPF, Yoshioka ETO, Mathews PD, *et al.* Anthelminthic efficacy of Cymbopogon citratus essential oil (Poaceae) against monogenean parasites of *Colossoma macropomum* (Serrasalmidae), and blood and histopathological effects. Aquaculture 2020; 528: 735500.

[107] dos Anjos ACP, Isaac A. The efficacy and dosage of *Mentha piperita* essential oil in the control of Monogenean parasites in *Oreochromis niloticus*. J Parasit Dis 2020; 44(3): 597-606.
[http://dx.doi.org/10.1007/s12639-020-01233-5] [PMID: 32801512]

[108] de Castro Nizio DA, Fujimoto RY, Maria AN, *et al.* Essential oils of *Varronia curassavica* accessions have different activity against white spot disease in freshwater fish. Parasitology Research 2018; 117(1): 97-105.
[http://dx.doi.org/10.1007/s12639-020-01233-5] [PMID: 32801512]

CHAPTER 13

The Role of Antimicrobial Peptides (AMPs) in Aquaculture Farming

Mohammed A. E. Naiel[1,*]**, Mohamed E. Abd El-Hack**[2,*]**, Amlan Kumar Patra**[3] and **Mahmoud Alagawany**[2]

[1] *Department of Animal Production, Faculty of Agriculture, Zagazig University, Zagazig 44519, Egypt*

[2] *Department of Poultry, Faculty of Agriculture, Zagazig University, Zagazig 44519, Egypt*

[3] *Department of Animal Nutrition, West Bengal University of Animal and Fishery Sciences, Kolkata 700037, India*

Abstract: Antimicrobial peptides (AMPs) are the vital constituents that stimulate the innate immune defense system against pathogens and perform several biological activities, which provide the first defensive line against infectious diseases. Owing to their unique structure, they can be utilized as a therapeutic strategy for infectious diseases in fishes. Several kinds of AMPs are reported in fishes with broad-spectrum antimicrobial properties. Besides, the bacterial cells cannot develop resistance strains against these cationic compounds with low molecular weight. Thus, AMPs may be considered an alternative to antibiotics to prevent or control infectious diseases in aquaculture. It is essential to provide sufficient knowledge about the mode of action of AMPs against fish pathogenic agents and their future applications.

Keywords: AMPs, Antibiotics, Antifungal, Antimicrobial, Antiparasitic, Antiviral.

INTRODUCTION

Antimicrobial peptides (AMPs) are small endogenous peptides (less than 13 kDa), having both hydrophilic and hydrophobic parts and a cationic charge. These low molecules are vital constituents, promoting the immune system against a wide range of pathogenic microbial agents [1]. The AMPs could be isolated from various organisms such as insects, plants, animals and humans, and they are identified to protect living organisms against pathogens [2]. The AMPs display

* **Corresponding author Mohammed A. E. Naiel and Mohamed E. Abd El-Hack:** Department of Animal Production, Faculty of Agriculture, Zagazig University, Zagazig 44519, Egypt and Department of Poultry, Faculty of Agriculture, Zagazig University, Zagazig 44519, Egypt; E-mails: mohammednaiel1984@gmail.com and dr.mohamed.e.abdalhaq@gmail.com

massive variations in structures and sequences but reveal some common properties such as cationic charge, amphipathic and hydrophobic structure [3, 4].

The AMPs have a positive charge constructed from nearly 20 to 50 amino acid sequences and have some essential amino acids, such as arginine and lysine [2]. The presence of disulfide bridges in AMPs structure is responsible for the positive charge and hydrophilic features [2].

Several types of AMPs, such as defensins and cathelicidins, have been identified and isolated from several mammals [6]. The variation in the structure of cationic AMPs delivers high antimicrobial activity to these residuals [5]. Besides the antimicrobial activity of AMPs, these cationic compounds have numerous biological effects like an immunomodulatory role against the virus, parasitic and fungus infections [7, 8].

Recent studies have identified several resources of AMPs and their importance as antibacterial, antiviral, antiparasitic and antifungal diseases. In this chapter, the common AMPs used against fish culture diseases have been discussed in detail, along with the mode of action of AMPs. Usage, advantage and disadvantage of these AMPs have also been illustrated.

Antimicrobial Peptides (AMPs) Types and Structure

The first AMP, cecropin, was isolated from diapausing pupae of the lepidopteran Hyalopholacecropia by Boman and Hultmark [9]. Since this discovery, multiple AMPs have been characterized and identified in many vertebrate and invertebrate species (Hertru *et al.*, 1994). The structural and functional characteristics of AMPs depend upon their amino acid sequences [10].

The AMPs could be categorized into three main classes: firstly, α-helical, with open and closed rings including disulfide-combined β-sheets, and prolonged features with a prevalence of a single amino acid (for instance, tryptophan, proline, or histidine) [11]. The α-helical AMPs include cecropins, magainins, pleurocidins, chrysophins, piscidins, moronecidins, misgurnin, pardaxin and cathelicidins have been isolated from insects, amphibian or teleost fish [12-18] Chang *et al.* 2006). The α-helical AMPs are rarely found in invertebrates, although styelins, dicynthaurin, halocidin, clavanins and clavaspirin were detected in tunicate hemocytes [19 - 22].

The second class of AMPs is β-sheet, which includes amino acids combined with disulfide peptides such as defensins (recently isolated from rainbow trout), mytilins, myticins, mitomycin, tachyplesins, polyphemusins, and hepcidins [23 - 30]. The third class of AMPs is defined as proline-rich compounds with an extended single amino acid structure. Astacidin, an AMP enriched with proline synthesized from hemocytes of crayfish *Pacifastacus leniusculus*, has its molecular weight similar to bovine bactenicin generated from shore crab, *Carcinus maenas* (6.5 kDa) [31, 32]. Also, callinectin isolated from blue crab, *Callinectes sapidus*, contains extensive quantities of arginine and proline with a 3.7 kDa molecular weight, but their sequences are typically proline-rich AMPs [33]. Moreover, penaeidins AMPs isolated from the small spider crab, *H. Araneus*, and shrimp, *Penaeus vannamei*, consisting of 37 amino acids dominate with proline and arginine followed by cysteines [34, 35].

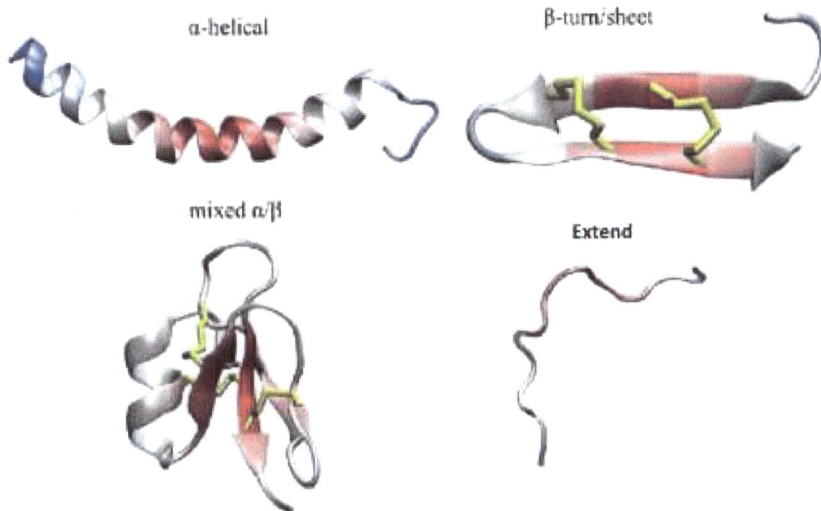

Fig. (1). The main classes of AMPs.

Resources of Antimicrobial Peptides

Naturally, AMPs have been discovered in all prokaryotic and eukaryotic creatures with a broad spectrum of antimicrobial effects [36]. The nisin was the first AMP isolated from the *Lactococcus lactis*, containing a 34-amino acid peptide belonging to the bacteriocins family [37]. Also, the melittin peptide was isolated from the Iranian honey bee (*Apis melliferameda*) venom by high-performance liquid chromatography, which has a 26-amino acid peptide [38] and showed anticancer properties besides its high antimicrobial effects. In 1939, the gramicidin was isolated from *Bacillus brevis* and identified as a leading AMP, which exhibited high effectiveness against various pathogenic bacteria and was commercially manufactured as safe antibiotics [38, 39]. Recently, AMPs have been isolated from different resources such as plants, mammals, marine invertebrates, insects, and other organisms. For instance, the purothionin compound was isolated from the wheat endosperm and categorized as AMP [40].

Lately, several peptides have been identified from invertebrate fish displaying numerous biological actions [36, 41, 42]. The most common AMPs isolated from fish belong to the defensin, cathelicidin, piscidin and hepcidin families [43]. For example, Primor and Tu [44] isolated and identified pardaxin AMPs (toxin AMPs) from the sole fish (*Pardachirus marmoratus*). The AMPs isolated from fish or called fish AMPs are effective against bacterial and virus diseases and play a vital role in the ion osmoregulation process [45 - 47]. The 1,2 and 3 iso-types of piscidins AMP were successfully isolated from mast cells of crossbreed striped bass [43] and, also had been discovered in phagocytic granulocytes of sea bream [48]. The AMP piscidin iso-type-2 showed higher effectiveness against parasitic infection [49].

Moreover, the newly discovered piscidin iso-type-4 in the gills of crossbreed striped bass exhibited a higher ability to prevent several fish pathogenic bacteria [50]. Peptides influences pathogenic bacteria *via* the toroidal pore mode of action. Briefly, it could adsorb, aggregate, and bind with a surface of the bacterial membrane and then penetrate the interior region. This may lead to wrapping the upper and lower bilayer and creating the pore form [51, 52]. The piscidins could be formed from about 22 amino acids in pre-form cleavage to produce the activated form responsible for the biological activity against the bacterial cells [53].

Antimicrobial Peptides Mechanisms, Advantages and Disadvantages

As an essential immune defense system in teleost fishes, the innate immunity system is more valuable than the specific immunity system against pathogenic bacterial infections. Antimicrobial peptides are classified as an important part of the innate immune responses. AMPs could be alternatives to antibiotics, especially against antibiotic or drug-resistant bacteria [54]. Antimicrobial peptides (AMPs) are a kind of short-chain polypeptides with wide biological properties and commonly exist in almost all living organisms [55]. Naturally, living organisms produce the AMPs for survival. They resist the infecting pathogens to the host *via* a wide range of antimicrobial effects that can differ according to the pathogen types [56].

The structure of AMPs differs from alpha-helix to beta strands [2]. The defensive mechanisms of AMPs against pathogens may be due to their selective toxicity effects. The AMPs could induce several changes in the lipid structure of the cell membranes between the eukaryotic and the prokaryotic species that operate as their target of action to avoid harmful effects on the host. Besides, the higher selective activity of AMPs is promoted under the influence of hydrophobicity and the net charge ability. The AMPs could carry a wide-range positive charging (+2 to +9), while few reports indicated they might be anionic molecules [57].

The AMPs mechanism may depend upon its cationic nature *via* allowing its binding capability to the negatively charged lipopolysaccharides, thus invading the pathogen membranes through attractive electrostatic forces. During this reaction, the AMPs obtain an amphipathic property that enhances the binding ability towards the pathogen's membranes. This type of AMPs is defined as membrane permeabilizing peptides due to their capability of inducing pores on the entire cell surface membrane after their electrostatic reaction, thus performing in the ejection of cellular constituents and sudden cell death [58]. Other types of AMPs categorized as non-membrane permeabilizing peptides, which could affect the internal biological process of target cells after their translocation across the cellular membranes [33, 59].

AMPs have several benefits: low cost of punctual synthesis, easy storing for a long time in large quantities, and being rapidly available after infection [1]. Despite the advantages of using the AMPs in fish culture systems, there are some critical drawbacks: restricted stability at specified pH levels, disruption of eukaryotic cell membranes triggering adverse hemolytic effects, and high technical and production cost [60 - 62]. Besides, there is a lack of information on their toxicological effects, pharmacodynamics, and pharmacokinetics influences, and inactivity in the attendance of some ions such as Ca^{+2}, Fe^{+2} and other serum constituents [55, 63].

The Application of AMPs Against Fish Diseases

Until now, more than 5,500 AMPs have been discovered, isolated, identified, characterized, or manufactured [64]. In fish, AMPs are produced in various tissues such as the skin epidermis layer and gill filaments exposed to the aquatic environment and are considered the most important defensive line against pathogens [65]. The AMPs contain up to 100 amino acids and most structures have amphipathic nature with aggregates of cationic and hydrophobic amino acids [66].

The mechanisms of AMPs are not totally comprehended. The selectivity of AMPs activity depends on the physicochemical characteristics of the outer bacterial membranes. The bacterial outer phospholipid membrane has negatively charged head groups, while the membrane of plants or animals has no net charge [67]. The mode of action may be attributed to the aggregation of AMPs at the outer cell bacterial membrane because of the differences between cell net charges alongside a-helices and binding with the hydrophobic clusters resulting in non-selective membrane permeation [68]. The antibacterial activity towards specific organisms may be induced by the destruction of the outer membrane permeation and activating plasmolysis, suppression of protein synthesis process and affecting the intracellular molecules [69]. Thus, the bacterial resistance to AMPs would not be developed because of the physical destruction of the outer cell membrane. AMPs do not aggregate in the outer membrane of multicellular organisms due to the absence of the net charge and high cholesterol content [57]. Therefore, the AMPs are more selective toward non-target organisms [70]. Hence, the AMPs have been considered a suitable strategy alternative to conventional therapeutics and possibly a valuable tool to prevent and control infection diseases (Table 1.) [41, 65, 71].

Table 1. The therapeutic role of several AMPs against fish pathogenic agents in aquaculture.

AMPs	Type	Resource	Activity	Pathogenic Agent	References
Chionodracine (Cnd-m3a)	piscidin-like AMP	the Antarctic ice fish *Chionodraco hamatus*	Antibacterial	*E. coli, Psychrobacter sp.*	Buonocore, Picchietti [40]
NK-lysin	The Saposinlike protein AMPs	The large yellow croaker (Larimichthys crocea).	Antimicrobial and antitoxic	*Staphylococcus aureus, Escherichia coli, Vibrio harveyi, Vibrio parahaemolyticus, Aeromonas hydrophila, Photobacteria damselae*	Zhou, Wang [55]
Penaeidin-5	Penaeidins	Black tiger shrimp (*Penaeus monodon*)	Antifungal, antibacterial	*Fusariumpisi and Fusarium oxysporum; Aerococcus viridans, Vibrio alginolyticus and A. viridans*	Hu, Huang [11]
Cyclotide	Cysteine AMPs	The Leaves of Bauhinia rufescens Lam (Fabaceae)	Antibacterial	*E. coli and A.salmonicida*	Nganso, Sidjui [118]
Defensin	Cysteine-rich cationic AMP	*M. edulis*	Antiparasitic, antibacterial, antiviral	*Escherichia coli, Salmonella typhimurium, Alteromonas carrageeno vora, Pseudomonas alginovora, Cytophaga drobachiensis, Micrococcus luteus, Aerococcus viridans, Bacillus megaterium, Enterococcus faecalis, Staphylococcus aureus*	Charlet, Chernysh [24]

(Table 1) cont.....

AMPs	Type	Resource	Activity	Pathogenic Agent	References
MGD-1,2	A defensin-like peptide	*Mytilis galloprovencialis*	Antifungal, antibacterial	*Albicans and C. glabrata*, *Staphylococcus aureus*	Boisard, Le Ray [119]
Mytilin	cysteine-rich cationic AMP	*M. edulis*	Antifungal, antibacterial, antiviral	white spot syndrome virus(WSSV), *Vibrio splendidus LGP32, Vibrio anguillarum, Micrococcus lysodeikticus and Escherichia coli*	Roch, Yang [120]
Myticin	cysteine-rich cationic AMP	*M. edulis*	Antifungal, antibacterial	Inhibits two fungal strains, Gram positive and negative bacteria, *E. coli*	Domeneghetti, Franzoi [121]
Mytimycin	A cationic rich cysteine-AMP	*M. edulis*	Antifungal	*F. oxysporum*	Sonthi, Cantet [122]
Penaeidins	Shrimp AMP	Many Crustaceans	Antifungal, antibacterial	Gram-negative strains, including the *Vibrio* species	Destoumieux, Munoz [123]
Carcinin	Type 1 AMP	*Carcinusmaenas*	antibacterial	*Planococcus citreus*	Brockton and Smith [124]
Callinectin	arginine-rich-AMP	*Callinectes sapidus*	antibacterial	the gram-negative bacterial cell	Johnson, Burnett [125]
Tachyplesin	a cationic β-hairpin AMP	*Tachyplesus tridentatus*	Antifungal, antibacterial, antiviral, antiparasitic	*Candida* spp., *Babesiamicroti, B. bigemina, B. caballi, B. equi, B. bovis, Plasmodium* spp., *Toxoplasma* spp., *Babesia* spp., *Leishmania* spp.	Dalhoff [126]
Big defensin	gene-encoded disulfide-rich AMP	*T. tridentatus*	antibacterial	*V. anguillarum*	Rosa, Santini [127]
Tachycitin	chitin-binding AMP	*T. tridentatus*	Antifungal, antibacterial	*grobacteriumtume faciens, Enterobacter cloacae, Erwinia amylovora, Pectobacterium carotovorum subsp., carotovorum,*	Al-Huqail, Behiry [128]
Polyphemusin	a cationic β-hairpin AMP	*Limulus polyphemus*	Antibacterial	Gram-positive and Gram-negative bacteria	Marggraf, Panteleev [129]

(Table 1) cont.....

AMPs	Type	Resource	Activity	Pathogenic Agent	References
Clavanin	alpha-helical AMP	*Styelaclava*	antibacterial	*E. coli and S. aureus*	Silva, Fensterseifer [130]
Styelin	phenylalanine-rich AMP	*S. clava*	antibacterial	gram negative and gram-positive bacterial pathogens	Lee, Cho [19]
Hepcidin	Hormone AMPs	Many fish species	Antifungal, antibacterial	*Vibrio vulnificus (204), Streptococcus agalactiae (SA)*	Bulet, Stöcklin [131]

Anti-parasitic and Antifungal Activity of AMPs

The parasitic infestation plays a vital role in limiting the final output, health, and economic returns of the global finfish aquaculture production. It could directly lead to high losses in fish production by decreasing final fish yield or indirectly by increasing the treatment costs to control the parasitic infestation. Despite low available studies on the adverse effects of fungal or parasitic infection on aquaculture, several AMPs are useful in preventing and controlling this type of disease. The piscidin-2 isolated from hybrid striped bass illustrated powerful antifungal activity against fungal pathogens such as *Candida albicans*, *Trichosporon beigelii* and *Malassezia furfur* [72]. Ofpis-1 and Oreoch-1, -2 and -3 showed high antifungal efficiency against *C. albicans* infection, but this effect is also towards *Saccharomyces cerevisiae* [73]. Besides, Kim, Yang [74] indicated that hepcidins (ecPis- 2S, -3S and -4S) isolated from orange-spotted grouper had powerful antifungal effects against *Pichia pastoris*. Furthermore, the synthetic hepcidin peptide from hybrid striped bass exhibited antifungal influence against *Aspergillus niger* [75], and AS-hepcidin-2 and -6 and PChepc isolated from large yellow croaker fish displayed antifungal activity against several fungus strains such as *Fusarium graminearum, C. albicans, Fusarium solani* and/or *P. pastoris* [76]. Moreover, numerous histone-like proteins employed higher antifungal effects. At the same trend, Zahran and Noga [77] reported that two types of histone-like proteins (HLP-1 and -2) or piscidin-2 isolated from channel catfish significantly reduced *Saprolegnia parasitica* development in infected fish. Also, synthetic histone-like proteins had synergistic antifungal effects against *Saprolegnia* spp. and *Tetrahymena pyriformis* combined with malachite green and copper sulfate treatment [77]. Atlantic cod diminished the development of *C. albicans* infection by exerting mucus content H2B and three types of ribosome-derived peptides (L35, L36A and L40) [78]. Moreover, isolated parasin I from catfish illustrated high efficiency against yeast such as *C. albicans, S. cerevisiae* and *Cryptococcus neoformans* [79], while the antifungal activity of myxinidin isolated from hagfish had been proved against *C. albicans* infection [80].

Misgurin could display antifungal efficiency against several kinds of fungus such as *C. albicans, C. neoformans* and *S. cerevisiae* [81]. Also, the epinecidin-1 isolated from grouper showed higher antifungal activity against many filamentous fungi and yeast (such as *C. albicans*, and *P. pastoris*) [82]. Furthermore, Yin, He [83] confirmed that epinecidin-1 had high effects against filamentous fungus (*Microsporosis canis, Trichophytonsis mentagrophytes* and *Cylindrocarpon* sp.).

Recent findings proved the antiparasitic efficiency of several types of AMPs against some parasitic infectious diseases. For instance, Colorni, Ullal [49] exhibited the antiparasitic activity of synthetic piscidin-2 isolated from hybrid striped bass against natural ectoparasites (*Trichodina* sp., *Cryptocaryon irritans, Ichthyophthirius multifiliis* and *Amyloodinium ocellatum*). Also, isolated ofpis-1 from rock bream possessed revealed high antiparasitic effects towards *Miamiensisavidus* infection [84]. Likewise, Atlantic cod piscidin-1, piscidin-2 and piscidin-2b in cationic or non-cationic form verified high antiparasitic activity against *Tetrahymena pyriformis* infection under low parasiticidal levels [85]. Moreover, synthetic ecPis-3 from orange-spotted grouper and HLP-1 from hybrid striped bass showed the ability to damage *C. irritants* infective stage [86]. In addition, the HLP-1 showed negative effects on the intensity and propagation of *A. occelatum* infection and reported high homology sequencing to histone H2B [87]. In *in vitro* study verified that the HLP-1isolated from rainbow trout (*Oncorhynchus mykiss*) and hybrid striped bass (female *Moron echrysopsx* male *M. saxatilis*) had antiparasitic activity against trophonts in early and mature stages [88]. Also, Colorni, Ullal [49] reported that the expression of AMPs was the level acquired for the anti-parasitic activity in the skin and gill epithelia in *in vitro*. These compounds could promote the fish-specific immune system against velvet disease, as they inhibit parasites especially in the early stage of the parasitic infection. The expression of the AMPs contains humic substances that might be responsible for the immunomodulatory effects of these molecules, but there is currently no prove about this attribution.

The Activity of AMPs Toward Bacterial Fish Diseases

The AMPs showed effective defensive mechanisms against a wide range of Gram-positive and Gram-negative bacteria [89]. The wide spectrum antimicrobial properties of AMPs are attributed to their capability to deactivate cell membranes and form pores [90, 91]. The AMPs could be stimulated in different fish tissues after exposure to pathogenic bacteria [92]. Furthermore, the AMPs could be an alternative to antibiotics and show higher effectiveness against antibiotic-resistant bacteria strains [93]. The AMPs have an amphipathic structure that may be responsible for their antibacterial properties [94]. Several expressed peptides contain a positively charged hydrophobic face which binds with negative charged bacterial membranes (anionic charge) [95]. The binding between AMPs and bacterial membrane leads to pore formation and depolymerization of cell mitochondria by influencing its elasticity and subsequently destroying the bacterial cells [96]. Several types of peptides could disrupt the bacterial cells by binding with the outer cellular membranes or even intracellularly [97]. The antibacterial activity of these AMPs depends upon their selectivity towards the specific pathogenic membrane, leading to the variations between the charge of lipids and the bacterial outer cell membrane charge [98]. Remarkably, the histone-like proteins isolated from fish showed a specific antibacterial action. Histones isolated from invertebrates or synthetic histones or fractioned exhibit a wide spectrum of antimicrobial properties through binding with the outer membrane, entering into the bacterial cell through the membranes, or inducing neutrophil extracellular traps (NETs) [99]. Given the basic pathways of produced histones, they remain unexplained as antibacterial agents by the induction of NETs in fish. Histones have been confirmed to play a critical role in the development of NETs traps, since their antibody obstruction prohibits their antimicrobial activity [100 - 102].

Many recent studies investigated the antibacterial activity of several AMPs. For instance, Gyan, Yang [103] investigated that the Pacific White-leg shrimp diets with 0.4% AMPs significantly promoted growth, serum antioxidant status, and innate immunity and reduced the mortality rate induced by *V. harveyi* infection. Likewise, the liver-expressed antimicrobial peptide-2 (LEAP-2) isolated from the golden pompano (*Trachinotus ovatus*) significantly inhibited the *E. tarda* and *S. agalactiae* development and stimulated the immune response against pathogenic infections in golden pompano [104]. At the same trend, and Wan [105] found that the chemically synthesized Spgly-AMP peptides (a glycine-derivatives AMPs) displayed high antibacterial activities against Gram-positive and -negative bacteria and greater thermal stability in the mud crabs reared under tropical conditions. Besides, Shan, Yang [106] investigated that the NKL-24 (a truncated peptide derived from zebrafish NK-lysin) could kill the *Vibrio parahaemolyticus* pathogenic bacteria and enhanced a non-specific immunity of Yesso scallop (*Patinopecten yessoensis*).

The Antiviral Effects of AMPs

Globally, the antiviral properties have been usually ignored compared to the antibacterial ability. This may be due to the motivation to find new effective antibacterial molecules as an alternative to antibiotics and resolve public health problems [107]. For the last two decades, the viruses have been widely spread within aquatic organisms, causing a big problem threatening the aquatic life in which suitable solutions have yet not been attained. Recently, several AMPs isolated from fish have been proved the antiviral properties. Also, several studies demonstrated the antiviral ability of AMPs against virus infections. Chinchar, Bryan [108] investigated that the piscidin-1N, -1H, -1 and -2 isolated from hybrid striped bass showed higher efficiency against the channel catfish virus (CCV) and frog virus 3 (FV3). The required concentrations of these AMPs were higher in the case of treating the FV3 compared with CCV, as well as the four types of piscidins proved the ability to inactivate the studied viruses. Also, beta defensin 2 (zfBD2) isolated from zebrafish caused a significantly decreased spring viraemia of carp virus (SVCV) infection by approximately 90%. This decrease in fish mortalities is dependent on the applicable dose, which was linked with the greater upregulation of Mx gene expression after infection [109]. The b-defensin isolated from the orange-spotted grouper liver documented the ability to diminish the pathogeneceity of two combined marine pathogens (Singapore grouper iridovirus, SGIV and nervous necrosis virus, NNV) [110] as also inspected with the ability of EC-hepcidin-1 or -2 towards SGIV infection [111].

Furthermore, isolated b-defensin from the orange-spotted grouper testis and pituitary proved higher effective towards the RGV iridovirus [112]. Additionally, the artificially synthetic spotted cat (*Scatophagus argus*) hepcidin, SA-hepcidin-1 or -2, seems to be active against siniperca chuatsi rhabdovirus (SCRV) and large mouth bass micropterus salmoides reovirus (MsReV), along with the epithelioma papulosum cyprini (EPC) and fin (GCF) cells of grass carp. The SA-hepcidin-2 revealed higher antiviral activity [113]. The recently isolated AMPs TO17 revealed antiviral activity against spleen and kidney necrosis virus (ISKNV) after injection with a virus and reduced the spleen viral load after 3, 5 and 7 days of injection [114]. Although the antiviral activity of many fish AMPs has been investigated, the entire cellular mechanisms are still unclear. Likewise, the antiviral activity of the pro-peptide Omhep1 exhibited higher efficiency towards the white spot syndrome virus (WSSV) infection in hematopoietic tissue cells and it downregulated the expression of WSSV mediate-early coding gene-1 [115]. The antiviral activity of CsNKL1-derived synthetic peptide NKLP27 was also verified against megalocytivirus RBIV-C1 [116]. Chang, Pan [117] established that the ability of the hepcidin isolated from tilapia (*Oreochromis mossambicus*) TH 1-5 towards NNV antiviral infection is dependent upon the applicable dose, and induces clumps by activating the adhering of viral particles. Therefore, the possible antiviral mechanisms of TH 1-5 may be dependent on the ability of NNV to enter the cells, thus decreasing the virus adsorption capacity and causing no stimulation of the Mx activity.

CONCLUDING REMARKS

The fish are always exposed to various pathogenic agents in the ecosystem, causing high mortality and economic losses. Hence, antimicrobial substances are very important to control infections, decrease fish mortality and promote general fish health status. The wide application of antibiotics in fish farms led to the increasing amount of antibiotic-resistant bacterial strains, which necessitates developing a new applicable strategy to produce safe antimicrobial compounds. Therefore, AMPs have great attention as immunomodulatory peptides that stimulate the immune defensive system against infectious diseases. The teleost fish depends on a direct innate immune system as a first defensive line against infection. Also, AMPs could be considered an alternative therapeutic molecule against infectious diseases. The AMPs have unique features due to the differences in base encoding sequences. The synthetic AMPs could be useful for developing pharmacological treatment and control of fish infectious diseases with high specificity and more effectiveness.

CONSENT FOR PUBLICATION

Not applicable.

CONFLICT OF INTEREST

The author declares no conflict of interest, financial or otherwise.

ACKNOWLEDGEMENTS

Declared none.

REFERENCES

[1]   Ravichandran S, Kumaravel K, Rameshkumar G, Ajithkumar T. Antimicrobial peptides from the marine fishes. Research journal of immunology 2010; 3(2): 146-64.

[2]   Hancock REW. Cationic peptides: effectors in innate immunity and novel antimicrobials. Lancet Infect Dis 2001; 1(3): 156-64.
[http://dx.doi.org/10.1016/S1473-3099(01)00092-5] [PMID: 11871492]

[3]   Hancock REW, Lehrer R. Cationic peptides: a new source of antibiotics. Trends Biotechnol 1998; 16(2): 82-8.
[http://dx.doi.org/10.1016/S0167-7799(97)01156-6] [PMID: 9487736]

[4]   Hancock REW, Rozek A. Role of membranes in the activities of antimicrobial cationic peptides. FEMS Microbiol Lett 2002; 206(2): 143-9.
[http://dx.doi.org/10.1111/j.1574-6968.2002.tb11000.x] [PMID: 11814654]

[5]   Vanhoye D, Bruston F, Nicolas P, Amiche M. Antimicrobial peptides from hylid and ranin frogs originated from a 150-million-year-old ancestral precursor with a conserved signal peptide but a hypermutable antimicrobial domain. Eur J Biochem 2003; 270(9): 2068-81.
[http://dx.doi.org/10.1046/j.1432-1033.2003.03584.x] [PMID: 12709067]

[6]   Cederlund A, Gudmundsson GH, Agerberth B. Antimicrobial peptides important in innate immunity. FEBS J 2011; 278(20): 3942-51.
[http://dx.doi.org/10.1111/j.1742-4658.2011.08302.x] [PMID: 21848912]

[7]   Mayer ML, Easton DM, Hancock RE. 12 Fine Tuning Host Responses in the Face of Infection: Emerging Roles and Clinical Applications of Host Defence. Antimicrobial peptides: discovery, design and novel therapeutic strategies 2010.

[8]   Steinstraesser L, Kraneburg U, Jacobsen F, Al-Benna S. Host defense peptides and their antimicrobial-immunomodulatory duality. Immunobiology 2011; 216(3): 322-33.
[http://dx.doi.org/10.1016/j.imbio.2010.07.003] [PMID: 20828865]

[9]   Boman HG, Hultmark D. Cell-free immunity in insects. Annu Rev Microbiol 1987; 41(1): 103-26.
[http://dx.doi.org/10.1146/annurev.mi.41.100187.000535] [PMID: 3318666]

[10]   Hetru C. Antibacterial peptides/polypeptides in the insect host defense: a comparison with vetebrate antibacterial peptides/polypeptides. Phylogenetic Perspectives in Immunity: The Insect Host Defense 1994.

[11]   Hu SY, Huang JH, Huang WT, *et al.* Structure and function of antimicrobial peptide penaeidin-5 from the black tiger shrimp *Penaeus monodon*. Aquaculture 2006; 260(1-4): 61-8.
[http://dx.doi.org/10.1016/j.aquaculture.2006.06.017]

[12]   Douglas SE, Gallant JW, Gong Z, Hew C. Cloning and developmental expression of a family of pleurocidin-like antimicrobial peptides from winter flounder, Pleuronectes americanus (Walbaum). Dev Comp Immunol 2001; 25(2): 137-47.
[http://dx.doi.org/10.1016/S0145-305X(00)00052-5] [PMID: 11113283]

[13]   Iijima N, Tanimoto N, Emoto Y, *et al.* Purification and characterization of three isoforms of chrysophsin, a novel antimicrobial peptide in the gills of the red sea bream, Chrysophrys major. Eur J Biochem 2003; 270(4): 675-86.
[http://dx.doi.org/10.1046/j.1432-1033.2003.03419.x] [PMID: 12581207]

[14]   Silphaduang U, Colorni A, Noga EJ. Evidence for widespread distribution of piscidin antimicrobial peptides in teleost fish. Dis Aquat Organ 2006; 72(3): 241-52.
[http://dx.doi.org/10.3354/dao072241] [PMID: 17190202]

[15]   Lauth X, Shike H, Burns JC, *et al.* Discovery and characterization of two isoforms of moronecidin, a novel antimicrobial peptide from hybrid striped bass. J Biol Chem 2002; 277(7): 5030-9.
[http://dx.doi.org/10.1074/jbc.M109173200] [PMID: 11739390]

[16]   Park CB, Lee JH, Park IY, Kim MS, Kim SC. A novel antimicrobial peptide from the loach, *Misgurnus anguillicaudatus*. FEBS Lett 1997; 411(2-3): 173-8.
[http://dx.doi.org/10.1016/S0014-5793(97)00684-4] [PMID: 9271200]

[17]   Shai Y, Fox J, Caratsch C, Shih YL, Edwards C, Lazarovici P. Sequencing and synthesis of pardaxin, a polypeptide from the Red Sea Moses sole with ionophore activity. FEBS Lett 1988; 242(1): 161-6.
[http://dx.doi.org/10.1016/0014-5793(88)81007-X] [PMID: 2462511]

[18]   Chang CI, Zhang YA, Zou J, Nie P, Secombes CJ. Two cathelicidin genes are present in both rainbow trout (*Oncorhynchus mykiss*) and atlantic salmon (*Salmo salar*). Antimicrob Agents Chemother 2006; 50(1): 185-95.
[http://dx.doi.org/10.1128/AAC.50.1.185-195.2006] [PMID: 16377685]

[19]   Lee IH, Cho Y, Lehrer RI. Styelins, broad-spectrum antimicrobial peptides from the solitary tunicate, Styela clava. Comp Biochem Physiol B Biochem Mol Biol 1997; 118(3): 515-21.
[http://dx.doi.org/10.1016/S0305-0491(97)00109-0] [PMID: 9467865]

[20]   Lee IH, Lee YS, Kim CH, *et al.* Dicynthaurin: an antimicrobial peptide from hemocytes of the solitary tunicate, Halocynthia aurantium. Biochim Biophys Acta, Gen Subj 2001; 1527(3): 141-8.
[http://dx.doi.org/10.1016/S0304-4165(01)00156-8] [PMID: 11479030]

[21]   Lee IH, Zhao C, Cho Y, Harwig SSL, Cooper EL, Lehrer RI. Clavanins, α-helical antimicrobial peptides from tunicate hemocytes. FEBS Lett 1997; 400(2): 158-62.
[http://dx.doi.org/10.1016/S0014-5793(96)01374-9] [PMID: 9001389]

[22]   In I-H, Zhao C, Nguyen T, *et al.* Clavaspirin, an antibacterial and haemolytic peptide from *Styela clava.* J Pept Res 2001; 58(6): 445-56.
[http://dx.doi.org/10.1034/j.1399-3011.2001.10975.x] [PMID: 12005415]

[23]   Saito T, Kawabata S, Shigenaga T, *et al.* A novel big defensin identified in horseshoe crab hemocytes: isolation, amino acid sequence, and antibacterial activity. J Biochem 1995; 117(5): 1131-7.
[http://dx.doi.org/10.1093/oxfordjournals.jbchem.a124818] [PMID: 8586631]

[24]   Charlet M, Chernysh S, Philippe H, Hetru C, Hoffmann JA, Bulet P. Innate Immunity. J Biol Chem 1996; 271(36): 21808-13.
[http://dx.doi.org/10.1074/jbc.271.36.21808] [PMID: 8702979]

[25]   Mitta G, Vandenbulcke F, Hubert F, Salzet M, Roch P. Involvement of mytilins in mussel antimicrobial defense. J Biol Chem 2000; 275(17): 12954-62.
[http://dx.doi.org/10.1074/jbc.275.17.12954] [PMID: 10777596]

[26]   Mitta G, Vandenbulcke F, Roch P. Original involvement of antimicrobial peptides in mussel innate immunity. FEBS Lett 2000; 486(3): 185-90.
[http://dx.doi.org/10.1016/S0014-5793(00)02192-X] [PMID: 11119700]

[27]   Nakamura T, Furunaka H, Miyata T, *et al.* Tachyplesin, a class of antimicrobial peptide from the hemocytes of the horseshoe crab (*Tachypleus tridentatus*). Isolation and chemical structure. J Biol Chem 1988; 263(32): 16709-13.
[http://dx.doi.org/10.1016/S0021-9258(18)37448-9] [PMID: 3141410]

[28]   Miyata T, Tokunaga F, Yoneya T, *et al.* Antimicrobial peptides, isolated from horseshoe crab hemocytes, tachyplesin II, and polyphemusins I and II: chemical structures and biological activity. J Biochem 1989; 106(4): 663-8.
[http://dx.doi.org/10.1093/oxfordjournals.jbchem.a122913] [PMID: 2514185]

[29]   Yang M, Wang KJ, Chen JH, Qu HD, Li SJ. Genomic organization and tissue-specific expression analysis of hepcidin-like genes from black porgy (Acanthopagrus schlegelii B.). Fish Shellfish Immunol 2007; 23(5): 1060-71.
[http://dx.doi.org/10.1016/j.fsi.2007.04.011] [PMID: 17574440]

[30]   Cho YS, Lee SY, Kim KH, Kim SK, Kim DS, Nam YK. Gene structure and differential modulation of multiple rockbream (*Oplegnathus fasciatus*) hepcidin isoforms resulting from different biological stimulations. Dev Comp Immunol 2009; 33(1): 46-58.
[http://dx.doi.org/10.1016/j.dci.2008.07.009] [PMID: 18761369]

[31]   Jiravanichpaisal P, Lee SY, Kim YA, Andrén T, Söderhäll I. Antibacterial peptides in hemocytes and hematopoietic tissue from freshwater crayfish *Pacifastacus leniusculus*: Characterization and expression pattern. Dev Comp Immunol 2007; 31(5): 441-55.
[http://dx.doi.org/10.1016/j.dci.2006.08.002] [PMID: 17049601]

[32]   Schnapp D, Kemp GD, Smith VJ. Purification and characterization of a proline-rich antibacterial peptide, with sequence similarity to bactenecin-7, from the haemocytes of the shore crab, Carcinus maenas. Eur J Biochem 1996; 240(3): 532-9.
[http://dx.doi.org/10.1111/j.1432-1033.1996.0532h.x] [PMID: 8856051]

[33]   Kondejewski LH, Jelokhani-Niaraki M, Farmer SW, *et al.* Dissociation of antimicrobial and hemolytic activities in cyclic peptide diastereomers by systematic alterations in amphipathicity. J Biol Chem 1999; 274(19): 13181-92.
[http://dx.doi.org/10.1074/jbc.274.19.13181] [PMID: 10224074]

[34]   Bachère E, Gueguen Y, Gonzalez M, de Lorgeril J, Garnier J, Romestand B. Insights into the anti-microbial defense of marine invertebrates: the penaeid shrimps and the oyster Crassostrea gigas.

Immunol Rev 2004; 198(1): 149-68.
[http://dx.doi.org/10.1111/j.0105-2896.2004.00115.x] [PMID: 15199961]

[35]   Stensvåg K, Haug T, Sperstad SV, Rekdal Ø, Indrevoll B, Styrvold OB. Arasin 1, a proline–arginine-rich antimicrobial peptide isolated from the spider crab, Hyas araneus. Dev Comp Immunol 2008; 32(3): 275-85.
[http://dx.doi.org/10.1016/j.dci.2007.06.002] [PMID: 17658600]

[36]   Shabir U, Ali S, Magray AR, *et al.* Fish antimicrobial peptides (AMP's) as essential and promising molecular therapeutic agents: A review. Microb Pathog 2018; 114: 50-6.
[http://dx.doi.org/10.1016/j.micpath.2017.11.039] [PMID: 29180291]

[37]   Rogers LA. The inhibiting effect of *Streptococcus lactis* on Lactobacillus bulgaricus. J Bacteriol 1928; 16(5): 321-5.
[http://dx.doi.org/10.1128/jb.16.5.321-325.1928] [PMID: 16559344]

[38]   Fischer FG, Neumann WP. [The venom of the honeybee. III. On the chemical knowledge of the principle active constituent (melittin)]. Biochem Z 1961; 335: 51-61.
[PMID: 13893130]

[39]   Dubos RJ. Studies on a bactericidal agent extracted from a soil bacillus: I. Preparation of the agent. Its activity in *in vitro.* J Exp Med 1939; 70(1): 1-10.
[http://dx.doi.org/10.1084/jem.70.1.1] [PMID: 19870884]

[40]   Buonocore F, Picchietti S, Porcelli F, *et al.* Fish-derived antimicrobial peptides: Activity of a chionodracine mutant against bacterial models and human bacterial pathogens. Dev Comp Immunol 2019; 96: 9-17.
[http://dx.doi.org/10.1016/j.dci.2019.02.012] [PMID: 30790604]

[41]   Rajanbabu V, Chen JY. Applications of antimicrobial peptides from fish and perspectives for the future. Peptides 2011; 32(2): 415-20.
[http://dx.doi.org/10.1016/j.peptides.2010.11.005] [PMID: 21093512]

[42]   Rathinakumar R, Wimley WC. Biomolecular engineering by combinatorial design and high-throughput screening: small, soluble peptides that permeabilize membranes. J Am Chem Soc 2008; 130(30): 9849-58.
[http://dx.doi.org/10.1021/ja8017863] [PMID: 18611015]

[43]   Noga EJ, Silphaduang U. Piscidins: A novel family of peptide antibiotics from fish. Drug News Perspect 2003; 16(2): 87-92.
[http://dx.doi.org/10.1358/dnp.2003.16.2.829325] [PMID: 12792669]

[44]   Primor N, Tu AT. Conformation of pardaxin, the toxin of the flatfish Pardachirus marmoratus. Biochim Biophys Acta Protein Struct 1980; 626(2): 299-306.
[http://dx.doi.org/10.1016/0005-2795(80)90124-5] [PMID: 7213649]

[45]   Pan CY, Chen JY, Cheng YSE, *et al.* Gene expression and localization of the epinecidin-1 antimicrobial peptide in the grouper (Epinephelus coioides), and its role in protecting fish against pathogenic infection. DNA Cell Biol 2007; 26(6): 403-13.
[http://dx.doi.org/10.1089/dna.2006.0564] [PMID: 17570764]

[46]   Chia TJ, Wu YC, Chen JY, Chi SC. Antimicrobial peptides (AMP) with antiviral activity against fish nodavirus. Fish Shellfish Immunol 2010; 28(3): 434-9.
[http://dx.doi.org/10.1016/j.fsi.2009.11.020] [PMID: 20004246]

[47]   Shi J, Camus AC. Hepcidins in amphibians and fishes: Antimicrobial peptides or iron-regulatory hormones? Dev Comp Immunol 2006; 30(9): 746-55.
[http://dx.doi.org/10.1016/j.dci.2005.10.009] [PMID: 16325907]

[48]   Mulero I, Noga EJ, Meseguer J, García-Ayala A, Mulero V. The antimicrobial peptides piscidins are stored in the granules of professional phagocytic granulocytes of fish and are delivered to the bacteria-containing phagosome upon phagocytosis. Dev Comp Immunol 2008; 32(12): 1531-8.

[http://dx.doi.org/10.1016/j.dci.2008.05.015] [PMID: 18582499]

[49]   Colorni A, Ullal A, Heinisch G, Noga EJ. Activity of the antimicrobial polypeptide piscidin 2 against fish ectoparasites. J Fish Dis 2008; 31(6): 423-32.
[http://dx.doi.org/10.1111/j.1365-2761.2008.00922.x] [PMID: 18471098]

[50]   Corrales J, Gordon WL, Noga EJ. Development of an ELISA for quantification of the antimicrobial peptide piscidin 4 and its application to assess stress in fish. Fish Shellfish Immunol 2009; 27(2): 154-63.
[http://dx.doi.org/10.1016/j.fsi.2009.02.023] [PMID: 19268546]

[51]   Mihajlovic M, Lazaridis T. Antimicrobial peptides in toroidal and cylindrical pores. Biochim Biophys Acta Biomembr 2010; 1798(8): 1485-93.
[http://dx.doi.org/10.1016/j.bbamem.2010.04.004]

[52]   Campagna S, Saint N, Molle G, Aumelas A. Structure and mechanism of action of the antimicrobial peptide piscidin. Biochemistry 2007; 46(7): 1771-8.
[http://dx.doi.org/10.1021/bi0620297] [PMID: 17253775]

[53]   Perrin BS Jr, Fu R, Cotten ML, Pastor RW. Simulations of membrane-disrupting peptides II: AMP piscidin 1 favors surface defects over pores. Biophys J 2016; 111(6): 1258-66.
[http://dx.doi.org/10.1016/j.bpj.2016.08.015] [PMID: 27653484]

[54]   Pasupuleti M, Schmidtchen A, Malmsten M. Antimicrobial peptides: key components of the innate immune system. Crit Rev Biotechnol 2012; 32(2): 143-71.
[http://dx.doi.org/10.3109/07388551.2011.594423] [PMID: 22074402]

[55]   Zhou QJ, Wang J, Liu M, *et al.* Identification, expression and antibacterial activities of an antimicrobial peptide NK-lysin from a marine fish Larimichthys crocea. Fish Shellfish Immunol 2016; 55: 195-202.
[http://dx.doi.org/10.1016/j.fsi.2016.05.035] [PMID: 27238427]

[56]   Zasloff M. Magainins, a class of antimicrobial peptides from Xenopus skin: isolation, characterization of two active forms, and partial cDNA sequence of a precursor. Proc Natl Acad Sci USA 1987; 84(15): 5449-53.
[http://dx.doi.org/10.1073/pnas.84.15.5449] [PMID: 3299384]

[57]   Zasloff M. Antimicrobial peptides of multicellular organisms. Nature 2002; 417(6870): 389-95.

[58]   Sengupta D, Leontiadou H, Mark AE, Marrink SJ. Toroidal pores formed by antimicrobial peptides show significant disorder. Biochim Biophys Acta Biomembr 2008; 1778(10): 2308-17.
[http://dx.doi.org/10.1016/j.bbamem.2008.06.007] [PMID: 18602889]

[59]   Brogden KA. Antimicrobial peptides: pore formers or metabolic inhibitors in bacteria? Nat Rev Microbiol 2005; 3(3): 238-50.
[http://dx.doi.org/10.1038/nrmicro1098] [PMID: 15703760]

[60]   Zhu S. Discovery of six families of fungal defensin-like peptides provides insights into origin and evolution of the CSαβ defensins. Mol Immunol 2008; 45(3): 828-38.
[http://dx.doi.org/10.1016/j.molimm.2007.06.354] [PMID: 17675235]

[61]   Oeemig JS, Lynggaard C, Knudsen DH, *et al.* Eurocin, a new fungal defensin: structure, lipid binding, and its mode of action. J Biol Chem 2012; 287(50): 42361-72.
[http://dx.doi.org/10.1074/jbc.M112.382028] [PMID: 23093408]

[62]   Semreen MH, El-Gamal MI, Abdin S, *et al.* Recent updates of marine antimicrobial peptides. Saudi Pharm J 2018; 26(3): 396-409.
[http://dx.doi.org/10.1016/j.jsps.2018.01.001] [PMID: 29556131]

[63]   Thackray PD, Moir A. SigM, an extracytoplasmic function sigma factor of Bacillus subtilis, is activated in response to cell wall antibiotics, ethanol, heat, acid, and superoxide stress. J Bacteriol 2003; 185(12): 3491-8.
[http://dx.doi.org/10.1128/JB.185.12.3491-3498.2003] [PMID: 12775685]

[64]   Zhao X, Wu H, Lu H, Li G, Huang Q. LAMP: a database linking antimicrobial peptides. PLoS One 2013; 8(6)e66557
[http://dx.doi.org/10.1371/journal.pone.0066557] [PMID: 23825543]

[65]   Lieke T, Meinelt T, Hoseinifar SH, Pan B, Straus DL, Steinberg CEW. Sustainable aquaculture requires environmental-friendly treatment strategies for fish diseases. Rev Aquacult 2020; 12(2): 943-65.
[http://dx.doi.org/10.1111/raq.12365]

[66]   Schulze M. Die Wirkung antimikrobieller Peptide (AMP) bei der Konservierung von Ebersperma. 2010.

[67]   Irazazabal LN, Porto WF, Fensterseifer ICM, *et al.* Fast and potent bactericidal membrane lytic activity of PaDBS1R1, a novel cationic antimicrobial peptide. Biochim Biophys Acta Biomembr 2019; 1861(1): 178-90.
[http://dx.doi.org/10.1016/j.bbamem.2018.08.001] [PMID: 30463701]

[68]   Matsuzaki K. Why and how are peptide-lipid interactions utilized for self defence?. Portland Press Ltd. 2001.
[http://dx.doi.org/10.1042/bst0290598]

[69]   Tornesello AL, Borrelli A, Buonaguro L, Buonaguro FM, Tornesello ML. Antimicrobial peptides as anticancer agents: Functional properties and biological activities. Molecules 2020; 25(12): 2850.
[http://dx.doi.org/10.3390/molecules25122850] [PMID: 32575664]

[70]   Zasloff M. Antimicrobial peptides of multicellular organisms: my perspective Antimicrobial Peptides. Springer 2019; pp. 3-6.
[http://dx.doi.org/10.1007/978-981-13-3588-4_1]

[71]   Rollins-Smith LA, Conlon JM. Antimicrobial peptide defenses against chytridiomycosis, an emerging infectious disease of amphibian populations. Dev Comp Immunol 2005; 29(7): 589-98.
[http://dx.doi.org/10.1016/j.dci.2004.11.004] [PMID: 15784290]

[72]   Peng KC, Lee SH, Hour AL, Pan CY, Lee LH, Chen JY. Five different piscidins from Nile tilapia, Oreochromis niloticus: analysis of their expressions and biological functions. PLoS One 2012; 7(11): e50263.
[http://dx.doi.org/10.1371/journal.pone.0050263] [PMID: 23226256]

[73]   Colombo AL, Padovan ACB, Chaves GM. Current knowledge of *Trichosporon spp.* and Trichosporonosis. Clin Microbiol Rev 2011; 24(4): 682-700.
[http://dx.doi.org/10.1128/CMR.00003-11] [PMID: 21976604]

[74]   Kim KH, Yang IJ, Kim WJ, *et al.* Expression analysis of interferon-stimulated gene 15 in the rock bream *Oplegnathus fasciatus* against rock bream iridovirus (RSIV) challenge. Dev Reprod 2017; 21(4): 371-8.
[http://dx.doi.org/10.12717/DR.2017.21.4.371] [PMID: 29354783]

[75]   Shike H, Lauth X, Westerman ME, *et al.* Bass hepcidin is a novel antimicrobial peptide induced by bacterial challenge. Eur J Biochem 2002; 269(8): 2232-7.
[http://dx.doi.org/10.1046/j.1432-1033.2002.02881.x] [PMID: 11985602]

[76]   Valero Y, Saraiva-Fraga M, Costas B, Guardiola FA. Antimicrobial peptides from fish: beyond the fight against pathogens. Rev Aquacult 2020; 12(1): 224-53.
[http://dx.doi.org/10.1111/raq.12314]

[77]   Zahran E, Noga EJ. Evidence for synergism of the antimicrobial peptide piscidin 2 with antiparasitic and antioomycete drugs. J Fish Dis 2010; 33(12): 995-1003.
[http://dx.doi.org/10.1111/j.1365-2761.2010.01205.x] [PMID: 21091726]

[78]   Bergsson G, Agerberth B, Jörnvall H, Gudmundsson GH. Isolation and identification of antimicrobial components from the epidermal mucus of Atlantic cod (Gadus morhua). FEBS J 2005; 272(19): 4960-9.

[http://dx.doi.org/10.1111/j.1742-4658.2005.04906.x] [PMID: 16176269]

[79]   Lamberty M, Bulet P, Latorse M-P, Hoffmann J. Antimicrobial peptides of the family of defensins, polynucleotides encoding said peptides, transformed vectors and organisms containing them. Google Patents 2004.

[80]   Subramanian S, Ross NW, MacKinnon SL. Myxinidin, a novel antimicrobial peptide from the epidermal mucus of hagfish, Myxine glutinosa L. Mar Biotechnol (NY) 2009; 11(6): 748-57.
[http://dx.doi.org/10.1007/s10126-009-9189-y] [PMID: 19330556]

[81]   Fernández de Ullivarri M, Arbulu S, Garcia-Gutierrez E, Cotter PD. Antifungal Peptides as Therapeutic Agents. Front Cell Infect Microbiol 2020; 10: 105.
[http://dx.doi.org/10.3389/fcimb.2020.00105] [PMID: 32257965]

[82]   Matejuk A, Leng Q, Begum MD, *et al.* Peptide-based antifungal therapies against emerging infections. Drugs Future 2010; 35(3): 197.
[http://dx.doi.org/10.1358/dof.2010.035.03.1452077] [PMID: 20495663]

[83]   Yin ZX, He W, Chen WJ, *et al.* Cloning, expression and antimicrobial activity of an antimicrobial peptide, epinecidin-1, from the orange-spotted grouper, Epinephelus coioides. Aquaculture 2006; 253(1-4): 204-11.
[http://dx.doi.org/10.1016/j.aquaculture.2005.10.002]

[84]   Umasuthan N, Mothishri MS, Thulasitha WS, Nam BH, Lee J. Molecular, genomic, and expressional delineation of a piscidin from rock bream (*Oplegnathus fasciatus*) with evidence for the potent antimicrobial activities of Of-Pis1 peptide. Fish Shellfish Immunol 2016; 48: 154-68.
[http://dx.doi.org/10.1016/j.fsi.2015.11.005] [PMID: 26549174]

[85]   Ruangsri J, Salger SA, Caipang CMA, Kiron V, Fernandes JMO. Differential expression and biological activity of two piscidin paralogues and a novel splice variant in Atlantic cod (Gadus morhua L.). Fish Shellfish Immunol 2012; 32(3): 396-406.
[http://dx.doi.org/10.1016/j.fsi.2011.11.022] [PMID: 22178249]

[86]   You X, Shan X, Shi Q. Research advances in the genomics and applications for molecular breeding of aquaculture animals. Aquaculture 2020; 526735357.
[http://dx.doi.org/10.1016/j.aquaculture.2020.735357]

[87]   Noga EJ, Fan Z, Silphaduang U. Host site of activity and cytological effects of histone-like proteins on the parasitic dinoflagellate *Amyloodinium ocellatum*. Dis Aquat Organ 2002; 52(3): 207-15.
[http://dx.doi.org/10.3354/dao052207] [PMID: 12553449]

[88]   Ullal AJ, Noga EJ. Antiparasitic activity of the antimicrobial peptide HbβP-1, a member of the β-haemoglobin peptide family. J Fish Dis 2010; 33(8): 657-64.
[http://dx.doi.org/10.1111/j.1365-2761.2010.01172.x] [PMID: 20561143]

[89]   Lei J, Sun L, Huang S, *et al.* The antimicrobial peptides and their potential clinical applications. Am J Transl Res 2019; 11(7): 3919-31.
[PMID: 31396309]

[90]   Kumar P, Kizhakkedathu J, Straus S. Antimicrobial peptides: diversity, mechanism of action and strategies to improve the activity and biocompatibility *in vivo*. Biomolecules 2018; 8(1): 4.
[http://dx.doi.org/10.3390/biom8010004] [PMID: 29351202]

[91]   Tincho MB, Morris T, Meyer M, Pretorius A. Antibacterial Activity of Rationally Designed Antimicrobial Peptides. International Journal of Microbiology 2020.
[http://dx.doi.org/10.1155/2020/2131535]

[92]   Masso-Silva J, Diamond G. Antimicrobial peptides from fish. Pharmaceuticals (Basel) 2014; 7(3): 265-310.
[http://dx.doi.org/10.3390/ph7030265] [PMID: 24594555]

[93]   Gupta A, Mumtaz S, Li CH, Hussain I, Rotello VM. Combatting antibiotic-resistant bacteria using nanomaterials. Chem Soc Rev 2019; 48(2): 415-27.

[http://dx.doi.org/10.1039/C7CS00748E] [PMID: 30462112]

[94]   Schmidt NW, Wong GCL. Antimicrobial peptides and induced membrane curvature: Geometry, coordination chemistry, and molecular engineering. Curr Opin Solid State Mater Sci 2013; 17(4): 151-63.
[http://dx.doi.org/10.1016/j.cossms.2013.09.004] [PMID: 24778573]

[95]   Dathe M, Wieprecht T. Structural features of helical antimicrobial peptides: their potential to modulate activity on model membranes and biological cells. Biochim Biophys Acta Biomembr 1999; 1462(1-2): 71-87.
[http://dx.doi.org/10.1016/S0005-2736(99)00201-1] [PMID: 10590303]

[96]   Cerón JM, Contreras-Moreno J, Puertollano E, de Cienfuegos GÁ, Puertollano MA, de Pablo MA. The antimicrobial peptide cecropin A induces caspase-independent cell death in human promyelocytic leukemia cells. Peptides 2010; 31(8): 1494-503.
[http://dx.doi.org/10.1016/j.peptides.2010.05.008] [PMID: 20493222]

[97]   Sato H, Feix JB. Peptide–membrane interactions and mechanisms of membrane destruction by amphipathic α-helical antimicrobial peptides. Biochim Biophys Acta Biomembr 2006; 1758(9): 1245-56.
[http://dx.doi.org/10.1016/j.bbamem.2006.02.021] [PMID: 16697975]

[98]   Malanovic N, Lohner K. Antimicrobial peptides targeting gram-positive bacteria. Pharmaceuticals (Basel) 2016; 9(3): 59.
[http://dx.doi.org/10.3390/ph9030059] [PMID: 27657092]

[99]   Hoeksema M, van Eijk M, Haagsman HP, Hartshorn KL. Histones as mediators of host defense, inflammation and thrombosis. Future Microbiol 2016; 11(3): 441-53.
[http://dx.doi.org/10.2217/fmb.15.151] [PMID: 26939619]

[100]   Kawasaki H, Iwamuro S. Potential roles of histones in host defense as antimicrobial agents. Infectious Disorders-Drug Targets (Formerly Current Drug Targets-Infectious Disorders) 2008; 8(3): 195-205.
[http://dx.doi.org/10.2174/1871526510808030195]

[101]   Delgado-Rizo V, Martínez-Guzmán MA, Iñiguez-Gutierrez L, García-Orozco A, Alvarado-Navarro A, Fafutis-Morris M. Neutrophil extracellular traps and its implications in inflammation: an overview. Front Immunol 2017; 8: 81.
[http://dx.doi.org/10.3389/fimmu.2017.00081] [PMID: 28220120]

[102]   Li T, Zhang Z, Li X, Dong G, Zhang M, Xu Z, *et al.* Neutrophil Extracellular Traps: Signaling Properties and Disease Relevance. Mediators of Inflammation 2020.

[103]   Gyan WR, Yang Q, Tan B, *et al.* Effects of antimicrobial peptides on growth, feed utilization, serum biochemical indices and disease resistance of juvenile shrimp, *Litopenaeus vannamei*. Aquacult Res 2020; 51(3): 1222-31.
[http://dx.doi.org/10.1111/are.14473]

[104]   Lei Y, Qiu R, Shen Y, Zhou Y, Cao Z, Sun Y. Molecular characterization and antibacterial immunity functional analysis of liver-expressed antimicrobial peptide 2 (LEAP-2) gene in golden pompano (*Trachinotus ovatus*). Fish Shellfish Immunol 2020; 106: 833-43.
[http://dx.doi.org/10.1016/j.fsi.2020.09.002] [PMID: 32891790]

[105]   Xie Y, Wan H, Zeng X, Zhang Z, Wang Y. Characterization and antimicrobial evaluation of a new Spgly-AMP, glycine-rich antimicrobial peptide from the mud crab Scylla paramamosain. Fish Shellfish Immunol 2020; 106: 384-92.
[http://dx.doi.org/10.1016/j.fsi.2020.08.009] [PMID: 32771609]

[106]   Shan Z, Yang Y, Guan N, Xia X, Liu W. NKL-24: A novel antimicrobial peptide derived from zebrafish NK-lysin that inhibits bacterial growth and enhances resistance against Vibrio parahaemolyticus infection in *Yesso scallop, Patinopecten yessoensis*. Fish Shellfish Immunol 2020; 106: 431-40.
[http://dx.doi.org/10.1016/j.fsi.2020.08.020] [PMID: 32810530]

[107] Zorofchian Moghadamtousi S, Abdul Kadir H, Hassandarvish P, Tajik H, Abubakar S, Zandi K. A review on antibacterial, antiviral, and antifungal activity of curcumin. BioMed research international 2014.
[http://dx.doi.org/10.1155/2014/186864]

[108] Chinchar VG, Bryan L, Silphadaung U, Noga E, Wade D, Rollins-Smith L. Inactivation of viruses infecting ectothermic animals by amphibian and piscine antimicrobial peptides. Virology 2004; 323(2): 268-75.
[http://dx.doi.org/10.1016/j.virol.2004.02.029] [PMID: 15193922]

[109] Verhelst J, Hulpiau P, Saelens X. Mx proteins: antiviral gatekeepers that restrain the uninvited. Microbiol Mol Biol Rev 2013; 77(4): 551-66.
[http://dx.doi.org/10.1128/MMBR.00024-13] [PMID: 24296571]

[110] Guo M, Wei J, Huang X, Huang Y, Qin Q. Antiviral effects of β-defensin derived from orange-spotted grouper (Epinephelus coioides). Fish Shellfish Immunol 2012; 32(5): 828-38.
[http://dx.doi.org/10.1016/j.fsi.2012.02.005] [PMID: 22343108]

[111] Burge CA, Hershberger PK. Climate change can drive marine diseases. Marine Disease Ecology 2020; p. 83.

[112] Jin JY, Zhou L, Wang Y, *et al.* Antibacterial and antiviral roles of a fish β-defensin expressed both in pituitary and testis. PLoS One 2010; 5(12)e12883
[http://dx.doi.org/10.1371/journal.pone.0012883] [PMID: 21188147]

[113] Yuan J, Yang Y, Nie H, *et al.* Transcriptome analysis of epithelioma *papulosum cyprini* cells after SVCV infection. BMC Genomics 2014; 15(1): 935.
[http://dx.doi.org/10.1186/1471-2164-15-935] [PMID: 25344771]

[114] Guo CJ, Wu YY, Yang LS, *et al.* Infectious spleen and kidney necrosis virus (a fish iridovirus) enters Mandarin fish fry cells *via* caveola-dependent endocytosis. J Virol 2012; 86(5): 2621-31.
[http://dx.doi.org/10.1128/JVI.06947-11] [PMID: 22171272]

[115] Verbruggen B, Bickley L, van Aerle R, *et al.* Molecular mechanisms of white spot syndrome virus infection and perspectives on treatments. Viruses 2016; 8(1): 23.
[http://dx.doi.org/10.3390/v8010023] [PMID: 26797629]

[116] Zhang M, Li M, Sun L. NKLP27: a teleost NK-lysin peptide that modulates immune response, induces degradation of bacterial DNA, and inhibits bacterial and viral infection. PLoS One 2014; 9(9)e106543
[http://dx.doi.org/10.1371/journal.pone.0106543] [PMID: 25180858]

[117] Chang WT, Pan CY, Rajanbabu V, Cheng CW, Chen JY. Tilapia (*Oreochromis mossambicus*) antimicrobial peptide, hepcidin 1–5, shows antitumor activity in cancer cells. Peptides 2011; 32(2): 342-52.
[http://dx.doi.org/10.1016/j.peptides.2010.11.003] [PMID: 21093514]

[118] Oscar Ditchou Nganso Y, Sidjui Sidjui L, Gabrielle A Ngnoung Amang A, *et al.* Identification of Peptides in the Leaves of <i>Bauhinia rufescens</i> Lam (Fabaceae) and Evaluation of Their Antimicrobial Activities Against Pathogens for Aquaculture. Science Journal of Chemistry 2020; 8(4): 81.
[http://dx.doi.org/10.11648/j.sjc.20200804.12]

[119] Boisard S, Le Ray A-M, Landreau A, Kempf M, Cassisa V, Flurin C, *et al.* Antifungal and antibacterial metabolites from a French poplar type propolis. Evidence-Based Complementary and Alternative Medicine 2015.
[http://dx.doi.org/10.1155/2015/319240]

[120] Roch P, Yang Y, Toubiana M, Aumelas A. NMR structure of mussel mytilin, and antiviral–antibacterial activities of derived synthetic peptides. Dev Comp Immunol 2008; 32(3): 227-38.
[http://dx.doi.org/10.1016/j.dci.2007.05.006] [PMID: 17628674]

[121] Domeneghetti S, Franzoi M, Damiano N, *et al*. Structural and antimicrobial features of peptides related to Myticin C, a special defense molecule from the Mediterranean mussel *Mytilus galloprovincialis*. J Agric Food Chem 2015; 63(42): 9251-9.
[http://dx.doi.org/10.1021/acs.jafc.5b03491] [PMID: 26444944]

[122] Sonthi M, Cantet F, Toubiana M, *et al*. Gene expression specificity of the mussel antifungal mytimycin (MytM). Fish Shellfish Immunol 2012; 32(1): 45-50.
[http://dx.doi.org/10.1016/j.fsi.2011.10.017] [PMID: 22037382]

[123] Destoumieux D, Munoz M, Bulet P, Bachère E. Penaeidins, a family of antimicrobial peptides from penaeid shrimp (Crustacea, Decapoda). Cell Mol Life Sci 2000; 57(8): 1260-71.
[http://dx.doi.org/10.1007/PL00000764] [PMID: 11028917]

[124] Brockton V, Smith VJ. Crustin expression following bacterial injection and temperature change in the shore crab, *Carcinus maenas*. Dev Comp Immunol 2008; 32(9): 1027-33.
[http://dx.doi.org/10.1016/j.dci.2008.02.002] [PMID: 18343497]

[125] Johnson NG, Burnett LE, Burnett KG. Properties of bacteria that trigger hemocytopenia in the Atlantic blue crab, Callinectes sapidus. Biol Bull 2011; 221(2): 164-75.
[http://dx.doi.org/10.1086/BBLv221n2p164] [PMID: 22042435]

[126] Dalhoff A. Antiviral, antifungal, and antiparasitic activities of fluoroquinolones optimized for treatment of bacterial infections: a puzzling paradox or a logical consequence of their mode of action? Eur J Clin Microbiol Infect Dis 2015; 34(4): 661-8.
[http://dx.doi.org/10.1007/s10096-014-2296-3] [PMID: 25515946]

[127] Rosa RD, Santini A, Fievet J, Bulet P, Destoumieux-Garzón D, Bachère E. Big defensins, a diverse family of antimicrobial peptides that follows different patterns of expression in hemocytes of the oyster Crassostrea gigas. PLoS One 2011; 6(9)e25594.
[http://dx.doi.org/10.1371/journal.pone.0025594] [PMID: 21980497]

[128] Al-Huqail A, Behiry S, Salem M, Ali H, Siddiqui M, Salem A. Antifungal, antibacterial, and antioxidant activities of *Acacia saligna* (Labill.) HL Wendl. flower extract: HPLC analysis of phenolic and flavonoid compounds. Molecules 2019; 24(4): 700.
[http://dx.doi.org/10.3390/molecules24040700] [PMID: 30781352]

[129] Marggraf M, Panteleev P, Emelianova A, *et al*. Cytotoxic potential of the novel horseshoe crab peptide polyphemusin III. Mar Drugs 2018; 16(12): 466.
[http://dx.doi.org/10.3390/md16120466] [PMID: 30486233]

[130] Silva ON, Fensterseifer ICM, Rodrigues EA, *et al*. Clavanin A improves outcome of complications from different bacterial infections. Antimicrob Agents Chemother 2015; 59(3): 1620-6.
[http://dx.doi.org/10.1128/AAC.03732-14] [PMID: 25547358]

[131] Bulet P, Stöcklin R, Menin L. Anti-microbial peptides: from invertebrates to vertebrates. Immunol Rev 2004; 198(1): 169-84.
[http://dx.doi.org/10.1111/j.0105-2896.2004.0124.x] [PMID: 15199962]

SUBJECT INDEX

A

Abdominal 26, 27, 87, 104
  fat deposition 87
  pain 26, 27, 104
Acid(s) 21, 23, 33, 34, 43, 55, 58, 66, 67, 69,
70, 82, 83, 84, 86, 88, 102, 104, 105, 112,
115, 125, 126, 128, 129, 141, 142, 143, 144,
145, 147, 163, 165, 173, 175, 196, 199, 200,
203
  ascorbic 43, 82
  bile 112, 115
  bioactive 126
  caffeic 21, 66, 67, 82, 125, 126, 165, 199
  carboxylic 55
  chlorogenic 82, 125, 126
  cichoric 126
  cinnamic 196
  citronellic 196
  conjugated linoleic 88
  coumaric 82
  crategolic 66, 67
  ellagic 66, 67, 70, 82, 83, 86
  ferulic 66, 67, 82
  galactosyluronic 105
  gallic 67, 70, 82, 83, 86
  gallotannic 66, 67
  hydroxybenzoic 67
  hydroxycinamic 67
  lactic 141, 142
  monocaffeoyltartaric 126
  oleanolic 70, 83
  organic 23, 33, 34, 58, 143, 144, 145, 203
  pantothenic 69, 102
  phenolic 67, 82, 129, 173
  punicic 82, 84
  rosmarinic 175, 199
  salicylic 66, 67, 104
  taraxinic 126
  thiobarbituric 128
  uric 144, 200
  uronic 105

  ursolic 83
  valeric 147
Additives 86, 98
  coffee 98
  pomegranate-derived 86
Adhesion 141, 145
  epithelial junction 141
Adipose tissue 88
Adjunctive therapy 1
Adrenal gland 131
Aerococcus viridans 220
Aeromonas hydrophila 173, 174
Agar diffusion method 44
Agent 71, 105, 112, 113, 129
  natural hepatoprotective 105
  plausible herbal 112
  traditional therapeutic 129
Agro-climatic conditions 68
Alanine 41, 98, 128
  aminotransferase 98, 128
  transaminase 41
Albumin protein 8
Alkaloids 167, 169, 171, 172
  dietary 172
  elegans 167, 169, 171
Allergic 1, 9, 10
  disorders 9
  reaction 1, 10
Allium sativum 204
Allochthonous probiotics 139
Aloe vera 20, 21, 25, 26
  ingesting 26
Alpha-glucosidase, inhibiting 86
Amino acids 142, 143, 165, 216, 217, 218,
  219
  essential 142, 216
  hydrophobic 219
Aminoglycosides 7, 9, 71
Amino transferase 163
Amoxicillin 7, 8
AMPs 216, 218, 220, 221, 224, 226
  cationic 216

chitin-binding 221
cysteine-rich cationic 220, 221
glycine-derivatives 224
synthetic 226
toxin 218
Amylase 143, 199
Amylase activity 142, 166
intestinal 166
Amyloodinium ocellatum 223
Angiogenesis 127, 129
inhibited 129
Angiosperms 100
Anthelmintic activity 178, 205
Anthocyanidins 82, 83
Anthocyanins 82, 85
Anthracenosides 129
Antibacterial 5, 34, 43, 44, 46, 70, 71, 74, 88,
106, 127, 129, 173, 174, 219, 223, 224
activity 43, 44, 70, 71, 88, 173, 174, 219,
224
drugs 129
effects 5, 44, 127, 129, 173
growth promoters 34
properties 74, 106, 174, 223
therapeutics 46
Antibacterial agents 33, 34, 70, 224
growth-promoting 33
natural 70
Antibiotic(s) 4, 6, 34, 66, 89, 99, 125
antibacterial 34
chlortetracycline 4
commercial 89
synthetic 5, 99
therapeutic 125
traditional 66
tylosin 6
virginiamycin 6
Antibiotic resistance 5, 7, 10, 20, 137, 172,
174
genes (ARGs) 7
Antibody 147, 224
obstruction 224
production 147
Anticancer agents 111
Antidiabetic activity 98

Antifungal 21, 22, 35, 52, 99, 172, 215, 220,
221, 222
Antifungal 20, 52, 59, 179, 180, 216, 222
activity 59, 179, 180, 222
diseases 216
influence 222
properties 20, 52
Antifungal effects 36, 222
synergistic 222
Antihelminthic 179
activity 179
effect 179
Anti-infective agents 5
Anti-inflammatory activities 86, 129, 167
Antimicrobial 2, 5, 8, 9, 27, 36, 81, 98, 102,
195, 217, 218, 126, 129, 195
agents 8, 98, 102, 126
effects 36, 129, 217, 218
growth promoters 2, 81
microflora 9
resistance 5, 9, 27, 195
Antimicrobial activity 81, 85, 86, 161, 162,
163, 164, 170, 172, 180, 216, 224
broad-spectrum 172
inhibiting 81
Antimicrobial properties 21, 34, 70, 73, 74,
172, 215, 224
broad-spectrum 215
Antioxidant 20, 21, 22, 38,41, 42, 66, 71, 72,
73, 74, 81, 82, 84, 85, 87, 89, 100, 127,
128, 129, 161, 162, 163, 164, 168, 169,
180, 201, 202, 203
activity 42, 71, 72, 74, 81, 84, 85, 129, 161,
162, 163, 164, 168, 169, 202
agent 100
capacity 85, 202, 203
effect 21, 89, 127
enzymes 22, 41, 74, 128, 169, 180, 201
intestinal 169
phytochemicals 21
properties 21, 74, 129, 163, 202
Antiparasitic 204, 205, 129, 223
activities 129, 223
agent 204
drug resistance 205

effectiveness 204
Anti-protozoan activities 81
Antiviral 57, 68, 70, 172, 175, 176, 177, 225
    activity 68, 70, 172, 175, 176, 177, 225
    agents 176, 177
    effects 57, 177
Antivirulence activity 174
Aortic tissue homogenate 42
Applications of chicory 104
Aquaculture 160, 161, 162, 163, 164, 167,
        168, 172, 175, 177, 178, 179, 180, 194,
        195, 220, 222
    ecosystems 172
    environment 164, 167, 168
    systems 179
Aquaculture production 160, 222
    global finfish 222
Arapaima gigas 179
Aspartate 98, 99, 163, 198
    aminotransferase 98, 99
    amino transferase activities 163
    transaminase 198
Aspergillus niger 36, 43, 44, 222
Aspirate aminotransferase 128
Assay 54
    reducing antioxidant power 54
Atherosclerosis 42, 45, 46
Azalea plant toxicity 27

B

Bacillomycin 141
Bacillus 44, 68, 70, 71, 89, 138, 142, 143,
        144, 145, 146, 147, 217, 220
    brevis 217
    cereus 68, 70, 71, 89
    lichenifermis 142, 143
    licheniformis 145
    megaterium 220
    pumilus 44
    subtilis 44, 138, 142, 144, 145, 146, 147
Bacteria 2, 3, 5, 6, 9, 59, 71, 73, 137, 138,
        139, 145, 147, 162, 170, 172, 175, 218
    drug-resistant 218

harmful 59, 170
    intestinal 137, 162, 170
    resistant 2, 5, 9
Bacterial 26, 86, 168, 223
    cell wall peptidoglycans 168
    fish diseases 223
    protein secretions 86
    reverse mutation 26
Balance gut microbiota 59, 127, 130
Bamboo leaf flavonoid (BLF) 59
Beluga juveniles 169
Bifidobacterium 143, 146
    bifidium 143
    thermophilum 146
Bile 115, 116, 127
    secretions 115, 127
    secretions fusion 116
Black cumin 99, 200
Blood 8, 45, 59,74, 88, 105,116,149, 167, 198,
        200 202
    antioxidant enzymes 45
    cholesterol 116
    haematology 149
    neutrophil 200
    plasma 74, 88
Body weight gain (BWG) 37, 38, 39, 41, 42,
        43, 56, 57, 58, 127, 128
Bone marrow 1
Boosting 19, 59,88
    health 88
    immune system modulation 19
    poultry birds 59
Breast muscles 7, 87, 144
Broilers 7,8, 20, 21, 22, 23, 24, 37, 40, 41,42,
        45, 46, 58, 59, 82, 87, 88, 105, 128,
        131,143, 144,146, 147,148
    breast meat 87
    chicken meat 45
    chickens 37, 42, 45, 46, 82, 87, 88, 128,
        131, 144, 146, 147, 148
    commercial 143
    dietary supplementing 128
    meat 7, 8, 87
Broiler chicks 38, 39, 42, 56, 58, 60
    female 56

Broiler diets 45, 57, 58, 72, 87, 88
  supplementing 87

C

Campylobacter jejuni 145
Cancers 26,105,118, 127, 128
  colon 118
  uterine 105
Candida albicans 43, 44, 68, 70, 222
Capacity 45, 53, 82, 163, 169,170, 176,225
  digestive 53
  hydrogen donor 169
  protein-binding 176
  virus adsorption 225
Capsicum 22, 23
  annuum 23
  oleoresin 22, 23
Catalase 42, 59, 87, 202
  activity 202
  enzyme activities 59, 87
  enzymes 42
Catechins 82, 83, 176
Cephalosporins 5
Chemical 84
  analysis of pomegranate 84
  constituents of pomegranate 84
Chicken immunity 89
Chloroquine-resistant *Plasmodium 70*
falciparum 70
Chlorotetaine 141
Cholagogic activity 128
Cholesterol 46, 59, 85, 86, 115
  absorption 86
  lowering activities 85
  metabolism 86, 115
  plasma 59
  predisposes 46
  synthesis 59
Chronic anemic conditions 198
Chrysophins 216
Cichlidogyrus 205
  halli 205
  thurstonae 205

tilapiae 205
Cichorium intybus 105
  root powder 105
  supplementation 105
Cinnamaldehyde 21, 22, 23, 54, 56, 58, 59,
  60, 200, 203, 205
  supplementation 56
Cinnamomum 55, 56, 57, 58
  bejolghota 56
  izeylanicum 56
  rivulorum 55
  sinharajense 55
  zeylanicum 56, 57, 58
Cinnamon 20, 55, 56
  extract 20
  leaf 55
  powder 56
Cinnamon oil 52, 54, 55, 56, 57, 58, 60
  production 55
Ciprofloxacin 1, 7, 8
Cistus plant 176
Citrus aurantium 164
Clostridium 6, 138, 144, 147
  butyricum 147
  perfringens 6, 138, 144
Clove 67, 71, 72
  antioxidant activity of 71, 72
  growth 67
Coccidiosis 20, 22, 81, 145, 148
Complement component chemokine 199
Composition 68, 83
  nutrient 68
  phenolic 83
Constipation 111, 113, 115, 117, 118
Consumption 26, 112, 113, 128,138, 142
  global meat 138
  of dietary fiber foods 112
Copaifera oleoresin 179
Coumarin 55, 176, 177, 196
Cryptococcus neoformans 222
Curcuma longa 23, 44, 162, 163
Curcumin 35, 36, 37, 162, 163, 196
Cyanogenic glycosides 21, 27
  toxic 27
Cysteine-rich peptides 173

Cytokines 4, 25, 138, 163, 201, 203
  anti-inflammatory 163, 203
    release 25
    secretions 201
Cytoplasmic 68, 70, 71, 173
  leakage 173
  membrane 68, 70, 71

D

Damage 9, 19, 23, 27,103, 104, 176, 168, 223
  cardiac 27
  economic 176
  hepatic 104
  mitigating intestinal 19, 23
Dandelion, dietary 167, 172
Defense systems 172
Defensins 216, 218, 220, 221
Development 9, 102, 164, 170, 171, 172, 194, 195, 222, 224
  agalactiae 224
  immune system's 170
  microbial resistance 172
Diabetic disorders 86
Diacetylactis 140
Diarrhea 26, 83, 113, 117, 118
Diarrheal-related disorders 118
Dichloromethane fraction 55
Dietary 22, 40, 43, 44, 46, 74, 103,112, 128, 143, 145, 148,163, 164, 165, 166, 167, 169
  aflatoxicosis 103
  antioxidants 22
  clove 74
  curcumin 163
  fiber foods 112
  ginger 164
  supplementation 40, 43, 44, 46, 128, 143, 145, 148, 164, 165, 166, 167, 169
Diets 23, 24,72, 87, 200
  bird 24
  broiler chicken 72
  curcuma 23
  fat-containing 87

  oil-based 200
Digestibility143, 147
  ileal amino acid 147
  nutrient 143
Digestion 6, 3, 20, 57, 60, 73, 125, 142, 143, 170
  coefficients 143
  nutrient 57, 170
  phospholipid 6
Digestive 125, 127
  disorders 125
  stimulant and prebiotic activity 127
Digestive enzyme(s) 34, 52, 57, 73, 74,142, 163, 199
  activities 74, 163
  endogenous 57
Disease(s) 1, 4, 5, 9, 21, 22, 35, 86, 99,113,115, 118, 161, 162,163, 166, 168, 170, 172,175, 179,180, 195,
  cardiovascular 86, 113
  coronary heart 115
  fungal 162, 179
  immunopathological 9
  inflammatory bowel 118
  intestinal 22
  parasitic 195
  protozoal 195
  respiratory 35
   resistance 166, 168, 172
  saprolegniasis andicthyophonus 179
  viral 175
DNA 70, 168
  damage 168
  polymerase enzyme 70
Dolops discoidalis 196
Drobachiensis 220
Drugs 3, 6, 8, 9, 34, 129, 160, 180
  growth-promoting 3
  veterinary 160, 180
Dysbacteriosis 138
Dysfunction 171

E

Economic income 195
Ecosystem 161, 164, 226
Edible tissue 7
Effects 9, 20, 34, 35, 44, 45, 82, 86, 87, 117,
    125, 126, 127, 128, 129, 130, 131,
    161,195, 205, 219
    anticoccidial 20
    anti-inflammatory 35, 86, 127, 129
    bacteriostatic 44
    cancer-causing 9
    cholesterol-lowering 117
    disease 205
    fungicidal 44
    gastrointestinal 125
    health-promoting 131
    hemolytic 219
    hypoglycaemic 130
    hypolipidemic 87, 127, 130
    immunostimulatory 87
    inflammatory 126
    neuroprotective 82
    radical-scavenging 45
    synergetic 161
    therapeutic 34, 128, 129, 195
Efficacy 34, 70, 98, 145,146, 176, 177
    anticoccidial 146
    antiviral 70
    economic 34
Efficiency 45, 57, 58, 59, 127, 128, 137, 138,
    194,195, 200, 203, 204, 223
    anti-oxidative 204
    antiparasitic 194, 223
    economic 45
Eggs 6, 8, 9, 10, 20, 24, 35, 38, 41, 44, 45, 46,
    112, 113, 119
    contaminated 9
    hypocholesterolemic 46
    producing organic 119
Electrostatic 219
    forces 219
    reaction 219
Elements, therapeutic 100
Energy 4, 24, 69, 101, 111

    metabolizable 24
    raising 111
Enteric pathogens 22
Enterobacteria 5, 44
Enterococcus 53, 71, 143, 146, 147, 220
    faecalis 53, 71, 220
    faecium 143, 146, 147
Environmental 7, 20, 171,
    antibiotic residues 7
    perturbations 171
    stress-related effects 20
Enzymes 20, 21, 34, 44,45, 59, 86, 127, 162,
    163, 199, 201, 202
    antioxidative 45
    liver function 163, 199
    plant proteolytic 44
Epithelioma papulosum cyprini (EPC) 225
Erwinia amylovora 221
Escherichia coli 5, 9, 43, 53, 71, 129, 138,
    220, 221
    resistance 5
Esculetin 104
Essential 5
    micro, protecting 5
Essential oil(s) 34, 36, 38, 39, 40, 41, 42, 43,
    44, 52, 53, 55, 58, 59, 61, 162, 164, 165,
    166, 173, 179, 194, 195, 196, 203, 205
    extract (EOE) 36, 41, 42, 43, 44, 55, 58, 61
    natural 194
    of cinnamon 173
    of oregano 165
    plant-derived 34
Eudesmanolides 130
Expression 86, 89, 162, 163, 164, 165, 167,
    170, 174, 198,199, 201, 203, 223, 225
    cytokine gene 203
    growth hormone 163
    immune gene 199
    immune-related genes 162, 198
    pro-inflammatory gene 203

F

Factors 8, 10, 45, 84, 99, 116,144, 161, 172,
    174,180, 195, 201,203

antiaging 84
essential virulence 174
hazardous 116
redox-sensitive transcription 203
Fat 35, 87
   digestibility 87
   peroxidation 35
Fatty acids 9, 23, 25, 58, 59, 67, 68, 74, 82,
     83, 129, 147, 173,174
   unsaturated 58, 74, 173
Feeding 1, 10, 46, 73,105, 143, 168, 171
   chicory root powder 105
   poultry antibiotic 1
Feed intake (FI) 19, 23, 24, 37, 38, 58, 142,
     143, 146, 163, 165
Fertilizers 9, 1
   residual 1
Fiber, psyllium 113
Fish 163, 166, 180, 195, 205, 215, ,218, 220,
     225, 226,
   diets 163
   disease resistance 180
   ectoparasites 195
   gold 205
   invertebrate 218
   immunity 166
   mortalities 225, 226
   parasite infections 205
   pathogenic agents 215, 220
Fish oil (FO) 163, 201, 203
   dietary 201
Flavonols, estrogenic 82
Food 34,89, 112, 138,
   natural 112
   additives 34, 138
   industry 89
Foodborne illness 160
Functions 81, 103, 124, 126, 127, 147, 201,
   hepatoprotective 103
   hypotensive 126
   immune 81, 147, 201
   macrophage 127
   stimulatory 124
Fusarium 36, 68, 70, 220, 222
   graminearum 222

moniliforme 68, 70
oxysporum 36, 68, 70, 220
solani 222

G

Garlic metabolites 22
Gas chromatography (GC) 35, 36, 37, 44, 56
   mass spectrometry 35, 56
Gas emission 87
Gastric juices 130, 139
   tolerance 139
Gastrointestinal 1, 4, 9, 40, 73, 98, 126, 130,
     144, 145, 170, 171
   system 126, 130
   tract 1, 4, 9, 40, 73, 98, 144, 145, 170, 171
Gastroprotective 99
GC-MS 36, 56,
   gas chromatography-mass spectrometry 56
   method 36
   techniques 36
GC techniques 36
Gelsemium elegans 165
Genes 2, 10, 22, 164, 167 169, 170, 175, 198,
     199, 201, 203,
   antimicrobial-resistant 2, 10
   antioxidant-related 169
   anti-oxidation-related 203
   cytokine-related 167
   immune-related 167
   mucosal immune 167
Genotoxic analysis 26
Genotoxicity 26
   evaluation tests 26
Gingerols 35
GIT 139, 104, 141
   disorders, minor 104
Glucose 45, 82, 86, 113, 198
   and lipid metabolism activities 86
   plasma 45
Glutamate pyruvate 104
Glutamic 131
   oxaloacetate transaminase 131
   pyruvic transaminase (GPT) 131

Glutathione 35, 201, 202
   reductase activities 202
Glutathione peroxidase (GP) 25, 42, 88, 128,
   163, 201
   enzyme 88
   plasma 25
Glycosides 126, 130, 175
  flavonol 175
Glycyrrhiza glabra 163
Growth 2, 73, 82, 89, 127, 144, 165,
  booster 73
   enhancer bacteria 144
   enhancing effects of in-feed antibiotics 2
   microbial 89
   promoter activity 165
   promoting effects 2, 82
  tumor 127
Growth promoters 1, 2, 3, 4, 5, 6, 10, 20, 21,
   23, 24, 26, 34, 72, 73, 99, 126, 137, 138,
   164
  herbal 24
  natural 23, 26, 34, 72, 126
Gumboro disease 148
Gut 20, 23, 169, 170, 171, 180
   dysbiosis 170
   homeostasis 170, 171, 180
   integrity 20
   lymphocytes 23
   microbes 20
   microbial ecology 170
   microbiome 169
Gut microbiota 81, 86, 161, 166, 169, 170,
   171
   punicalagins modulate 86
   functions 166
   homeostasis 171
Gyrodactylus turnbulli 204

H

Health, gastrointestinal 146
Hemagglutination 176
Hemagglutinin 27
Heme oxygenase 24

Hemocytes 216, 217
  tunicate 216
Hemoglobin 42, 131, 163
Hemorrhoids 111, 118
  bleeding internal 118
  internal bleeding 118
  progression 118
  symptomatic 118
  therapy 111
Hepatic 103
  androsterone 103
  enzymes 103
Hepatitis 70, 124
Hepatocytes 99, 103, 128
Hepatotoxicity 9
Herbal 20, 23, 26, 39, 81, 99, 112, 127, 163,
  196
  mutagenicity 26
  plants 23, 26, 39, 81, 99, 112, 127, 163,
   196
  preparations 20
  toxicity 26
Herbs 19, 20, 22, 23, 24, 25, 26, 27, 44, 53, 74,
  124, 125, 126, 127, 128, 129, 195 196
  medical 22
  medicinal 20, 44, 129, 195
High density lipoprotein (HDL) 113
Hippophae rhamnoides 175
Hormones 4, 6, 9, 115, 131
  catabolic 4
Hydrodistillation 44
Hydrophila 163, 174, 175
  disease 163,
  induced branchial bioenergetics 174
  infection 174, 175
Hydrophobicity 218
Hypercholesterolemia 112, 113, 115, 116, 118
  prevention 116
Hyper-cholesterol-related syndromes 113
Hyperlipidemia 112, 116

I

Immune 130, 146, 226

communications 130
  defensive system 226
  reactions 130
  response genes 146
Immune activity 200, 201, 203
  innate 203
Immune responses 22, 23, 25, 87, 88, 147,
    148, 163, 166, 167, 168, 169, 171, 198,
    199, 201, 218
  and disease resistance 168
  humoral innate 167
  innate 22, 23, 166, 218
  systemic 167
Immune system 4, 20, 147, 161, 166, 167,
    170, 175, 215, 223
  mucosal 147, 167, 175
  responses 166
Immunity 19, 20, 21, 22, 23, 24, 59, 98, 99,
    137, 138, 141, 147, 148, 166, 170, 199,
    200, 218, 224
  innate 22, 141, 166, 170, 224
  modulated 59
  promoted 199, 200
  system 218
Immunogenic 25, 35, 57
  activity 57
  reaction 35
Immunoglobulins 141, 164, 175, 198, 199
Immunomodulation 149
Immunomodulatory 20, 21, 52, 61, 81, 82, 90,
    124, 125,130, 162, 163, 164, 166, 167,
    168, 173, 195, 216, 223, 226,
  activity 162, 163, 164, 166, 173
  effects 52, 124, 130, 167, 223
  peptides 226
  properties 21
  role 168, 195, 216
Immunostimulants 98, 161, 166, 167, 180,
    204
  dietary 167
Immunostimulatory 172, 204
  activity 172
Immunosuppression 160, 161
Impact 35, 52, 198
  hypo-cholesterolaemic 52

immunostimulatory 198
  of ginger oil 35
Incidence, down-regulated disease 66
Industries 66, 81, 99
  cosmetics 81
Infection 2, 3, 4, 5, 6, 21, 22, 35, 83, 174, 175,
    177, 194, 195, 201, 216, 218, 219, 222,
    223, 225, 226
  albicans 222
  antibiotic-resistant 2
  antiviral 225
  bacterial 5, 174, 218
  bacteriological 35
  drug-resistant 2
  fungal 83
  fungus 216
  parasitic 177, 194, 201, 218, 222, 223
Infectious bronchitis 148
  and Gumboro disease 148
Infectious diseases 4, 137, 139, 194, 195, 201,
    215, 223, 226
Inflammation 89, 113, 115, 127, 141,162, 201
  chronic 141
  intestinal 113
  soothing 115
Inflammatory cytokine 89
Insulin sensitivity 86
Intestinal 66, 115, 144, 147, 165, 171, 172
  cavity 115
  health 144, 147, 172
  histomorphology 165
  histomorphometry analysis 171
  microbiota population 66
Intestinal microbes 21, 90, 147
  reducing harmful 147
Irritable bowel syndrome 118

K

Kidney enzymes activity 199

L

Lactate dehydrogenase 199
Lactic acid bacteria 44, 143, 144, 145, 171, 175
Lactic acid bacteria production 6
lactobacillus 142, 143, 145, 146, 147
    acidophilus 142, 143, 145, 146, 147
    casei 143, 146
Lactococcus lactis 217
Leishmania 70
    amazonensis 70
    donovani 70
    tropica 70
Licheniformis 141
Lipid peroxidation 41, 59, 104, 106, 168
Liquid 55, 139, 145, 217
    aromatic oily 55
    chromatography, high-performance 217
Listeria monocytogenes 60, 68, 70, 71, 86, 145
Live 72, 98, 103, 104, 105, 128
    body weight (LBW) 72, 128
    dysfunction 103
    enzymes 98, 103, 104, 105, 128
Livestock 1, 4, 6, 10, 34, 73, 112
    diseases 6
    growth 73
    industry 1, 4
    production 112
    productivity 112
    systems 112
Luteinizing hormone 25
Lymphocytes 25, 26, 141, 199, 200, 201
    activation 25
    activity 26
    intestinal intraepithelial 141
Lysozyme 88, 162, 163, 164, 167, 168, 173, 175, 198, 200
    activity 162, 163, 164, 167, 168, 173, 200
    mucus 167
    promoted tilapia plasma 198

M

Macrolactin 141
Macronutrients 5
Macrophages 35, 141, 201
Malnutrition 160
Malondialdehyde 38, 41, 149, 202
Mannan-oligosaccharide 39
Meat, refrigerated 87, 89
Mechanisms 3, 4, 98, 137, 138, 141, 166, 169, 172, 175, 176, 179, 180, 203, 225,
    antiviral 176, 225
    cellular defense 203
    of probiotic action 141
Medicinal plants 81, 104, 111, 125, 160, 167, 169, 172, 177
    traditional 81
Medicinal products 66, 161
    veterinary 161
Medicines 6, 7, 19, 22, 26, 86, 112, 125, 126
    ayurvedic 86
    traditional 112, 125
    traditional herbal 112
Mediterranean basin 80
Melaleuca alternifolia 204
Metabolic 24, 86, 170
    disorders 86, 170
    pathway 170
    profile 24
Metabolism 7, 66, 71, 73, 74, 103, 117, 124
    carbohydrate 74
    nutrient 73
Metabolites 9, 19, 21, 24, 26, 27, 112, 169, 173, 194
    secondary 24, 112, 173, 194
    toxic microbial 21
Methanolic extracts and essential oils 179
Microbial 3, 5, 43, 170
    dysbiosis 170
    Infections 3, 43
    population growth 5
Microbial balance 144, 171
    intestinal 144

Microbiota 59, 138, 147, 166, 169, 171, 175, 204
  intestinal 59, 147, 171, 204
  intestinal pathogenic 175
Micrococcus luteus 220
Microflora 40, 142, 145
  intestinal 142, 145
Microsporum 68, 70
  canis 68, 70
  gypseum 68, 70
Modulating 60, 81, 89, 161, 169, 172, 198, 202, 203
  gut microbiota 161
  mRNA expression 169, 172
Monocytogenes 44
Monogeneans 177, 178, 179, 196, 197, 205
Monogenoideans 197, 204
Moringa oleifera 20, 21
Morphology, intestinal 171
MS technique 44
Multiple antibiotic-resistant bacteria 6 (MARBs) 6
Mutagenicity 1, 9
Myeloperoxidase activities 166
Mytimycin 221

N

Natural 144, 168, 196, 223
  antioxidant agents 168
  defence 144
  ectoparasites 223
  monoterpene 196
Nephropathy 9
Neutrophil 200, 224
  extracellular traps (NETs) 224
  production process 200
Newcastle disease 147, 148
Non-esterified fatty acid (NEFA) 105, 201
Nutrient(s) 57, 138, 142, 171
  absorption 57, 138, 142
  food-derived 171

O

Ocimum sanctum 26
Oil(s) 20, 35, 36, 37, 39, 40, 41, 42, 43, 44, 52, 53, 55, 54, 55, 56, 57, 58, 59, 60, 68, 70, 71, 72, 74, 88, 163, 196, 198, 199, 200, 202, 203, 204
  Bhaisa 36
  chemical composition of ginger 37
  cinnamon essential (CEO) 53, 55, 56, 58, 59, 60, 68, 70
  cinnamomum bejolghota bark 56
  cinnamomum verum leaf 54
  clove 71, 72, 74
  corn 163
  dietary vegetable 203
  garlic 40
  ginger 35, 37, 39, 40, 42, 43, 200
  linseed 88, 200
  menthol 203
  oregano 39
  root 55
  rosemary 200
  sesame 200
  soybean 88
  sunflower 36
  thyme 39, 42, 198, 199, 202
  turmeric 44
  vegetable 163
  Virgin coconut 163
Olea europaea 163
Oregano essential oil (OEO) 59, 148, 164, 165, 171
Oxygen radical absorption capacity (ORAC) 71

P

Parahemolyticus 53
Parasites 70, 179, 194, 196, 197, 198, 204, 223
  monogenean 179
  monogenoidean 204
Parasitic 201, 204, 205, 222

cytotoxic effect 205
  infestation 201, 204, 222
Parasitosis 195
Pastoris 222, 223
Pathogenic bacteria 10, 53, 86, 99, 138, 141,
    144, 171, 174, 175, 217, 218, 223
  multidrug-resistant 86
  reducing gastrointestinal tract 99
Pathogenic 215, 224
  infections 224
  microbial agents 215
Pathogens 52, 53, 56, 74, 162, 163, 164, 166,
    167, 168, 170, 171, 174, 175, 176, 179,
    194, 200, 215, 222
  biofilm-forming 174
  fungal 56, 179, 222
  harmful 52, 53
Penicillium auregenosa 71
Peptides 216, 217, 218, 221, 222, 224
  disulfide 216
  melittin 217
  ribosome-derived 222
Peroxidase 169, 201
Phagocytic 5, 200, 218
  granulocytes 218
  index 200
  processes 5
Phagocytosis 168, 173, 198
  activity 198
Phytogenic 126, 164, 166, 168, 169, 180, 202,
    203
  agent 126
  constituents 202
  feed additives 164, 166, 168, 169
  molecules 180, 203
  plants 203
Phytonutrients 22, 23
  dietary 23
Phytotherapeutic agents 172, 173
Phytotherapy 172, 195
Plants 21, 23, 24, 34, 52, 99, 100, 102, 105,
    130, 131, 162, 173, 174, 179, 215, 217
  aromatic 52, 174
  dicotyledonous 105
  monocotyledonous herbal 34

Plasma protein 198
Plasmolysis, activating 220
Polyethylene glycol preparations 118
Polyphemusins 216, 221
Polyphenol oxidase 169
Polysaccharides 113, 115, 117, 126, 130
  psyllium 113
Pomegranate 82, 87, 88, 89
  molasses 82, 87, 88, 89
  peel powder (PPP) 82, 87, 88
  peel powder meal (PPPM) 87
Population 44, 130, 139
  anaerobic bacterial 44
  caecal coliform 139
  lymphocyte 130
Poultry 8, 23, 34, 81, 82, 90, 112,
  disease 23
  farming disorders 112
  muscle doxycycline 8
  nourishment 82
  nutritionists 81, 82, 90
  productivity 34
Poultry meat 10, 34, 44, 99
  antibiotic-free 99
Prebiotic activity 127
Probiotics 34, 124, 125, 137, 138, 139, 141,
    142, 143, 144, 145, 146, 147, 148, 149
  commercial 139, 145
Production 5, 6, 10, 60, 81, 103, 127, 141
  bile hydrolase 6
  inflammatory molecule 81
  methane 10
  mucus 60
  nitric oxide 127
  pro-inflammatory cytokines 141
  reduced toxic 5
  reducing ceramide 103
Products 3, 5, 6, 19, 20, 23, 33, 35, 40, 43, 44,
    46, 53, 57, 66, 139, 140, 194, 195
  cinnamon-derived 57
  microbial 140
  natural 53, 66
  on egg quality and production 40
Pro-inflammatory cytokine 201
Proliferating cell nuclear antigen (PCNA) 202

Properties 24, 52, 56, 66, 73, 86, 125, 219,
    224, 225,
  amphipathic 219
  anti-inflammatory 66, 86
  antiviral 224, 225
  growth-promoting 73
  health-promoting 125
  immunomodulating 52
  sensory 24
  therapeutic 56
Proteases 143, 175, 199
Protein(s) 9, 22, 24, 68, 69, 74, 102, 125, 142,
    168, 176, 222, 224
  degradation 168
  efficiency index (PEI) 24
  immunogenic 22
  viral-binding 176
Protein synthesis 5, 6
  inhibiting 5
Proteobacteria 170
Proteolysis 44
Pseudomonas 43, 44, 53, 68, 70
  aeruginosa 43, 44, 53, 68, 70, 220
  alginovora 220
Psidium guajava 163
Psychotherapy 113

R

Radical(s) 35, 43, 104
  oxidative 104
  scavenging action (RSA) 43
Reactive oxygen species (ROS) 22, 37, 39, 71,
    72, 81, 146, 168, 202
Reducing parasitic infection 196
Resistance 10, 138, 172, 174, 201
  drug 174
  genes 10
  microbial 138, 172
  reduced fish 201
Resources of antimicrobial peptides 217
Respiratory distress 27
Responses 40, 60, 68, 88, 148, 163, 164, 166,
    180, 202

humoral antibody 148
  immunogenic 60
  immunological 88, 164, 166
  inhibitory 68
Risks 6, 10, 19, 21, 142, 161,
  environmental safety 161
  lowering cardiovascular disease 142
Rosmarinus officinalis 21

S

Sacchalomyces 44, 138, 141, 142, 143, 145,
    147, 148, 222
  cervisiae 142, 143
  boulardii 138, 148
  cerevisiae 44, 138, 141, 145, 147, 148, 222
Salmonella 68, 70, 71, 138, 145, 220
  enteritidis 68, 70, 145
  typhimurium 68, 70, 71, 138, 220
Salvia officinalis 21
Saprolegniasis 179
Schistosoma mansoni 70
Scutellaria baicalensis 26
Secretion 73, 74, 115, 116, 129, 130, 142
  bile acid 116
  bile salt 74
Soil 2, 6, 7, 10, 67, 125, 139
  communities 2
  ecosystem 2
  organic matter-rich loam 67
Staphylococcus 9, 43, 53, 68, 70, 71, 86, 138,
    220, 221
  aureus 9, 43, 53, 68, 70, 71, 86, 138, 220,
    221
  epidermis 53, 68, 70
Stimulation 168, 201
  of lysozyme activity 168
  protective 201
Streptococcus 68, 70, 142, 143
  faecium 142, 143
  pneumonia 68, 70
  pyogenes 68, 70
Streptomycin 8

Stress 21, 22, 27, 38, 88, 131, 148, 161, 162, 169, 201
  breathing 27
  heat 38, 88, 148
  oxidative 88, 162, 169, 201
Stressors 161, 198, 201
  environmental 161, 198, 201
Superoxide dismutase 128, 169, 201, 202
Synergistic effect of phytochemicals 22
Synthesis 45, 115, 176
  bile acid 115
  viral RNA 176
Synthetic antioxidants 36

T

Taraxacum officinale 124, 125
Tea, green 20, 85, 176
Techniques 7, 43, 113
  radical scavenging 43
Tetrahymena pyriformis infection 223
Therapeutic activities 84, 113
Therapeutic herbs 34, 39, 42, 99, 124, 125, 128, 131
  traditional 125
Thymus 21, 39, 196
  glandulosus 196
  hyemalis 196
  vulgaris 21, 39, 196
Tissues 6, 8, 26, 41, 168, 202, 219
  edible poultry 8
  germinal 41
TLC 37
  supported bioautography procedure 37
  techniques 37
Total protein concentrations 43
Toxicity, bone marrow 9
Toxic metals 168
Transaminases 41, 42
  serum aspartate 41
Trans-cinnamaldehyde 55
Treatments 5, 9, 112, 116, 118, 138, 143, 162, 163, 164, 174, 179
  allopathic 112

cholesterol-reduction 116
Trichophyton 68, 70
  mentagrophytes 68, 70
  rubrum 68, 70
Trichophytonsis mentagrophytes 223
Trichosporon beigelii 222
Trypanosoma cruzi 70
Tumor necrosis factor (TNF) 130, 199, 201

U

Ulcerative colitis 118

V

Vaccines 148, 167
  coccidia 148
Viral neuraminidase activity 176
Virus 68, 70, 142, 148, 176, 177, 216, 221, 225
  herpes simplex 68
  infections 225
  kidney necrosis 225
  nervous necrosis 225
  viral haemorrhagic septicaemia 177
  white spot syndrome 176, 221, 225

W

Water 9, 6, 7, 45, 69, 117, 118, 145, 196, 197, 198, 199, 205
  contaminated 9, 199
Western blot analysis 176
White spot syndrome virus (WSSV) 176, 221, 225
World population growth 161

X

Xanthine dehydrogenase 105

Y

Yeast 4, 138, 139, 141, 222, 223
   microbial 4
Yersinia enterocolitica 71, 86

Z

Zataria multiflora 162
Zingiber officinale 34, 35, 39, 40, 164, 197